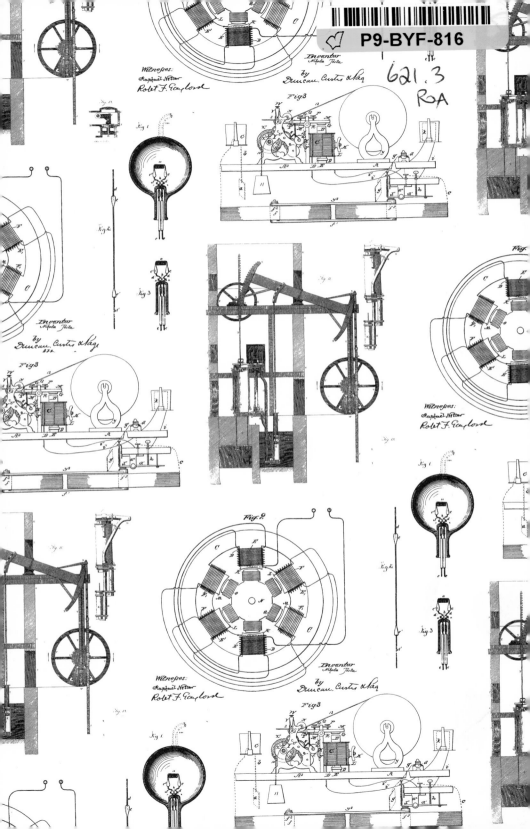

PRAISE FOR *SIMPLY ELECTRIFYING*

"Ten years in the making, *Simply Electrifying* is an inspiring book about the pioneers and visionaries who brought the transformative technology of electricity to all of society. Thoroughly researched, with insightful commentary, the book provides a thought-provoking and comprehensive examination of the challenges and evolving opportunities surrounding the electrification of our economy. The book also underscores the critically important role that electricity plays in powering our lives and our future."

—Tom Kuhn, president, Edison Electric Institute

"*Simply Electrifying* is as impressive as it is engaging and informative. Documenting the comprehensive history of one of humanity's greatest achievements is no small feat, but Craig R. Roach has given us an elegant book that provides an excellent review of electricity and the utility industry in this country."

—Joshua W. Martin III, director, Southwest Power Pool

"Craig Roach is a very accomplished economist, so as I sat down to read his *Simply Electrifying*, I expected another economic treatise replete with graphs and equations geared more toward the academic community than the general public. Just the opposite: As an economist myself, I was captivated as he filled in the blanks of history on a subject that has governed most of my career. The perseverance of entrepreneurs, the politics, the government, and the competitions—all of which gave us the miracle of electricity—is written beautifully in the genre of historical fiction. But this is not fiction, and Craig does not stop in the present nor with power alone, but rather with sensitivity to man's responsibility in future pursuits of scientific achievements. This book should fill a prominent place on anyone's bookshelf as a reference to one of man's greatest achievements."

—Alan R. Schriber, PhD, former chair,
Public Utilities Commission of Ohio

"Dozens of times a day we flip a switch and lights, TVs, computers, washing machines, and all the other devices we rely upon in our modern lives come on. We simply take for granted that this magical thing we call electricity will be available on-demand, 24/7, 365 days a year. Rarely do we even wonder about how that happens, where electricity comes from, and how it gets to us, instantly ready when we need it. In this well-researched, well-written, and engaging book, Dr. Roach presents an accessible history of this simply electrifying topic: electricity itself. Starting with Benjamin Franklin and continuing through to Elon Musk and Bill Gates, Dr. Roach gives us the inside story, unpacking scientific breakthroughs and connecting physics and engineering to the economic, regulatory, legal, business, and environmental policies that shape this crucial industry. Both newbies to the field and experienced policy wonks will learn much from this important book."

—Julie Simon, energy industry expert, Washington, DC

"A history buff's delight! Step into the incredible story of the origins of electricity like never before. It's fascinating to see how far we've come and consider what the future may hold. *Simply Electrifying* is a gem and a must-read for anybody that has ever reached for a light switch."

—Kent Parsons, energy and electricity lawyer, Baton Rouge, LA

"If you have any interest in learning about the roles that science, governments, and private enterprise can play in determining the economics and choice of new technologies for social benefit or harm, read Craig's book on the US electricity market. It is a comprehensive, balanced, and very informative analysis of the history of this essential product. I am already recommending *Simply Electrifying* to my students."

—Ken Hendricks, Laurits R. Christensen Distinguished Chair, Department of Economics, University of Wisconsin–Madison

"Dr. Roach's talent for identifying key issues and explaining complex ideas in straightforward terms enriches his 'wide-angle' review of the origin and development of the electricity industry and the business of supplying power. *Simply Electrifying* brings into vivid focus both the inventors responsible for the turning points in the discovery, production, and commercialization of electricity, and the executives and investors who shaped the electricity business. This incisive work connects advances in electricity technology and uses to the broader sweep of scientific progress, while also framing the transformative consequences of the democratization of electricity in economic, regulatory, and environmental terms. Dr. Roach's outstanding book, by explaining how electricity became fundamental to modern life, allows us to better understand where the industry will go next."

—Kenneth M. Simon, partner (ret.), Latham & Watkins LLP

Simply
Electrifying

Simply Electrifying

THE TECHNOLOGY THAT TRANSFORMED THE WORLD, *from* BENJAMIN FRANKLIN *to* ELON MUSK

Craig R. Roach

BenBella Books, Inc.
Dallas, TX

BenBella Books, Inc.
10440 N. Central Expressway, Suite 800
Dallas, TX 75231
www.benbellabooks.com
Send feedback to feedback@benbellabooks.com

Printed in the United States of America
10 9 8 7 6 5 4 3 2 1

Library of Congress Cataloging-in-Publication Data
Names: Roach, Craig.
Title: Simply electrifying : the technology that transformed the world, from
    Benjamin Franklin to Elon Musk / Craig R. Roach.
Description: Dallas, TX : BenBella Books, Inc., [2017] | Includes
    bibliographical references and index.
Identifiers: LCCN 2017004508 (print) | LCCN 2017006347 (ebook) | ISBN
    9781944648268 (trade cloth : alk. paper) | ISBN 9781944648275 (electronic)
Subjects: LCSH: Electricity—History. | Electrical engineering—History. |
    Inventors—Biography.
Classification: LCC QC527 .R63 2017 (print) | LCC QC527 (ebook) | DDC
    621.309—dc23
LC record available at https://lccn.loc.gov/2017004508

Editing by Stephanie Gorton Murphy
Copyediting by Miki Alexandra Caputo
Proofreading by James Fraleigh
    and Laura Cherkas
Indexing by Debra Bowman
    Indexing Services

Text design by Aaron Edmiston
Text composition by PerfecType, Nashville, TN
Cover design by Pete Garceau
Author photo by Dupont Photographers
Jacket design by Sarah Avinger
Printed by Lake Book Manufacturing

Distributed by Perseus Distribution
www.perseusdistribution.com

To place orders through Perseus Distribution:
Tel: (800) 343-4499
Fax: (800) 351-5073
Email: orderentry@perseusbooks.com

**Special discounts for bulk sales (minimum of 25 copies) are available.
Please contact Aida Herrera at aida@benbellabooks.com.**

To all those who agree with Benjamin Franklin that "when Men differ in Opinion, both Sides ought equally to have the Advantage of being heard by the Publick; and that when Truth and Error have fair Play, the former is always an overmatch for the latter."

CONTENTS

PART THREE: THE AGE OF BIG: BIG GOVERNMENT, BIG POWER PLANTS

PART FOUR: THE AGE OF HARM: CRISIS, CHANGE, AND SCANDAL

PART FIVE: THE AGE OF UNCOMPROMISING BELIEF

ACKNOWLEDGMENTS

I could not have written **Simply Electrifying** *without the tenacious and insightful* research of John Garnett; look for great things from this top-shelf historian out of George Mason University.

I also owe a great debt to all the biographers, historians, scientists, journalists, corporate executives, and public servants who have written about electricity over these many decades. As evidenced by the hundreds of endnotes, I have built upon the great work of others, and I hope I have given credit where credit is due; any errors, of course, are my own.

Many thanks to my agent, Jeff Ourvan, of the Jennifer Lyons Literary Agency in New York. Jeff saw the compelling narrative in *Simply Electrifying* early on and has worked tirelessly since then to make sure that readers had a chance to see it, too.

I am grateful to Glenn Yeffeth, publisher of BenBella Books, for putting together a great team ready to pursue success in this new era of book publishing. I could not have had a more engaged and skillful editor than Stephanie Gorton Murphy. Thanks also to Leah Wilson, BenBella's editor-in-chief, for her care and insight in the editing process.

A decade is a very long time to take to write a book. My good friend Julie Simon offered encouragement and advice over that entire time. She was the one who told me to put the wow in every chapter—I hope I have met her standard. Another friend, Mark Pacelli, is one of the best-read people I know. His advice on what to read and how to make the story work was a great help to me.

Thanks to my wife, Joanne Lepanto, for her encouragement and patience over the past decade of writing—and, of course, for all the other things that really matter.

PROLOGUE

Just Imagine

[Franklin] found electricity a curiosity and left it a science.
—Carl Van Doren (1938)[1]

Imagine your world without electricity: No morning alarm clock or coffee maker.
No electric lights, heat, or air conditioning. No whirling electric motors.
And no computers or cell phones.

This was the world 265 years ago in Philadelphia, the birthplace of
the modern science of electricity. The morning alarm was the sun rising or the cock crowing. Coffee and heat required a simple woodstove.
Muscle power and perhaps literal horse power drove businesses. And any
computing going on was surely done with pencil and paper. No one could
imagine a world with electricity because at that time the notion of such
a force was fantastical; man had not yet captured and harnessed such
power.

Without electricity the urban night was dimly lit at best and often dangerous. Imagine sitting around a candle trying to read or sew or do close
work of any kind; it would have taken over a hundred candles to give the
same amount of light produced by a single 100-watt incandescent electric

lightbulb. Getting close to the open flame could mean a shirtsleeve, lock of hair, or curtain could be set on fire. The dangers continued outside and ranged from tripping along a dark street to being mugged or worse. The one partial consolation might have been that, without electric lights, there was nothing to obstruct the beauty of a star-filled night sky.[2]

Amazingly, a few bold minds were eager and able to imagine electricity—to *see the unseen*. None was more important than Philadelphia's Benjamin Franklin, who allowed the world to "see" electricity one day in June 1752 by flying his kite in the midst of a storm. Because of the obvious danger, doubts persist to this day whether Franklin actually flew his kite in a storm, but there is no reason to doubt that he conducted his kite experiment, perhaps in less dramatic weather.[3]

While Franklin's unique intellect was surely central to his success, so too was the fact that he lived at a watershed moment for science and technology. The Scientific Revolution had already transformed man's understanding of the universe, and the Industrial Revolution would soon begin to transform the world's economy. Newton and others had already begun to explain in precise mathematics the mechanics of the universe in terms of the force of gravity. Eventually, Franklin and others would do the same for electricity, as well as for magnetism and light.

At the same time, there were several common scientific and medical beliefs that seem primitive or even cruel today. One example is the practice of bloodletting as a remedy for fever and infection. It is more likely that George Washington, among many others, died from such "remedies" than from the ailment he contracted toward the end of his life. Great shifts in science and technology sometimes spread progress unevenly across the many facets of life.[4]

Benjamin Franklin and his kite represent only the first step in a fascinating journey from an era when life with electricity was barely imagined by a few to a time in which life without electricity was unimaginable for most. That journey was powered by the bold actions of many famous (and some infamous) figures who were sometimes propelled forward, and as often pushed back, by the demands and mores of their times.

These innovators built the powerful, pervasive electric infrastructure on which everyone in the developed world now depends. The first cornerstone of that infrastructure was remarkable scientific advances made by Benjamin Franklin, Michael Faraday, James Clerk Maxwell, Nikola Tesla, and Albert Einstein. Indeed, Einstein believed that the discovery of electricity was not just a stop along the path to modern physics, it was the path itself. Man's understanding of electricity led to many of the major breakthroughs in all of physics.

The second cornerstone was the technology based on this great science. Samuel Morse's telegraph was the first fruit of the hard-won scientific knowledge linking electricity and magnetism. The fact that the purpose of the telegraph was for communication is significant. The electricity business and the telecommunications business were born of the same science, and yet the latter has far outpaced the former in technological advances in modern times.

The first war between technology titans in electricity featured Thomas Edison against George Westinghouse, with Nikola Tesla by Westinghouse's side. Their clash was the original Battle of the Currents. In the end, Westinghouse's alternating-current system bested Edison's direct-current system. Alternating current allowed for economies of scale because electric power plants could be linked to customers across a vast distance. Interestingly, few people today are fully aware of the key roles played by Westinghouse and Tesla.

The third cornerstone was the business practices that governed the nationwide spread of electric service. Innovative business magnate Samuel Insull set the standard here with a full understanding of and belief in economies of scale. His aim was to spread the fixed cost of power plants and transmission lines over an increasing number of customers, thus causing the per-unit price of electricity to fall. Eventually, electricity became available and affordable to all, and, in this way, Insull implemented the first democratization of electricity.

The pursuit of the big, however, is also what got Insull into trouble and why his reputation is still infamous to some today. In 1932 Insull owned one of just three holding companies that together controlled almost half of all power generation in America. When his holding company failed in the Great Depression, none other than President Franklin D. Roosevelt blamed Insull for the severity of the ensuing economic collapse.[5]

The fourth cornerstone, closely linked to the third, was government regulation. In the early 1900s, Insull had actually invited state regulation of utilities; with such regulation, he won the exclusive areas of service (utility monopolies that would allow economies of scale) in exchange for giving the states the right to enforce cost-based rates for electric service. Federal regulation, however, came about during the Great Depression. During his presidency, Franklin D. Roosevelt aggressively broke up the large holding companies—one of the largest government interventions ever outside of wartime. That set up a jurisdictional tension between states and the federal government that remains in full force today.

Environmental regulation also is part of the regulatory cornerstone—an increasingly important part. The deadly air pollution episodes in Donora, Pennsylvania, and New York City were an important impetus. The landmark book *Silent Spring* (1962) by Rachel Carson is often credited with creating the modern environmental movement, with its combination of Carson's compelling documentation of the detrimental effects of pesticides and her personal courage. She continues to have a major impact on the electricity business despite the fact that she never directly wrote or spoke about that industry. Carson's template for environmental action can be seen today in efforts ranging from former vice president Al Gore's *An Inconvenient Truth* in 2006 to former president Barack Obama's Clean Power Plan in 2015, initiatives meant to address widespread concern about global climate change.

Today, fundamental changes in both scientific knowledge and cultural values have led to the question of whether an entirely new electric infrastructure must be built. The fact that environmental concerns—especially global climate change—drive and dominate the electricity business is one of those changes. Add to that the growing concern about electric-service reliability at the customer's doorstep, rather than at the grid, due to more frequent storms and cyber warfare. And, further, add a demand for choice in electricity supply no different than the demand for choice in transport reflected in the rise of Uber. With such fundamental changes there is the possibility of a second Battle of the Currents. This time, the opposition would pit the well-established electrical grid against "personal power" from new industry leaders like Elon Musk, who often deploy smaller-scale, renewable technologies. A second democratization may also lie on the horizon, in which individual consumers can "vote" with their wallets for more environmental protection, more doorstep reliability, and more choice.

The ambitious narrative summarized here reflects the view that the only way to truly understand the past and anticipate the future is to use a wide-angle lens. All 265 years of the history of electricity must be taken into consideration. If the story starts with Edison, for example, it will have missed much of the first cornerstone—the great science of Franklin, Faraday, and Maxwell. Likewise, it is important to capture the full range of factors driving the evolution of the electricity industry. Only with assessments of the wider fields of science, military strategy, and geopolitics—particularly, the geopolitics of the Cold War—can one explain the rise of nuclear power. Only with assessment of the strategic opportunities in the auto business plus a deep dive into the culture of cars can one anticipate the future of electric vehicles. And only with the wide view can we judge whether the seemingly confrontational cabinet appointments of President Donald Trump might actually set the stage for constructive compromise on climate.

Electricity is at the core of all modern life, yet the full story of this revolutionary force has remained untold—until now.

PART ONE

The Age of Franklin

A Time of Awe and Discovery

CHAPTER 1

Benjamin Franklin's Kite

He snatched the lightning from the sky
and the scepter from the tyrants.
—Anne-Robert-Jacques Turgot (1779)[1]

"Death and Slaughter of a Man and his Son at Plough together with Four Horses"
read the headline of a news bulletin in 1680. The culprit? Lightning. The
bulletin went on to explain in graphic detail that the boy's clothes were
"rent from his body, and his Hat torn in two or three Pieces," and that his
body was burned from his neck down his back. So fearsome was lightning
at the time that a 1766 sermon predicted that on Judgment Day God's
wrath would be carried out with thunder and lightning so powerful that
"with one clap the earth will split to the very center, and the whole globe
crumble." In an earlier sermon of 1746 a different preacher expressed little
hope that man could understand this great mystery of lightning, concluding that "humane Capacities cannot fully comprehend it."[2]

Benjamin Franklin refused to concede that lightning was incom-
prehensible. Indeed, by flying his kite in the midst of a storm one day,
Franklin revealed that lightning could be understood by man as an

electrical phenomenon and not just as God's wrath. Imagine Franklin's courage when he flew his kite; this was not just intellectual courage with the gravest consequence being public embarrassment for a failed experiment. What he proposed was no safe lab trial. Franklin took on lightning, one of the most mysterious and dangerous forces then known.

With his kite experiment, Franklin enabled man to *see the unseen*. Franklin inspired an era of science in which electricity was the essence of scientific inquiry—the Age of Franklin. Through all this, Franklin showed the world the physical and intellectual courage that would continue to propel him throughout his remarkable life.

In his day, Franklin was recognized as one of the world's foremost scientists, with some of his contemporaries referring to him as the "Newton of electricity." Yet today there is a tendency to downplay his scientific contribution. Take a look at America's hundred-dollar bill: Franklin's face is on one side and Independence Hall on the other. Fair enough, but why are there no images related to his experiments, his kite, or his lightning rod? Some have argued that without his diplomatic success few would have paid attention to his science. Increasingly, however, modern historians are beginning to believe the reverse is more accurate. Scholars have recently argued that his reputation as a scientist created the opportunity for his success as America's original diplomat. His electrical experiment and theories made him a celebrity around the world, and that celebrity gained him access to the dinner tables of the well-to-do and influential in France. From that perch, he negotiated the alliance with France that gave the American colonies the troops, ships, and supplies needed to beat the English at Yorktown and win the Revolutionary War. Electricity's greatest contribution to the world is generally thought to be light, but, to America, it may quite well have been freedom.[3]

The lasting meaning of Franklin's story is that science and scientists matter greatly to the history of electricity. This makes electricity fundamentally different from the other widespread forms of energy, like coal and oil. In Franklin's day, scientists were thought to bring wisdom to any issue of the day, whether that be technology, politics, law, or culture.

Just think of the progress that could be made in now-deadlocked debates on global climate change and other challenging issues if that were still true today.

THE ORIGINS OF HIS SUCCESS

It is said that success happens when opportunity and talent collide. For Franklin, it is arguably more accurate to say his scientific success came when context and personal ambition collided. Born in Boston in 1706, Franklin was a product of his time; he lived in the wake of the Scientific Revolution. His was an era when, for those with the money and time, science was central to both entertainment and social conversation. It also was a time of economic progress for the American colonies, and Franklin would rapidly become a wealthy man.[4]

His early years molded his character in ways that contributed to his great success. At the age of twelve, Franklin was apprenticed to his brother James as a printer for a term that was expected to be nine years. Let's pause a minute to think of the importance of printing at the time: a printer was a skilled worker and a print shop might be thought of as the Google of the day, a place where information was routinely collected, updated, and disseminated to those with access. So powerful was this dissemination of information that historians identify printing as an essential tool for huge cultural shifts ranging from the Protestant Reformation to the Scientific Revolution. As an apprentice, Franklin spent hours reading and learning from what he printed. Printers were exposed to a unique education; another great scientist, Michael Faraday, would later enjoy the same education in the print shop of his mentor. Franklin came to see the printed word as something of an arbiter of the truth, writing that "printers are educated in the Belief that when Men differ in Opinion, both Sides ought equally to have the Advantage of being heard by the Publick; and that when Truth and Error have fair Play, the former is always an overmatch for the latter."[5]

At seventeen Franklin breached his apprenticeship and ran away from Boston. He eventually settled in Philadelphia, where he started his own printing business with another former apprentice, Hugh Meredith. During his time as a printer, Franklin enjoyed a series of successes that revealed his broad talents, depth of intellect, and abundance of energy. One such success was a new fireplace (the Franklin stove), which conserved wood. He called his invention "another sun" because of the light and warmth it provided. Reflecting an innate altruism, Franklin refused to patent his inventions; this to-patent-or-not-to-patent question became a recurring theme in the history of electricity.[6]

Franklin stepped away from his printing business in 1748, at age forty-two. The printing trade had given Franklin the wealth that allowed him to retire in order to serve the world with his scientific genius and to serve America with his talent for diplomacy. Franklin had a confidence born of hard-won successes, and he believed he could achieve anything he put his mind to. He had a breadth of curiosity made manifest by a long list of achievements that would be hard to match in any age, and he was savvy enough to let practical and political reality guide him to what worked. Finally, Franklin had a depth of commitment to serve mankind in general and America in particular.

It would be wrong, however, to see Franklin as a perfect man. Franklin was inclined to pranks and hoaxes. One example involved his brother James, who published a newspaper called the *New England Courant*. James was intrigued one day to find an article under his door signed "Silence Dogood," which he liked enough to print in his newspaper on April 2, 1722. Over the next six months he would find fifteen of these articles. James did not know that Mrs. Dogood was actually the nom de plume of his sixteen-year-old brother, Benjamin. In addition to playing tricks, Franklin became a scofflaw when he breached his apprenticeship. He also sired a son out of wedlock, perhaps with a prostitute.

Failure is sometimes a necessary step on the road to success, and this was the case with Franklin's first trip to London. The governor of Pennsylvania, William Keith, falsely promised to provide letters of introduction and credit for Franklin. At age eighteen, Franklin found himself abandoned and far from home with little money. He got a job as a printer and spent eighteen months learning about printing and about London. While cited far less than his later visits to London, the first journey helped forge a character of confident self-reliance.[7]

THE SCIENTIST

The Right Time and a Plan

The historian I. Bernard Cohen, a professor emeritus of the history of science at Harvard, declared that Franklin became "one of the best known living scientists" of his time by providing the "first satisfactory theory" of electricity and was "the primary creator of this new science." Astonishingly, in an era when simple communication meant sending a letter that could take days or weeks to arrive, Franklin won much of his fame in a short five-year span, from about 1747 to 1752, primarily by sending a series of letters to Peter Collinson in London concerning Franklin's own experiments with electricity.[8]

What did it take to transform Franklin from successful printer to internationally famous scientist? Clearly it took intellectual genius, but, more than that, Franklin had an intuition for the needs and preoccupations of his time. In Franklin's day, the term *science* meant "wisdom," while *natural philosophy* coincided with what is called physics today. In the eighteenth century, reports on the latest experiments were commonly published even in provincial newspapers. As an apprentice printer, Franklin had taught himself to write by studying journals like the popular *Spectator*, in which Joseph Addison boasted that he had "brought Philosophy out of Closets and Libraries, Schools and Colleges, to dwell in Clubs and Assemblies, at Tea-Tables and in Coffee Houses."[9]

It was the right time, and Franklin had a solid plan to get noticed. He would become what today would be called a public intellectual, and he understood that distant and lowly colonists needed the support of London in order to be recognized as high-profile scientists. To this end, Franklin built on his relationships with botanist and Royal Society fellow Peter Collinson in London, particularly when Franklin learned that Collinson was searching for a "colonial protégé." It is Collinson who would take Franklin's letters, which reported on his experiments with electricity, and have them read to the Royal Society in London.[10]

Science in Newton's Shadow

Isaac Newton's famous *Principia* was published in 1687, just nineteen years before Franklin's birth. Newton's other classic, *Opticks*, was published in 1704. These two works were fundamentally different, and it was *Opticks* that had a greater influence on Franklin. *Opticks* was accessible to him because it was written in English rather than Latin and, more important according to Cohen, was "devoted to the description of beautiful, simply performed experiments." Franklin's work was consistent not only with *Opticks*' focus on experimentation but also with the fundamental nature of the effort. In *Principia*, Newton theorized that gravity was a function of mass and that the type of material did not matter. *Opticks* contrasted sharply with this view because it focused deeply on types of material, including specific materials that would conduct electricity or were magnetic. Franklin's work is in the same vein as *Opticks* since Franklin was, as Cohen notes, "an experimental Newtonian scientist."[11]

Franklin's scientific work lived up to the motto of the Royal Society: "On the Word of No One." That is, the laws of nature should be discovered through experimentation, not taken for granted. This commitment to experimental proof was the foundation of the Scientific Revolution launched by Newton and others. The study of electricity was a great test of that commitment, since there was likely a temptation to stick with the celebrated findings of Newton, rather than develop a different view. But the boldest "electricians" did not succumb to that temptation; they let themselves be confronted and confounded by the differences between gravity and electricity. Newton's gravity was all about attraction of celestial bodies, but electricity was different: it concerned *both* attraction *and* repulsion. The goal was to explain the coexistence of attraction and repulsion, something that had never been attempted before. Franklin would do just that in his letters to Collinson.[12]

Franklin's Letters to London: An Exquisite Experiment

In two letters to Collinson dated May 25 and July 28, 1747, Franklin explained his basic theory of electricity. It depended on the concept of balance, or equilibrium. When an object—such as a piece of amber or a glass tube—had its normal amount of "electrical matter" or "electrical fluid," that object was in a normal state and did not attract or repel other objects. However, if the object lost or gained electrical fluid and became unbalanced, the object became "charged." Franklin established definitions for these imbalances: when an object gained electrical fluid, it was positively charged; when the object lost electrical fluid, it became negatively charged. Franklin's theory focused on the redistribution of a single type of electrical matter or electrical fluid.[13]

Most important, Franklin's theory explained attraction and repulsion. Attraction occurs when two objects or materials are brought together and one is positively charged and the other negatively charged; so, with electrical fluid, as in romance, you could say that opposites attract. In contrast, if two objects are brought together and both are either positively charged or negatively charged, the objects will repel one another—like repels like. Franklin's theory was sometimes called the single-fluid theory. The experiments in which he proved his theory were elegantly simple, and they sometimes had the feel of a parlor game because real people were the ones positively or negatively charged. In one experiment Person A rubs a piece of glass with silk to create static electricity. When A passes the glass to B, sparks fly.[14]

Franklin's Letter on the Leyden Jar

In 1746 a university researcher in Holland conducted what he described as "a new but terrible experiment." The researcher's name was Pieter van Musschenbroek, and the device he developed through the experiment became known as a Leyden jar, after Leiden University where Musschenbroek worked. The device is best envisioned as an ordinary, partially filled glass jar of water with a brass wire dipped into it. Noted Berkeley historian J. L. Heilbron says the "standard form" also has "metal foil inside and out." The foil on the inside of the Leyden jar is given an electrical charge by touching the brass wire with a metal rod; the rod got its charge from a glass tube rubbed with silk, which created static electricity. A charge of opposite sign and equal magnitude would then form on the outer metal foil. Because glass is a good insulator, the two charges hold each other in a standoff until a path is created to let the charges come together.[15]

Why was the Leyden jar a "terrible experiment"? When Musschenbroek put his right hand on the outside of the Leyden jar and his left hand on the iron bar that conveyed static electricity to the inside of the jar, his body linked two oppositely charged objects. That opposite charge was the source of the jolt with which his "whole body quivered just like someone hit by lightning."[16]

The world took sides on explaining the Leyden jar, with Franklin in one corner and the Frenchman Jean-Antoine Nollet in the other. Though Nollet's name is nowhere near as recognizable as Franklin's today, Heilbron concluded that from 1745 to 1752 Nollet's theory "enjoyed the widest consensus any electrical theory had yet received."[17]

Franklin's concept of positive and negative charges allowed him to explain the properties of the Leyden jar, and thus his theory won the day. Heilbron wrote that the electrical fire inside the Leyden jar accumulated "on the inner surface of the jar because it is unable to pass" through the glass; in other words, Franklin got the role of the glass right—it is a great insulator. This excess of electrical charge on the inside meant that an equal amount of electrical fire must be repelled, and that force can be neutralized only by an equal but opposite charge on the outside of the glass. Nollet conducted the terrible experiment himself, and it "bent him double and knocked out his wind." Ultimately, he was unable to explain precisely why the Leyden jar did what it did; he concluded that it was an "exception." In a letter to London in July 1747, Franklin explained the science behind the Leyden jar, and, with that feat, increased his international fame as a scientist.[18]

Franklin's Kite: Lightning Is Electricity

Franklin's victory in the battle of the theories was impressive. A lowly American went head-to-head with the best-educated and best-funded scientists Europe had to offer, and he bested them all. Not only did he win personal fame but he helped the striving American colonies gain recognition in the international scientific community. Franklin's success showed that Americans had the mettle to equal if not surpass their European counterparts.

The impact of this scientific victory on his international reputation, however, pales in comparison to the impact of the experiment he suggested that showed that lightning is best explained as the equivalent of electrical force. Franklin was not the first to suggest the equivalence, but he was the first to successfully propose and inspire an experimental test to illustrate it. In July 1750 Franklin proposed what became known as the sentry-box experiment in a letter to Collinson. His proposal started with the idea of an enclosed sentry box that would be big enough for a man standing on insulation. The sentry box would have an iron rod that extended twenty or thirty feet into the sky. Franklin's proposed experiment was an attempt to understand the nature of lightning by drawing it down from the sky to an iron rod.[19]

What Franklin wanted to test was the theory that lightning began with an electrical charge in the storm clouds, and that lightning discharges would descend to the highest object around, whether that was a church steeple or a mighty oak. The inspiration to others began with the publication of Franklin's letters in April 1751 in a book titled *Experiments and Observations on Electricity, Made at Philadelphia in America by Mr. Benjamin Franklin*. A Frenchman, the Comte de Buffon, read and admired Franklin's work, and he asked that Jean-François Dalibard translate the work into French, with the help of an electrical experimenter named Delor. Buffon's interest also reflected his intense dislike for Nollet and his colleagues, for a variety of reasons, including a personal feud with René-Antoine Ferchault de Réaumur, Nollet's patron. Buffon and Réaumur had traded insults directly and through others. Since Franklin's work opposed Nollet, and by association Réaumur, Buffon wanted to promote Franklin's views.[20]

Buffon, Dalibard, and Delor gained their first opportunity to publicize Franklin by re-creating some of his experiments for none other than the king of France, Louis XV. The three Frenchmen then sought to conduct Franklin's sentry-box experiment. In May 1752 Dalibard set up the needed apparatus at Marly-la-Ville, a village about twenty miles outside Paris. On May 10 two other men were instructed to conduct the dangerous experiment by touching a pole to a brass wire and "saw the first spark man ever intentionally drew from the sky." Dalibard reported the Marly experiment to the Academy of Sciences in Paris. The experiment confirmed "the kinship of lightning and electricity" and that Franklin was right about the lightning rod: "an elevated, *grounded*, pointed rod" offered protection from storms. The success at Marly proved that the study of electricity was not "toy physics" but rather a significant path of inquiry that could explain a major physical phenomenon.[21]

Franklin's experiment to prove that lightning was electricity—flying a kite during a storm—was a variant of the sentry-box experiment. Franklin conducted the kite experiment in June 1752—which was before he heard of the successful sentry-box experiment at Marly, but not before the French experiment had actually been conducted. It was the kite experiment, rather than the sentry box, that caught the public imagination and led the philosopher Immanuel Kant to refer to Franklin as a "modern Prometheus."[22]

Franklin described his apparatus in his letter to Collinson in October 1752. The body of the kite was a silk handkerchief stretched over a cross made of two pieces of cedar wood. Extending from the top of the kite was a thin metal wire. A long string ran from the bottom of the body of the kite. At a point close to Franklin's hand a metal key was tied to the string, and below that key was another piece of silk that Franklin held in his hand. This simple apparatus would tame the power of nature.[23]

With the kite flying high, he invited lightning to strike the tip of the metal wire extending above his silk kite and then send the captured electric current down the string to the waiting key. That key would then allow Franklin to transfer and store the electrical charge in a Leyden jar. Once in the Leyden jar, the electrical charge from the lightning was shown to be the same as the static electricity that scientists used at that time to experiment with electricity—a much less fearsome proposition. Since the extra piece of silk Franklin used as a handle was crucial to his safety—this is why the charge supposedly stopped at the key—he suggested that it was best for future experimenters to stand inside a small building to keep themselves and the silk handle dry.[24]

The danger is clear: Franklin's experiment put him in the perfect position to be struck by lightning and killed. This dire consequence was brought home by the death of George Richman, a scientist attempting Franklin's experiment in July 1753 from his apartment in St. Petersburg. An obituary for Richman tells of both his notable scientific stature and the danger of experiments with electricity. Richman's death was a loss to science: he was well educated, a member of the Imperial Academy of Sciences and Arts at St. Petersburg, and a professor of experimental philosophy. While well aware of the "dangerous consequences," as the obituary notes, he "neglected no opportunity" to measure the strength of the charge in a "thunder-cloud." Richman's death and the obvious danger of the experiment provoke doubts to this day that Franklin actually flew his kite in a storm.[25]

The Harvard historian I. Bernard Cohen makes clear his view that Franklin did indeed conduct the kite experiment, while acknowledging the debate around it. In an article from 1952, Cohen starts by pointing to a famous depiction of the kite experiment by Currier and Ives and then points out all that is wrong with the depiction. In the end, Cohen concludes that there is no "legitimate doubt" that Franklin conducted his kite experiment. Cohen also says there is good evidence that others replicated the experiment, so there is no reason to doubt Franklin's claim. Still, doubts persist. Author Tom Tucker makes his conclusion quite obvious in his 2003 book, *Bolt of Fate: Benjamin Franklin and His Electric Kite Hoax.* Tucker frames the kite experiment by concentrating on Franklin's propensity for trickery, and he sees it as just another classic Franklin prank. Yet if Franklin did not conduct his kite experiment as he said he did, this would have gone well beyond a good-natured hoax in the name of fun. It would have been a deception on a grand scale. There was no reason for Franklin to lie; he would have won international fame by simply suggesting the sentry-box experiment and having Dalibard conduct it. As Cohen suggests, Franklin probably did not conduct the experiment with the drama emphasized by the Currier and Ives depiction, but there is no reason to conclude he did not do it at all.[26]

Franklin's Invention: The Lightning Rod

Not only was Franklin an innovative experimental scientist, he was also a master at the practical application of his science. Franklin described his most important invention, the lightning rod, in his *Poor Richard's Almanac* of 1753 under the heading "How To Secure Houses, & C. From LIGHTNING." He first said, "It has pleased God in his goodness" to provide some level of protection from "mischief by thunder and lightning." The protection comes in the form of a "small iron rod" with "one end being three or four feet in the moist ground."[27] The other end should be "six or eight feet above the highest part of the building." He adds that a foot-long brass wire should be attached to the upper end. And he promised that with such a device the house "will not be damaged by lightning" because the lightning would be attracted "by the points, and passing thro the metal into the ground without hurting any thing."[28]

Although the lightning rod seems intuitive and even quaint today, it represents a hard-won achievement in both science (discovery that lightning is an electrical phenomenon) and technology (the consequent invention of a lifesaving tool available to all). Similarly, it is important not to underestimate the possible philosophical advance it represented in the battle of man versus nature—with the balance tipping toward man.

The lightning rod also demonstrated how politics can enter the debate surrounding any technology. In England a controversy arose over the shape of the lightning rod: should it be "pointed" as Franklin proposed or should it be a blunt knob as suggested by the English scientist Benjamin Wilson and others? Heilbron calls it a "comical battle" that started in 1772 when the Royal Society was asked how to protect an arsenal stocked with gunpowder from lightning. The Royal Society's committee, despite Wilson's dissent, concluded that pointed lightning rods were the best protection. In 1777 the town of Purfleet suffered damage from lightning despite the pointed lightning rods near its arsenal and other buildings. By then, of course, Franklin's America had begun its revolution. King George III himself entered the fray, with a request to the Royal Society to reverse its decision. Cohen paraphrased the president of the Royal Society Sir John Pringle's reply thus: "His Majesty might change the laws of the land but could not reverse or alter the laws of nature." Suffice it to say that Pringle lost his job at the Royal Society. It was a "comical" and unnecessary public debate since, as Heilbron puts it, when viewed from the same height "on nature's scale both [a knob and a point] appear to be points." This, of course, was not the last time politics would interfere with scientific inquiry.[29]

THE CONSEQUENCES OF HIS SUCCESS

His Fame in Science

Franklin earned renown as America's diplomat in France, but, as mentioned earlier, modern historians increasingly find that Franklin's international fame as a scientist gave him advantages as a diplomat. Harvard American history professor Joyce Chaplin more boldly states, "Franklin was not a statesman who did science. He became a statesman because he had done science. And he was able to do so because, in the eighteenth century, science became part of public culture."[30]

Franklin's international scientific fame is well-documented. His book *Experiments and Observations on Electricity* (1751) was printed in five English editions and was translated into three languages, all before the American Revolution and Franklin's assignment to France. Franklin also received many awards and honorary degrees for his scientific contributions. In 1753 he was awarded the Sir Godfrey Copley Gold Medal, the highest award for scientific research from the British Royal Society. Before Franklin's award, the medal had been given only to British Fellows of the Royal Society. His other honors during his lifetime included honorary degrees from Harvard, Yale, the College of William and Mary, Scotland's University of St. Andrews, and England's Oxford University. Franklin was elected as a Fellow of the British Royal Society, a member of the Royal Society of Göttingen in Germany, and a member of the Royal Academy of Sciences in Paris—only one of eight foreigners to be so honored at the time.[31]

Two Missions in London

Scientific fame, however, does not guarantee diplomatic success, as evidenced by Franklin's two diplomatic missions to England. He went one for two, with his loss being far more consequential than his win. On his first mission to England, from 1757 to 1762, Franklin was sent to London as the representative of the Pennsylvania Assembly to negotiate with William Penn's heirs. Franklin's awkward goal was to ask the Penn family as proprietors to allow themselves to be taxed to generate revenue for the colony. Chaplin reports that the Pennsylvania Assembly correctly guessed that Franklin was "famous enough to shine" in London, and Franklin did take "his place in London as a somebody—as the electrical genius, the winner of the Copley Medal, and the famous American philosopher." Still, despite

Franklin's glowing reputation, by the end of his first mission in 1762 one of the Penn family called him a "malicious villain." In August 1760, however, "Franklin won the day" as the "Privy Council ruled mostly in favor of the Pennsylvania Assembly and against the Penns."[32]

Franklin's second mission to England covered an even longer period, the eleven years between 1764 and 1775. This time, the issues at hand were more delicate and dangerous; they would lead to the Revolutionary War. Franklin's second mission ended, as did the first, with a session of the privy council. Unlike the first, however, Franklin won nothing—in fact, he lost a great deal. The unstated agenda of this session of the privy council was to rebuke Franklin for releasing private letters written by the appointed governor of Massachusetts. Meetings of the privy council were not often well attended, but most of the political elite were present for this session. The prime minister, Lord Frederick North, was there, as were Lord Dartmouth, the secretary of state for the colonies; the archbishop of Canterbury; the bishop of London; and other leading figures. The solicitor general himself, Alexander Wedderburn, was asked to conduct the examination of Franklin. H. W. Brands, a noted Franklin biographer, writes that through it all, the packed audience in the gallery "laughed and cheered at the solicitor's slashing assault on Franklin's behavior and character." He notes that Wedderburn struck out at the fame Franklin had earned for his work on electricity when he said Franklin was "the prime conductor" of the affair. He ridiculed Franklin's position as postmaster, stating that, going forward, he would "esteem it a libel to be called a *man of letters*."[33]

The ridicule was meant to put Franklin and the colonies in their place. In England's view, America was not its equal on any grounds. Franklin, however, was clearly the equal of any of their great minds; indeed, his scientific achievements were judged to be as exciting and innovative in England as elsewhere. Moreover, Franklin had made his own success; nothing had been given to him by title or birthright. Perhaps on that day, the ideal that all men are created equal began to form in his head. Surely it was on that day that Franklin turned firmly toward revolution.

The Alliance with France

Soon after the scathing session in front of the privy council, Franklin returned to America and was asked to take on another great challenge for his country. It was one that would be both physically and professionally demanding: at age seventy, Franklin was asked to make an arduous journey to France. He would travel across hostile waters, in difficult weather, with only his two young grandsons to accompany him. The professional challenge lay in the task he was given: to convince King Louis XVI of France to give monetary and military support to America so that it could assert its independence from King George III of England. Franklin, of course, accepted the challenge and achieved his goal. Thus, the scientist behind the discovery of electricity became the metaphorical "lightning rod" to the independence of the United States—the nation in which the development and application of electricity would be nurtured for centuries to come.

How did he do it? To start, the French knew him. Not only had Franklin bested the French scientist Nollet's theories on electricity but the first successful test of Franklin's conjecture that lightning was electricity—the sentry-box experiment—was done in France by the French.

However, while fame or reputation got Franklin in the door, it by no means guaranteed success. Franklin's persuasiveness as a diplomat must be due in equal part to the fact that he was savvy enough to understand what France wanted from an alliance with America. Franklin also understood what the influential figures in France wanted from him personally. He knew that the small group of powerful French were open to being convinced "at the dinner table," where he was "different enough to be interesting but familiar enough not to be frightening." Franklin's identity as a diplomat was not restricted to certain hours or meetings: it governed his approach to life.[34]

Unusually for an American, Franklin learned to fit perfectly into the European culture of his day. He spent much of the twenty-eight-year period from 1757 to 1785 in Europe. Historian Morris Bishop finds that Franklin's scientific fame gave him "access to the scientific and cultural world of Europe" in which he finally found "stimulating intellectual companionship." The science of electricity was not relegated to the laboratory; rather, it often was trotted out at dinner parties, as a topic of interest for the elites. Moreover, though Franklin had few intellectual equals in America, Bishop concludes that "Franklin's experience in France was the homecoming of his mind." The scientist and diplomat was stimulated by the other curious minds around him, regardless of nationality.[35]

A triumphant career did not come without personal pain. Franklin lost his son, William, when William chose to support England. Franklin said, "Nothing has hurt me so much and affected me with such keen sensations as to find myself deserted in my old Age by my only Son." William would be imprisoned throughout the war and lived out his days in London, exiled from both father and country.[36]

FRANKLIN'S LEGACY

Franklin's place in history is best captured by the caption of a famous 1779 engraving: "He snatched the lightning from the sky and the scepter from the tyrants." Snatching the lightning made him an important scientist; snatching the scepter made him an astonishing statesman. His dual fame makes him unique and somewhat difficult to place in a single category in the narrative of history. Was he a man of his time, or a man who created his time?[37]

He was a man of his time in that he lived in a country where freedom was blossoming and the social ladder was lowered for him to climb. He was a man of his time because the Scientific Revolution had begun, and so an interest in science was pervasive. His writings put him center stage as a scientist, while the American Revolution put him center stage as a diplomat. Still, more than most, he created his time when he left the middle-class comfort of his printing business to pursue his curiosity around science and politics.

Franklin's core theory of electricity from 265 years ago is consistent with modern theory. Today, scientists believe all matter is made up of

atoms. In the basic (but incomplete) picture of an atom, there are electrons orbiting around a nucleus that contains other particles including, but
not limited to, protons. Both the electrons and the protons carry an electrical charge, with electrons arbitrarily assigned a negative charge and
protons said to carry a positive charge. Inside that atom, Franklin's old
rules apply: opposites attract. Therefore, orbiting electrons are attracted
to protons, and that attraction serves as the force holding the different particles in an atom together. In most materials, atoms have exactly offsetting
positive and negative charges, so they are electrically neutral. However,
electrons in the atoms of some materials can be "taken" by other atoms.
When that happens, the atoms that have done the taking have an excess
of electrons, so the balance or equilibrium among the atoms is disturbed.
Electricity itself is a feature of the "movement" of these electrons in the
face of disequilibrium.

Franklin's stature was unmatched in colonial America. His contemporaries, whether they knew him or had only heard about him, recognized
his genius. In a letter to Franklin in his later years, George Washington
wrote, "If to be venerated for benevolence, if to be admired for talents. If
to be esteemed for patriotism, if to be beloved for philanthropy, can gratify
the human mind, you must have the pleasing consolation to know, that
you have not lived in vain."[38]

When Franklin died in April 1790, more than twenty thousand people, the equivalent of half Philadelphia's population, attended his funeral.
In France, where the National Assembly decreed an unprecedented three
days of mourning, a secretary of the French Royal Society summarized the
feelings of the nation: "A man is dead, and two worlds are in mourning."[39]
The two worlds that suffered the loss might be seen not only as the Old
World and the New but also as the worlds of science and statesmanship.

CHAPTER 2

James Watt's Steam Engine

Tamer of lightning and tamer of steam, Franklin and
Watt . . . how much further the steam engine is to
be the handmaid of electricity cannot be told.
—Andrew Carnegie (1913)[1]

If Franklin placed the first cornerstone for a world to be built on the pervasive production and use of electricity, James Watt placed the second cornerstone by perfecting the technology for the steam engine in the midst of the Industrial Revolution.

At first glance, electricity and steam seem unrelated. This reflects the uncertain dance of science and technology. Only the long view of history reveals the perfect fit between them. This is evident in two ways. First, for much of the electric power generated in the world, even today, the generation of steam is the first step. Steam is produced by burning fuels that range from coal to natural gas, or even by producing heat from a nuclear reaction. This steam is then used to drive a turbine that spins a magnet inside a coil of wire, which, in turn, generates electricity—the "flow" of electrons that Franklin "saw" and Michael Faraday would later prove.

Second, the ubiquity of steam engines in factories set the stage for the equally widespread use of electric motors because the steam engine inexorably linked factories to the use of mechanical power. This shift signaled the birth of the Industrial Revolution. Had manufacturers not become addicted to mechanical power with the steam engine, there may not have been the same opportunity to replace older systems with the electric motor—and to reap the enormous economic benefits of doing so.

More broadly, steam generation is a technology that provides us with an early template for technological change. Since technological change is offered as a solution to almost every problem in modern times, it is worth examining that template. Its evolution is a tale of perseverance, partnership, patents, and productivity.

PERSEVERANCE

In a letter to James Watt's son, well after his father's death, Dr. James Gibson, who was married to Watt's granddaughter, wrote up a family legend describing Watt's early fascination with steam. Watt was frequently ill as a child and had been sent to the countryside to recover. Gibson wrote, "While the elders were at table the subject of this éloge [Watt] was found sitting by the parlour fire holding a plate . . . to the spout of the boiling tea kettle noticing carefully the distances at which the steam condensed when it came in contact with the cold plate." Gibson wondered if this could have given Watt "the first idea of the power and elasticity of steam." Watt's own experimental notes discuss his use of a teakettle to understand the science underlying his first major breakthrough—that is, the science behind the theory of latent heat. It would be difficult to conclude that his encounter with the teakettle as a child set him off on an inevitable path to invention, but the recollection indicates an innate curiosity, the patience for observation, and an interest in precise measurement that served him well as a scientist.[2]

Born in 1736 in Greenock, Scotland, James Watt enjoyed a comfortable childhood. His father was an established merchant, shipwright, and carpenter, and became renowned for his ability to make delicate mechanical instruments used for precise measurement. Watt first revealed his precocity through an aptitude for math, reading widely, and at his workbench where he made models of mechanical devices such as pumps and pulleys. In 1753 Watt suffered two serious hardships when his mother died and his father lost his wealth. In the face of his father's financial loss, Watt needed a trade and set his sights on becoming a mathematical instrument maker. In the next year, Watt moved to Glasgow, where he was mentored by John Anderson and Robert Dick, both professors of natural philosophy at the University of Glasgow. Watt then went to London in 1755 to look for a shortened apprenticeship. John Morgan, a mathematical instrument maker, agreed to grant him a one-year instruction. In 1756, at age twenty, Watt returned to Glasgow. The protective guilds there would not let him work as an instrument maker because he had not completed a seven-year apprenticeship, but the University of Glasgow did. At that moment, the university was filled with great men, including Adam Smith, the famous economist, and Joseph Black, a noted professor of medicine and chemistry. Watt became the instrument maker for the university and was given his own shop. The door to a life of invention was opened.[3]

How does an instrument maker become a world-class inventor and scientist? The answer starts with the link between science and instrument makers during Watt's lifetime. Instrument makers were the "unsung heroes" of the Scientific Revolution. Scientists of all fields depended on the instruments that these craftsmen provided. Meteorologists needed barometers, chemists needed tools for laboratory analysis, and surgeons needed forceps and medical devices. Newton had brought on the Scientific Revolution by focusing on measuring and quantifying relationships, and instrument makers were central to giving precision to those measurements. In essence, the instrument makers provided the cutting-edge technology that enabled the explosion of scientific advancement.[4]

The University of Glasgow had a model of a valuable steam engine, designed by early inventor Thomas Newcomen, that needed repair. In

1764 professor John Anderson asked Watt to fix it. The engine was very simple: Steam was injected into a cylinder, and it pushed a piston upward. When the steam cooled, the piston fell. The movement of the piston lowered and raised a beam that performed work. In normal use, the engine pumped water out of mines; the need to pump water out of coal and other mines drove a great demand for improved steam technology. These efforts to improve and fully exploit mines helped ignite the Industrial Revolution in England before it began anywhere else.

Watt found the Newcomen engine to be terribly inefficient. He had to burn substantial quantities of coal to boil water in order to produce enough steam to get the piston to move even a few strokes. Through his own experiments, however, Watt discovered Joseph Black's theory of latent heat. Independently, Black and Watt found that when heat was initially applied to water, it increased the water's temperature; we all observe this every time we heat water for a cup of tea. However, the temperature of water doesn't increase as one might predict. Once the temperature reaches 212 degrees Fahrenheit, the boiling point of water, it stubbornly remains the same as the water turns to steam. The heat applied to increase the temperature of the water is called *sensible heat*, and the heat used to transform a liquid like water to a vapor like steam is called *latent heat*. Black termed it "latent" or "hidden" because it could not be detected by measuring temperature. Watt concluded that by condensing or cooling the steam in the cylinder of the Newcomen engine with each stroke latent heat was lost and this undermined the engine's efficiency.[5]

Watt's great breakthrough was the separate condenser. To keep the engine's cylinder hot, he would draw the steam over to a separate vessel where it could condense back into water. Watt did this by creating a vacuum in the separate condenser so steam rushed into it from the cylinder and was cooled there without cooling the cylinder. Professor John Robison, speaking in 1796, captured the productive mix of talents that enabled Watt to achieve this great breakthrough. When visiting Watt's instrument shop, Robison said, "After first feasting my Eyes with view of fine instruments . . . I saw a Workman and expected no more—but was surprized to find a philosopher." Further, Robison wrote that "every thing became Science in his hands." Science came naturally to Watt; fortune, meanwhile, was more elusive.[6]

PARTNERS AND PATENTS

Watt was neither the first nor the last inventor to need an outside investor to transform his invention into a commercial success. It was a long road from a model steam engine to a full-scale engine that could pump water from a real mine. Once again, Joseph Black, his friend and mentor, helped Watt get to the next stage, first with his own money, and then by introducing him to John Roebuck, a prospective investor. Flooding was a big problem at Roebuck's coal mines, and the Newcomen engine he used to pump water out did not solve his problem efficiently enough.[7]

Roebuck gave Watt a place at his Kinneil estate near Barrowstoness, a coastal area in the Central Lowlands of Scotland, to build a full-scale steam engine. But ultimately, Roebuck was not ready to support Watt financially. Watt encountered a series of setbacks as he pushed forward, and in 1766 he had to sell his instrument maker's shop and do surveying work in order to support his family. Around this time, Watt met Matthew Boulton, a successful businessman who would eventually become central to his success. Boulton was already famous for developing the first great manufacturing facility at Soho outside Birmingham, England.[8]

In 1769 Watt won his first patent for the steam engine. As with any law, patents have benefits and shortcomings. In contemporary society, we are more likely to think of the good side of patents: they guarantee an inventor the rights to his or her invention, and this serves as both protection and incentive. In Watt's day, the view of patents tended to be more negative: it was thought that a patent could allow a monopoly, and that patents should be few and far between. Boulton and others advised Watt to seek a broad patent for his invention, to keep others from profiting from his idea by developing a slight variation.[9]

Watt and Boulton spent thirty years struggling against rivals under the patent law of the time. Some say it was because their early patent was too vague; it covered a principle, not the application of that principle. The issue is more complicated and, once again, reflects this uncertain dance between science and technology. Did the new scientific awareness of latent heat or, more broadly, thermodynamics justify this principle-based patent? Or should Watt and Boulton have filed a patent on the specific technology of the separate condenser? Must the technology be described in enough detail that once the patent expires others can use it to advance the technology even more? The debate on whether and how to patent hindered Watt, became an affliction for Samuel Morse, and persists even today. The *Economist* observed in 2015, "Today's patent regime operates in the name of progress. Instead, it sets innovation back."[10]

A series of setbacks followed the development of and the patent for the full-scale engine in 1769. About four years later, history repeated itself for Watt when, once again, life dealt him two serious setbacks at a crucial moment. He suffered a financial loss when his investor, John Roebuck, went bankrupt, and he suffered a personal loss when his wife, Peggy, died. Watt had married Peggy in 1764, and she died nine years later when Watt was thirty-seven years old. They had six children, though not all reached adulthood.

In 1774, nine years after his breakthrough with the separate condenser, Watt moved his steam engine to Boulton's Soho Manufactory. Birmingham was a place where people "came on foot and left in chariots," a place known for free enterprise, upward mobility, and freedom of religion—the same freedoms, at the same time, that inspired Franklin and other American colonists. Historian Arnold Toynbee says that when "Adam Smith wrote his bitter criticism of the corporations, he was probably thinking of the particular instance of Glasgow, where Watt was not allowed to set up trade." This move to Birmingham, a "free town," would allow him to flourish from that freedom. In 1775 Watt and Boulton won some breathing room when Watt's patent was extended for twenty-four years by the British government.[11]

Watt and Boulton then turned to the task of making the steam engine work. For this they recruited others, including John "Iron-Mad" Wilkinson. Wilkinson was "iron mad" because he thought the iron he made could and should be used for everything. The tin cylinder Watt had been using was problematic. Wilkinson had patented a technique to bore out a smooth iron cylinder for use with cannons, and that technology, along with switching from tin to iron, helped to fix some of the problems in Watt's steam engine. In addition, Boulton recruited William Murdoch, who became Watt's right-hand man in the mining district of Cornwall, the major location where steam engines were built and used at the time. It was not long before Watt would proclaim that "the business I am here about has turned out rather successful; that is to say, the fire-engine I have invented is now going, and answers much better than any other that has yet been made." Cross-pollination of ideas and skills from one industry to another was key to technological progress throughout the history of electricity. It still is today.[12]

Watt, always the scientist and inventor, continually improved the steam engine. Boulton, the businessman, filed more patents. Granting patents was a major way for the government to influence technological advances at the time. Contrast that with today's cash and tax subsidies, which give money directly to finance inventions. Still, at least one private investor, namely Boulton, did step in to finance and direct the development of this soon-to-be-pervasive technology. And Boulton brought much more than money to Watt's invention; he had confidence and vision. When he was first approached about a joint effort with Roebuck and Watt, Boulton dismissed the initial offer to share in sales from just three counties. Instead, Boulton boldly proposed, "We would serve all the world with Engines of all sizes."[13]

PRODUCTIVITY

The impact of Watt's steam engine on industry was unprecedented. In 1913, 144 years after his first patent, Andrew Carnegie, the famed steel magnate, published a biography of James Watt. Carnegie initially declined the publisher's request, but later agreed, saying, "Why shouldn't I write the Life of the maker of the steam-engine, out of which I had made fortune?"

This statement alone, by one of America's towering industrialists, indicates the significant and long-lasting impact of Watt's steam engine on the world's industrial development.[14]

Historian Eric Robinson emphasizes that "Watt's invention, after all, has been generally recognized as the key-invention of the Industrial Revolution." Also impressive is the range of industries that were transformed; there were revolutionary changes in copper mining and fundamental changes in the cotton industry. The iron industry benefited significantly, too. The steam engine revolutionized factories by freeing them of locational constraints (e.g., proximity to water) and allowing economies of scale to be realized. After additional improvements, it also revolutionized transportation with its use in trains and steamboats.[15]

In his biography of Watt, Carnegie presented a back-of-the-envelope calculation of his subject's economic impact that illustrated how the steam engine increased the world's industrial productivity. Carnegie reckoned that by 1905 the world had about 150 million effective horsepower of steam capacity at work; presumably, his estimate included steam power both in factories and in transportation. To complete his calculation, Carnegie concluded that, since each horsepower was the equivalent of twenty-four men's labor, the world's steam capacity in 1905 did the work of 3.6 billion men. This was ten times his estimate of the number of adult men in the world at that time.[16]

Carnegie was right to judge the economic impact of invention in terms of the effect on productivity; that is, in terms of enabling one man or woman to do the work of many. Modern economic historians, while more precise in their analyses, have similarly identified and quantified the steam engine's impact on productivity. They also go one step further to estimate how that productivity growth increased economic growth. In this growth-accounting mentality, increasing the number of people and the number of machines can only move economic growth in a linear fashion: a 2 percent increase in the same kind of worker using the same kind of tools would result in a 2 percent increase in economic output. It is the productivity growth from invention that lets a nation get ahead of the game. Only with productivity growth (broadly defined) can economic growth outpace population growth so that income per person increases and the standard of living is improved.

Although more analytically diligent than Carnegie, modern economic historians still agree on one point: they credit Watt's invention with having a transformative impact. That impact can be assessed in at least two ways. In the first, there is the direct impact: the cost of steam power fell dramatically and the amount of steam power used increased dramatically. By one account, about a hundred years after Watt's breakthrough, cost had fallen almost 80 percent and steam had beaten its rivals to account for 90 percent of total power from steam, water, and wind in the United States.[17]

The second way is to assess indirect impact. This is often termed *spillover effects*, and the impact is much greater and more surprising than the direct impact. Spillover effects mean that an invention induces other changes beyond the invention itself. For example, the manufacturers who used the steam engine could adapt their technologies to each new, more efficient version. Nathan Rosenberg, an economic historian from Stanford University, and Manuel Trajtenberg from Tel Aviv University point out an example in the textile industry. In the mid-1800s George Corliss introduced important advances to the steam engine that allowed precise control over the amount of power delivered, so that a uniformity of power delivery could be ensured. In support of a patent application by Corliss, one textile manufacturer from Pittsburgh reported that "the avoidance of thread breakage attending the irregular motion of his old engine brought a savings probably equal to that from reduced fuel consumption." The steam engine, then, enabled textile manufacturers not only to stop wasting coal but also to stop wasting thread. Moreover, the uniformity of power delivery enabled some manufacturers, when customer demands warranted, to shift to higher-quality textile products because there was less risk that material would be wasted.[18]

The most compelling spillover effect of the steam engine, however, was that it removed the severe geographic constraint of using water power. If an industry used water power, it needed to be located next to a river, without any exceptions. With steam power, especially after coal use was cut dramatically thanks to Watt's innovations, manufacturers could set up anywhere. This new freedom led to other benefits. For example, an industry could get closer to its customers or to a larger source of skilled workers. Today, this liberation from geographical constraints is cited by historians as the key link between the invention of the steam engine and rising urbanization. More broadly, it shows the connection between a general-purpose technology and economic growth.[19]

WATT'S LEGACY

There is no better evidence of the importance of Watt to electricity than that found by simply looking at monthly electric bills. Those bills show the charges in terms of kilowatt-hours, or kWh. That capital W, of course, stands for Watt; the watt is the unit of measurement for electric power.

Watt's direct legacy for the history of electricity is that even today steam is a fundamental ingredient for much of the production of electricity in the world. Watt's legacy also consists of a template for technological change. The template recognizes the perseverance it takes to go from an idea, typically a scientific idea, to a successful, scalable technological innovation. Boulton was rich and powerful, and that surely helped, but that is not what makes the partnership between him and Watt worthy of attention. Boulton brought the vision and the confidence of an entrepreneur, along with political connections and a steady source of money.

James Watt died on August 19, 1819, at the age of eighty-three. Carnegie writes that, fittingly, Watt was "laid to rest" next to Matthew Boulton, his investor, partner, and friend.[20] It comes as no surprise, then, that in March 2011 the Bank of England issued a new fifty-pound note featuring portraits of Matthew Boulton and James Watt, along with images of both the steam engine and Soho Manufactory.[21]

As we have seen, patents, lawyers, and courtrooms were of significant concern in the early years of science and invention; litigiousness is not just a modern ailment. Patent fights were common from the beginning. Boulton and Watt reported that they could list thirty-four patents related to the steam engine between 1698 and 1793 beyond those of Watt himself; they later said there were probably a dozen more. Watt himself gave a blunt response as to why there were so many patents when he said, "I do not think we are safe [from patent challenges] a day to an end in this enterprizing age. Ones thoughts seem to be stolen before one speaks them. It looks as if Nature had taken up an adversion to Monopolies and put the same things in several heads at once to prevent them." Watt was not alone in his obsession with the steam engine at a time when the rise of steam-dependent industry was reaching a "fever pitch."[22]

Ultimately, one cannot measure the impact of advances in the science and technology of electricity only by the change in the cost of electricity. The impact includes truly revolutionary change in broad-scale economic ways and means. And it is the impact on industrial productivity that made Watt a man of his time as well as a man of the future. Watt was a man of his time because his perfected steam engine drove the great manufacturing surge of the Industrial Revolution. He was a man of the future because the widespread adoption of his steam engine set the stage for the rapid and pervasive adoption, more than one hundred years later, of the electric motor. That electric motor was beyond Watt's horizon, but it would be created and perfected by Watt's inheritors in science, Michael Faraday and Nikola Tesla.

CHAPTER 3

Faraday's Faith
(and Maxwell's Equations)

From this simple laboratory toy was to come
the whole of the electric power industry.
—Pierce L. Williams (1965)[1]

Michael Faraday built ambitiously upon the cornerstones set by Franklin and Watt. His scientific experimentation was central to the discovery of the link between two of nature's most important forces: magnetism and electricity. This link led Faraday to develop the two fundamental technologies for the nascent electric power industry. He invented the first electric motor in 1821, and ten years later he invented the first "dynamo," or electric generator.[2]

A further triumph came later in his career, when he laid the foundation for modern theories about electromagnetic waves. Wave theory is still central to modern science and underlies much of modern telecommunication. If Franklin had the genius to see the unseen, then Faraday had the genius to see that the empty space between us is not empty; it is filled with electromagnetic waves.

Faraday, however, has not enjoyed the sustained fame of Benjamin Franklin. He did not have the immediate technological impact of James Watt, either. While his scientific breakthroughs were undeniably transformative, he is a somewhat unrecognized hero in the history of science. Two biographers have seen a need to resurrect Faraday's reputation. Writing in 1965, L. Pearce Williams hoped "to reveal the full dimensions of his genius and to place him more accurately in the mainstream of the history of science." James Hamilton, with his more recent biography in 2002, said he wanted "to look at Faraday in a way that gives a new and I hope urgent priority to his role as a central figure in the history of western culture."[3]

With the long-standing debate on the presumed conflict between science and religion as a backdrop, Faraday opens yet another important dimension to this story: he combined his deep faith in God with his deep commitment to science and created a seamless whole. Faraday's working relationship with James Clerk Maxwell opened still another dimension. With great respect, Maxwell took Faraday's thinking about the link between electricity and magnetism, brought to it the discipline and imagination of a new mathematics, and thereby revolutionized the science of physics. Maxwell's equations embody one of man's greatest scientific advances.

THE KINDNESS OF STRANGERS

Like Franklin, Faraday was trained as a printer. In 1804, at the age of thirteen, Faraday was apprenticed to a London bookbinder by the name of George Riebau. This was formative for Faraday. Riebau gave him access to books and, later, introductions to some of the most important scientists of the day. As Franklin did before him, Faraday took advantage of his job to read many of the books he handled. As he bound the *Encyclopaedia Britannica*, Faraday read a 127-page article on electricity by James Tytler. Tytler's article recounted the history of electricity substantially based on Joseph Priestley's then-famous book *The History and Present State of Electricity*, published in 1767, but Tytler went well beyond what Priestley had to say in his own view on the science of electricity.[4]

Instead of thinking of electricity as a fluid, as in Franklin's theory, Tytler thought of it as a "vibration," like the "waves produced by a stone being cast into a pool of still water." In his scientific worldview, Tytler promoted the idea of the unity of major forces; light, heat, and electricity were "identical in substance." Faraday did not buy into the Tytler theory completely when he first read it, but it is remarkable how many raw elements of Tytler's theory emerge later in Faraday's work. Tytler's article inspired Faraday to conduct his first experiment, and even though the "lab equipment" was nothing more than two bottles, Faraday had to wait until their price dropped before he could afford them.[5]

When Faraday's father died in 1810, Riebau became a father figure to him. The printer encouraged Faraday to attend public lectures by scientists of the day. Faraday first went to the lectures of John Tatum, a teacher, philosopher, and silversmith. Tatum's lectures covered a wide range of topics, including electricity, optics, geology, experimental mechanics, astronomy, and chemistry.[6]

Faraday worked hard to understand what he heard at these lectures and to gain a broad knowledge of scientific fields. He took exhaustive notes as Tatum spoke, and then, after he returned home, he would rewrite the notes to deepen and expand his comprehension. The Tatum lectures paid off in terms of giving Faraday a basic scientific education but resulted in another spectacular advantage, too. Riebau showed a copy of Faraday's bound notes to an acquaintance, who was impressed enough to get Faraday one of the hottest tickets in town: an invitation to the Royal Institution to hear the most popular English scientist of the day, Humphry Davy.[7]

THE UNIVERSITY OF EUROPE

The impact of Davy's lectures on Faraday was substantial. Davy, a chemist, was famous for his work with electrolysis, the decomposition of a substance by running electricity through it; Napoleon had even awarded Davy a French science medal for this work. In 1800 chemists and other students of electricity employed a powerful new tool that provided a steady source of electricity. The tool was called a "voltaic pile," a rudimentary but, for the time, powerful battery. The invention of the voltaic battery began in 1794 when Luigi Galvani, an Italian biologist, discovered what he believed to be a new form of electricity; he called it "animal electricity." Galvani had been conducting a series of experiments on the effects of electricity on animals. He noted that when his steel scalpel came into contact with the brass fixture holding a frog's leg the frog's leg twitched, and he attributed the twitch to electricity inherent in the frog—hence the term "animal electricity."[8]

Galvani's colleague Alessandro Volta admired Galvani's experiments, but reached entirely different conclusions as to why the frog's leg twitched. In a series of experiments, he showed that the source of the electricity was actually the two dissimilar metals coming into contact, and that those metals were the *active* source of electricity. Volta invented the voltaic pile by stacking plates made of these dissimilar metals with a wet conductor between each layer of metal. This voltaic pile was based on Volta's "contact theory," which found that "when dissimilar materials come into contact, one becomes positively and the other negatively charged." With this, Volta gave those who studied electricity a great source of electric current flowing between the positive and negative charges—and also gave the science of electrochemistry a great boost.[9]

Faraday took copious notes at Davy's lectures, and, once again, those notes won him recognition. Boldly, Faraday presented Davy with a copy of his bound lecture notes, and Davy soon hired Faraday as the lab assistant at the Royal Institution. Being a lab assistant was a lowly post, but it gave Faraday a chance to work with Davy on a variety of experiments. Davy decided to travel to France to receive the science medal he had won and asked Faraday to accompany him. Faraday happily agreed, but it became a wearying experience. At the last minute, Davy's valet backed out. Faraday was asked to serve as Davy's valet as well as his lab assistant. He had not expected having to take out Davy's clothes or deal with hotel employees or other such tasks. Moreover, Davy's wife looked down on Faraday because of his humble start, and treated him badly.[10]

Still, the journey with Davy was truly an education. It was, as biographer L. Pearce Williams suggests, a degree earned at the "University of Europe." Faraday got to see places that typically would be seen only by the rich. More importantly, he got to meet famous scientists, including André-Marie Ampère, one of the early contributors to the understanding of electromagnetism. Faraday even participated in Davy's lab analysis showing that iodine was a new element. After eighteen months of travel, Davy and Faraday returned to London. While he still worked with Davy and other scientists, Faraday was ready to strike out on his own.[11]

FARADAY'S FIRST TRIUMPH: THE ELECTRIC MOTOR

In 1820 Davy, Faraday, and the world learned of Hans Christian Ørsted's discovery of the interaction of electricity and magnetism. Ørsted, a Danish scientist, took his first step toward this discovery by accident while conducting a lecture and using a battery. When he connected the battery to create an electric current in a wire, he saw that the needle of a compass on a nearby table moved. In his words, the electricity in the wire "disturbed" the compass needle. With this, Ørsted showed that there was attraction (or repulsion) between the electricity in the wire and the magnetism in the compass needle. Scientists at the Royal Institution replicated Ørsted's discovery and pushed for explanations. William Wollaston, another scientist at the Royal Institution, spoke widely about it. The French scientist Ampère also jumped into the fray. But a complete explanation was not immediately clear to anyone.[12]

In 1821 Richard Philips, editor of the *Annals of Philosophy*, asked Faraday to sum up the state of knowledge on electricity in a series of articles. The work Faraday did for these articles led him to the discovery of "electromagnetic rotation"—or, in technological terms, the invention of the electric motor. Ørsted had shown that electricity has an effect on magnetism, and Faraday would go further to show that there is a mutual effect. When Faraday brought a permanent magnet toward a wire carrying an electric current, that wire rotated around the pole of the magnet. This meant that electricity, in the presence of a magnet, could be converted into mechanical energy (it would make a wire rotate) and thereby do mechanical work. All the electric motors that have been used in factories and homes since that day are based on Faraday's discovery of the basic relationship between electricity and magnetism.[13]

Faraday wrote two papers that were published in 1821. The first appeared in the *Annals of Philosophy* under the title "Historical Sketch of Electromagnetism." There he attempted to explain Ørsted's and Ampère's theories. It was as he finished the first paper that he conducted the experiments that would lead to his first triumph. Faraday then wrote a sixteen-page article about his original research titled "On Some New Electro-magnetical Motions, and on the Theory of Magnetism." It was published in the *Quarterly Journal of Science* and he signed it "M. Faraday, Chemical Assistant in the Royal Institution." Almost immediately, as biographer James Hamilton writes, Faraday was "crucified by the scientific world" because he had not acknowledged Davy and Wollaston, with the implication being that the "servant ran off with the silver."[14]

A sad consequence of this affair was that it led to the open betrayal of Faraday by his mentor, Humphry Davy. The Royal Society wanted Faraday as a member, but Davy tried to block his membership. Davy alleged that Faraday had stolen his ideas, not just on electromagnetism but also on the liquefaction of chlorine. But Davy was alone in his attempt to keep Faraday out of the Royal Society. In 1824 Faraday was voted in with only one no vote. The episode shows that since science is done by human beings, it is not always as objective as might be thought or hoped.[15]

FARADAY'S SECOND TRIUMPH: THE ELECTRIC GENERATOR

Ten years after he invented the first electric motor, Faraday invented its mirror image—the electric generator. The electric *motor* uses electricity to create mechanical action. The electric *generator* does the reverse; it uses mechanical action to create electricity. Both technologies have at their core the intimate link between magnetism and electricity. In a series of papers published from 1831 onward, Faraday described his lab work and announced, "These considerations, with their consequence, the hope of obtaining electricity from ordinary magnetism, have stimulated me at various times to investigate experimentally the inductive effect of electric currents. I lately arrived at positive results."[16]

Positive results indeed. Through his simple and systematic experiments, he proved that the movement of his ordinary magnet inside a coil of metal wire would generate electricity. That is the principle on which almost all electric power plants work today and the principle that links Watt to electricity—his steam engine is what moves the magnet back and forth.[17]

To fully perceive the simple brilliance of Faraday's insight, it is worth delving into a few of his experiments. Faraday first wrapped two sets of wires around a cylinder of wood. One of the wires was connected to a battery, and when the battery was turned on it would send an electric current through that wire; Faraday called that component the *inducing wire*. The other wire was connected to a galvanometer, a device used to detect electric current. Faraday called this part the *wire under induction*. Faraday wanted to see the electricity in the inducing wire create electricity in the wire under induction. With a small battery (and therefore a small electric current), Faraday reported that he found "not the slightest sensible deflection of the galvanometer needle." This failure was a necessary part of his success because it led him to define two new experiments.[18]

Faraday then turned to experiments with what might best be pictured as an iron doughnut—a welded ring of soft iron. Faraday wound separate wires around each half of the doughnut. The wire on the left side was connected to a galvanometer and the wire on the right side was connected to a battery. The hoped-for effect was that the inducing wire on the right (the one with the electric current from the battery) would magnetize the iron doughnut and that in turn would create an electric current in the wire under induction on the left side. Faraday did indeed find the effect on his galvanometer "to a degree far beyond what has been described when with a battery of tenfold power helices *without iron* were used." Importantly, only a temporary effect was detected with the galvanometer; that is, it detected electric current in the wire under induction only when the battery was started and then again when the battery was stopped.[19]

Finally, Faraday turned to an ordinary magnet. He pushed this magnet into a hollow coil of wire and his galvanometer detected the creation of an electric current. There was no battery involved here, as he was looking for the magnet to induce electricity in the coil. As in his other experiments, however, the induced current was temporary; if he had left the magnet sitting idle inside the coil of wire, the galvanometer needle would have returned to its original position. When he pulled the magnet out of the coil of wire, again his galvanometer needle was deflected, in the opposite direction.[20] Faraday concluded, "The various experiments of this section prove, I think, most completely the production of electricity from ordinary magnetism." Indeed, moving a magnet within a coil of wire is the essence of generating electricity at power plants to this day.[21]

There was little fanfare for this silent revolution created with the invention of the first electric generator. On one occasion, the prime minister, Sir Robert Peel, visited Faraday's lab. Apparently, Peel asked Faraday what the ramifications of his invention might be, to which Faraday replied, "I know not, but I wager that one day your government will tax it." Today we know that the experiments in Faraday's lab have been of enormous importance, even beyond being a source of tax revenue. Faraday discovered magneto-electric induction and invented the first electric generator. As Williams concludes, "From this simple laboratory toy was to come the whole of the electric power industry."[22]

FARADAY'S THIRD TRIUMPH: ELECTROMAGNETIC WAVES

Even after two triumphant experiments, Faraday was not finished with his contribution to science. Through the experiments described earlier, Faraday showed that magnetism could induce electricity; that is, a magnet moved back and forth inside a coil of wire created a sustained, detectable electric current in the wire. Faraday later showed that electricity itself could induce electricity; an electric current in one wire could induce a detectable electric current in a nearby wire. This electric-electric induction would lead to the technology of transformers, which are used widely today to raise and lower the voltage of electric currents for more efficient transmission of electricity and safer use. For his work on induction Faraday was awarded the highest prize of the Royal Society, the Copley Medal, which Benjamin Franklin had also won.[23]

To achieve his third triumph, Faraday's research was aimed at another essential question of science: How is force transmitted? Put simply, how does magnetic force get from the magnet to the wire, and how does the electric force in one wire get to the other wire? In Newtonian science, the answer can be summarized in the phrase *action at a distance*. Gravitational force was thought to act at a distance. Gravity as a force depended only on the mass of the objects affected and on the distance between them. For Newtonians, empty space, often referred to as *aether* or *ether*, filled the distance between these objects. The science of electricity reflected this Newtonian view in Coulomb's Law, which said that electrical force between objects depended on the quantity of the electric charge in the object and the distance between the objects. (Charles-Augustin de Coulomb was a French physicist born in 1736.)[24]

Faraday's theory broke with the mainstream and revolutionized science. He refuted action at a distance, a core of Newtonian science. He claimed that force was not transmitted through empty space but rather was transmitted particle to particle in that space. His theory eventually became mainstream through James Clerk Maxwell's and Albert Einstein's respective works on modern field theory. Through the work of these later scientists, Faraday's theories laid the foundation for the study of electromagnetic waves which came to underpin all modern telecommunication technology.[25]

Faraday's path to these groundbreaking theories is fascinating for anyone interested in scientific discovery and innovation, in any age. For example, his path began with chemistry, not physics; cross-pollination like this across sciences or businesses continues to be an avenue for important progress.

Faraday's most "spectacular" experiment concerning action at a distance happened in January 1836, in the theater at the Royal Institution. He built "a great cube with an edge of 12 feet" and a covering made up of good "conducting material." The "cage" was sitting on insulated "supports" so that "it could be charged with electricity until sparks flew from it." Inside that cage sat Faraday himself, "in perfect electrical peace" with equipment that showed no sign of electricity inside the cage. The point of the cage, and many other less dramatic experiments, was that the power of attraction or repulsion in "electrified bodies or in the electric spark does not lie in those bodies." Rather, "it is seated in the surrounding space," that is, in the empty space between the electrified objects. It took supreme confidence to perform this seemingly dangerous experiment, but Faraday was no stranger to belief in the unseen in his scientific or personal life.[26]

A SEAMLESS LIFE

Religion was "one of the first and foremost influences" on Faraday. He belonged to a sect founded by Robert Sandeman, a preacher out of Scotland. The Sandemanians, as his followers were called, were part of a wave of change in a time of splinter groups that ran through England. The biographer James Hamilton says, "The teachings of the Bible were literally and strictly true in Sandemanian belief," and believers sought balance under one power: God. While Faraday claimed that he kept science and religion separate, Williams concludes, "His deepest intuitions about the physical world sprang from this religious faith in the Divine origin of nature." Since there was one God, Faraday hoped there was one force in nature; that is why finding the link between the forces of nature was central to Faraday's science.[27]

Faraday was never hesitant to talk about God and science at the same time. He spoke of the forces he discovered as "the second causes," as compared to the "one Great Cause" who "works his wonders and governs this earth." Williams opined that Faraday believed that there was no higher calling for a man than to seek "knowledge of God's creation." In a speech on the importance of education in the sciences, Faraday summed up both his scientific and religious beliefs. He said that, yes, electricity can be called "wonderful—beautiful," but "only in common with the other forces of nature." He expressed his view of man's dominion over nature when he said that while nature's mystery is compelling, "it is under *law*" and the "taught intellect can even now govern it largely." Finally, he reflected his faith when he said education in science "is rendered supereminent in dignity" because "it conveys the gifts of God to man."[28]

Humility was a by-product of Faraday's faith. Faraday thought himself blessed to do the work he did. This attitude is evident throughout his correspondence and work. Faraday once said, "In all kinds of knowledge I perceive that my views are insufficient, and my judgment imperfect. In experiments I come to conclusions which, if partly right, are sure to be in part wrong." His acceptance of human error, including his own, gave him persistence as a scientist.[29]

As Faraday aged, however, he increasingly—and tragically—struggled with memory loss. Faraday described how his memory "both loses recent things and sometimes suggests old ones as new" and how "the past is gone, not to be remembered, the future is coming not to be imagined or guessed at, the present only is shaped in my mind." By 1860, when he was near seventy, he wrote, "When I try to write of science, it comes back to me in confusion." A German physicist of that era, G. C. Lichtenberg, describes the sad experience of memory loss: "As long as memory lasts . . . a crowd of people are working together as a unity: the twenty-year-old, the thirty-year-old and so on. As soon as it fails, one begins increasingly to be alone and the whole generation of selves stands back and mocks the helpless old man." In 1861 Faraday offered his partial resignation to the Royal Institution. In his letter he looked back on his forty-nine years of membership and thanked the managers of the Royal Institution, saying that he had "been most happy in your kindness." Faraday concluded that he was in a time of "gentle decay," but still felt that the "evening of life" was a "blessing."[30]

Faraday had not been made rich by his discoveries, since, unlike Watt, he never concerned himself with patents and was not driven by a desire for material gain. Fortunately, his contribution to science was recognized in a modest way by the British government, and in 1858 Queen Victoria rewarded his lifetime of devotion to science by giving Faraday the use of a house at Hampton Court Palace. It would become his residence for the remainder of his life.[31]

Taken on its own, Faraday's science is both impressive and important, but, in truth, his science cannot be fully understood without consideration of his spiritual faith. His faith motivated him to look for an invisible force, and it simultaneously compelled him to believe in what he could not see. It also gave him courage in both his science and his life, especially in his final illness. In a letter of 1861 Faraday wrote of death, "Such peace is alone in the gift of god, and as it is he who gives it, why shall we be afraid." Faraday died on August 25, 1867.[32]

MAXWELL'S EQUATIONS

Maxwell's deepening of Faraday's understanding of the link between electricity and magnetism was enough to earn him lasting fame among scientists, but his work went even a step further. He found that light waves were also electromagnetic waves, and that all electromagnetic waves travel at the speed of light. Thus, Maxwell combined electricity, magnetism, and light into a single unified theory. He is among the few scientists from the earliest days of science whose star only shines brighter today, as scientific advances continue to be based on his theory. Even Albert Einstein gave fulsome credit to Maxwell when tracing the source of his own great work, saying, "One scientific epoch ended and another began with James Clerk Maxwell."[33]

A Prodigy

Born in Edinburgh, Scotland, in 1831, Maxwell entered the world of science at the age of fourteen. He started with the simple game of drawing circles using a pencil attached by a piece of string to a pin poked into a fixed point on a piece of paper. He then added a second pin to draw an ellipse and then added more pins to guide his drawing. Then, taking the game a step further, Maxwell developed a mathematical equation to describe the curves he drew. His father showed the work to a friend, James Forbes, who was a professor of natural philosophy at Edinburgh University and would later become Maxwell's mentor. Forbes and a colleague in the mathematics department researched whether any such work had been done before, and they found that it had been covered by the famous French mathematician and philosopher René Descartes. Shortly thereafter, Professor Forbes read Maxwell's paper to the Royal Society of Edinburgh, since Maxwell was too young to attend.[34]

Maxwell was given ample exposure to the cutting-edge science and technology of the mid-nineteenth century. His uncle took him to the workplace of a noted optician, William Nicol, and Maxwell also visited his cousin who was married to Hugh Blackburn, a professor of mathematics at Glasgow University. Blackburn was friends with William Thomson, a professor of natural philosophy who would become one of the best-known scientists in England. Thomson and Faraday would become two of Maxwell's greatest influences.[35]

In 1847, at the age of sixteen, Maxwell began taking classes at Edinburgh University. There he learned various approaches to the study of science and read widely on mathematics. Among the scientific approaches was one espoused by David Hume, who believed that mathematics should be the foundation of all scientific truths and that any theory that was not backed by math was conjecture. Two of Maxwell's mathematical investigations were produced as papers and, once again, Maxwell was too young to read his work to the Society, so it was presented for him. Three years later, in 1850, Maxwell left Edinburgh for Cambridge University, where he thrived at Trinity College. In his university studies, Maxwell took a keen interest in the science of electricity and magnetism.[36]

"The Man Who Changed Everything"

Albert Einstein, writing for the centenary of Maxwell's birth, made it clear that Maxwell and Faraday together brought about radical change in the science of physics. Einstein said, "The greatest change in the axiomatic basis of physics—in other words, of our conception of the structure of reality—since Newton laid the foundation of theoretical physics was brought about by Faraday's and Maxwell's work on electromagnetic phenomena." Einstein went on to explain that Newton's system characterized reality as the interaction of objects ("material points" and the "forces" between those points). Maxwell came along and changed the mainstream view of nature, from the interaction of objects to the pervasiveness of electromagnetic waves; Faraday's and Maxwell's ideas thereby revolutionized the science of physics.[37]

Over a ten-year period from 1855 to 1865 Maxwell wrote three papers that greatly advanced man's knowledge of the relationship between electricity and magnetism, and fundamentally changed the science of physics. In the words of his biographer Basil Mahon, Maxwell can justifiably be looked at as "the man who changed everything." In 1855 and 1856, Maxwell published, in two parts, the first of these great papers on electricity and magnetism, titled *On Faraday's Lines of Force*. In 1861 and 1862, he published his second paper, *On Physical Lines of Force*, in parts, and in 1865 he published *A Dynamical Theory of the Electromagnetic Field*. Eight years later, in 1873, Maxwell published a book-length masterpiece titled *A Treatise on Electricity and Magnetism*.[38]

Maxwell's first argument was directly allied with Faraday's, as its title suggests. Faraday had said that the space between two objects was filled with what he termed "lines of force"; most scientists opposed the idea. Maxwell, an avowed follower of the "experimental genius" Faraday, thought that there had to be easy-to-find physical evidence of this phenomenon, and there was. To see these lines of force, one need only sprinkle iron filings around a magnet and observe the clear pattern in which those iron filings array themselves around the magnet. When Maxwell wrote his first paper, *On Faraday's Lines of Force*, Faraday wrote to thank him and to say he had been impressed that the lines of force theory had withstood mathematical scrutiny.[39]

A Story Well Told

It is tempting to say that Maxwell simply translated Faraday's ideas into math and that he did so to make Faraday's ideas acceptable to mathematical scientists of the day. Scholar Thomas K. Simpson, in his unique take on the *Treatise on Electricity and Magnetism* (*Figures of Thought*, published 2006), found that Maxwell's math was much more than just another interpretation; it was a "new way of looking at physical processes . . . a truly poetic conception." Math drives Maxwell's story forward, deepens it, and in the end disciplines the story and lends it added credibility. Indeed, it is math that carries the story to its surprise ending—that light itself is an electromagnetic wave.[40]

In his treatise Maxwell proved that Faraday was right in believing that the "empty" space between objects was not actually empty. To explain the importance of space not being empty, picture two people on opposite ends of a trampoline. The first person flexes her legs, leaps into the air, and comes down hard on the trampoline. The person at the other end of the trampoline bounces. Why does the second person bounce? Was it Newton's action at a distance—does the weight of the first person and distance from the second person somehow explain that bounce? Or was it the first person's effect on the fabric of the trampoline—the "empty space" or "ether" in between—that made its way to the second person and caused that bounce? Faraday and Maxwell would have concluded that the second person bounced because of the first person's effect on the fabric of the trampoline; that is, the action takes place in a seemingly empty space, which in fact is not empty.

To focus this concept on electricity and magnetism, Faraday and Maxwell believed that the current in the wire was not the direct cause of the movement in the magnetic needle. Rather, it was the current's impact on the space in between that caused the needle to move. So what is in that empty space? What is the equivalent of the fabric of the trampoline? The empty space is in fact filled with an intricate fabric of interwoven electric and magnetic fields. To see that, Maxwell made another fundamental change from the Newtonian paradigm: he thought in terms of energy. The first person flexing her legs and jumping created the energy that affected the fabric of the trampoline, and that effect made the other person bounce. It was not force—not the Newtonian notion of attraction or repulsion—that caused the other person to bounce.[41]

One of the unsolved mysteries in Faraday's work showed how important it was for Maxwell to focus on energy rather than force. Recall that Faraday got the effect he wanted—electricity created by a magnet—only with *movement* of the magnet inside the coil of wire. Maxwell wanted to explain this, and he did so by distinguishing between *potential* energy—the stored energy in the person's legs at the position when she was poised to leap—and *kinetic* energy—energy associated with her motion. It is kinetic energy that has the "power" to do work and cause force. That is why Faraday could only create electricity while the magnet was in motion.

Maxwell's perspective on electromagnetic waves revealed a fundamental difference between electricity and other forms of energy. Think of electricity in comparison to gasoline for cars. The owner of the car fills up the tank and drives that energy away. Visualize gasoline flowing out from the nozzle at the gas station into a car's gas tank. If an electric socket is like that nozzle, why doesn't electricity continually spill out?

It is because electricity is all about the movement of electrons, not about the movement of fluid. It is better to think of a long row of the electrons lined up in close rank like dominoes. Flick a finger to knock down the first and all the others soon follow. The energy is not transferred solely by moving the first domino but rather by jostling all the dominoes already standing in a row. Only when a lamp at a home or an electric motor at a business is plugged into the wall socket does the "flow"—the jostling—of electrons occur. Only when there is a source and a use at the same moment is electric power consumed.[42]

Lagrange Offers a Perfect Fit

Maxwell did not start from scratch with the math that enabled him to "change everything"; he drew on the mathematical work of others. Most important, Maxwell was led to Joseph-Louis Lagrange, a mathematician in France. Lagrange had developed a complete mathematical depiction of a system ideally suited to the theory Maxwell wanted to propose. With Lagrangian math, the goal is to measure what happens to total energy when a small change is made—for example, a small change in the electric current put into a loop of wire. How total energy changes with a small change in one quantity is captured in what mathematicians call a *partial derivative*. Maxwell acknowledged that the effect of a single change depends on the configuration of the whole system at that moment.[43]

Maxwell moved past Newton with Lagrangian math because the mass of an object became far less important as a principal way of comprehending the physical world. Indeed, Simpson concludes that Newton's dependence on mass—his knowledge of the objects that caused force—was no longer necessary for Maxwell. Motion could be explained by the changes in objects, but Maxwell didn't need to know which objects; he could abstract from their actual nature.[44]

By using math that moved beyond Newton, Maxwell also moved past the smart and curious lay reader. To help explain these revolutionary views without the math, Maxwell provided a "thought experiment" involving bell ringers. Imagine a room filled with bell ringers, each with their hands on their own rope that controls a bell above. All the ropes pass through a ceiling with holes in it, so the bell ringers cannot see the actual bells they are ringing, or whether the bells are linked.[45]

How does the bell ringer metaphor help the layperson to understand Maxwell's astonishing science? What do the bells and the bell ringers represent? First, it tells us that we do not need to know the complete intricate mechanics of the interconnected bells to understand the music played. Just as we do not need to know the intricate mechanics of nature to understand the total potential and kinetic energy produced in Maxwell's system.

Second, through experiments we can know the music played when the ropes are at particular positions and pulled with particular velocities. The same knowledge can be gained about the changes in total energy, when the quantities such as electrical charge have particular positions and velocities. Third, however, the effect of one bell ringer pulling on a rope depends on the status of all other bells at that moment in turn. That's true for a change in total energy caused by an electrical charge, too.[46]

Fourth, it is a dynamic system—the bell ringer must pull on the rope to have an effect, just as Faraday had to push and pull a magnet inside a coil of wire to produce electrical energy. In this way, Faraday was computing Maxwell's partial derivations.[47]

Maxwell Makes Waves: His Surprise Ending

The revolutionary impact of Maxwell's work with Lagrangian concepts cannot be overstated. For Maxwell, the most relevant implications were that his dynamic system view was the only way to understand electromagnetic waves. To clearly understand Maxwell's wave theory, one can start by recalling the Ørsted and Faraday effects: Ørsted found that electricity had an effect on magnetism, and Faraday found that magnetism had an electrical effect. More specifically, Ørsted found that a changing electric charge was surrounded by a magnetic field, and Faraday found that a changing magnetic field "gives rise to an electric field." If it is true that an electric field gives rise to a magnetic field and a magnetic field gives rise to an electric field, then electric and magnetic fields must extend out into space in an endless game of leapfrog. It is this weaving together of the Ørsted and Faraday effects that allows a full conception of electromagnetic waves.[48]

Maxwell pictured the electromagnetic wave as a smoothly oscillating, or sinusoidal, wave of the electric field coupled with the sinusoidal wave of the magnetic field. Picture the sinusoidal wave as a person sliding down a firehouse pole; instead of coming straight down, the person twirls around the pole. This vision of the way an electromagnetic wave propagates into space became the inspiration for the surprise ending. Maxwell believed that all such electromagnetic waves should propagate at the same speed. However, speed is a matter of observation, not theory, so speed had to be determined through experimentation. Maxwell determined that the speed of propagation of an electromagnetic wave was the speed of light, and thus light itself must be an electromagnetic wave. With this great insight, in 1861 Maxwell linked three of the most important phenomena in the physical world—electricity, magnetism, and light.[49]

FARADAY'S AND MAXWELL'S LEGACIES

Faraday's and Maxwell's legacies rest on the importance of science to electricity, the importance of character to scientists, and the importance of the study of electricity to scientific advance. Regarding the importance of science to electricity, Faraday's discoveries of the basic electric motor in 1821 and the basic electric generator in 1831 added significantly to the scientific and technologic cornerstones for the electricity business originally placed by Franklin and Watt. This dependence on science distinguishes electricity from most other types of routinely used energy, including the petroleum used to make gasoline to run cars, the natural gas used to heat homes and as feedstock for chemicals, and the coal used to fuel power plants. All are essential sources of energy and all have driven important advances in technology. None of these other forms of energy, however, in their creation and use, require a scientific understanding so intimately connected with the fabric of nature.

As to the importance of character to scientists, it is interesting that Faraday and Maxwell could not have been more different in their respective paths to scientific achievement. Faraday had little formal education and relied instead on public lectures and humble experimentation. In sharp contrast, Maxwell grew up with the great universities and royal societies of England and Scotland available to him for knowledge and support. Still, Maxwell, the young mathematical prodigy, sought out Faraday, the workbench experimenter with vision. Indeed, Maxwell seemed to give Faraday credit for his surprise ending, the electromagnetic theory of light. Maxwell said, "The electromagnetic theory of light, as proposed by him [Faraday], is the same in substance as that which I have begun to develope [*sic*] in this paper, except that in 1846 there were no data to calculate the velocity of propagation." So, too, both showed courage in their willingness to go in a direction that was at odds with the accepted theories of the day. They proved that empty space is not empty but, rather, is filled with electromagnetic waves—this undermined respected Newtonian theory. The scientific process should be celebrated in history only when it is squarely focused on finding the truth, unimpeded by other aims. Faraday and Maxwell had the character to do it right.[50]

At that time, the study of electricity was the core of scientific study. It was not driven purely by a narrow interest in understanding electricity but by a broad interest in understanding all the forces of nature. To put it another way, the study of electricity was not just a stop along the path to scientific advance; it was the path itself. It is in this sense that, when it came to science, the Age of Franklin was a time of awe and discovery.

Faraday and Maxwell were inspired, dedicated scientists whose discoveries are very much still with us. The technologies they developed or inspired have supported the breadth of the electricity and wireless telecommunication businesses from their very start through the present day. The scientific theories they grappled with revolutionized physics and, more broadly, man's fundamental understanding of nature. All the major advances in physics that came after, up to and including those by Albert Einstein, can be traced back to Faraday and Maxwell. And the ethical and humble manner in which they conducted themselves should inspire all scientists and all people to selflessly pursue the truth.

PART TWO

The Age of Edison

Let There Be Light (and Power)

CHAPTER 4

Samuel Morse's Telegraph

And Science proclaimed, from shore to shore,
That Time and Space ruled man no more.
—Rossiter Johnson (1872)[1]

Samuel Morse reaped the first fruits of man's discovery of the science behind
nature's electromagnetic wonders. Morse's telegraph was a simple, elegant application of Faraday's findings on the link between electricity and magnetism. That link is an electric current sent through a coil of wire that would magnetize an iron bar placed inside the coil. At the receiving end of the telegraph, on-and-off magnetizations drove a lever, and a pencil attached to that lever "wrote" the resulting dots and dashes on scrolling paper. Morse's code translated the dots and dashes into letters and numbers. With the widespread adoption of the telegraph in the late 1800s, electricity began to power modern communication and drive us toward instantaneous communication, bringing us all the way to today's smartphones.[2]

As did Watt's experience with his steam engine, Morse's invention of the telegraph shows the importance of persistence, partners, patents, and productivity. This is true for any new technology, but Morse's story involves a heightened intensity and perversity. Morse was exceptionally persistent in bringing together all the elements needed for the first commercial success with the telegraph. And yet he pursued that success despite the fact that creating the telegraph was not at all his original ambition. At first he wanted to be a famous painter.

Morse's lifelong defense of his patents is a much bigger part of his story than it was for Watt, and it puts the spotlight on the role of the courts. In 1854 the United States Supreme Court ruled that Morse "was the original and first inventor" of the telegraph, regardless of competitors who disputed the claim. It was an early instance of the court playing an outsized role in determining who received the fame and fortune from a new technology.[3]

Economy-wide productivity mattered as much for Morse's telegraph as it did for Watt's steam engine. The impact of the telegraph went far beyond the cleverness of the invention itself. The author Tom Standage aptly refers to the telegraph as the "Victorian Internet"; like today's internet, the telegraph connected the world. So, too, by enabling markets and businesses to increase in scope and scale, the telegraph had an economic impact well beyond that measured by the money spent to build the system.

Morse was the first to show us that electricity is the invisible force driving communication technology. Morse's innovation shows us, further, that modern communication is the foundation on which a global economy is built. And his life shows us that great men are not always great, nor always right on crucial issues of the day.

RELIGION, PAINTING, DISAPPOINTMENT

Morse was not a genius—scientific or practical—as were Franklin, Faraday, and Maxwell. His life was a hard slog forward as he staked a claim on a technology that many others claimed to have invented before him. Watt confronted this kind of challenge, too, but he had the good fortune of having Boulton as a partner. Still, it is hard to feel sorry for Morse. He was born into a well-educated, well-to-do family—unlike Franklin and Faraday. He ultimately won the great patent war over the telegraph. And it would be hard to muster sympathy in any event for a person choosing to take the positions he took on slavery and religious freedom.

On April 30, 1789, Morse's father, Jedidiah Morse, became the pastor of the Congregationalist Church in Charlestown, Massachusetts. In due course, Jedidiah married, and two years later, on April 27, 1791, Jedidiah and his wife, Elizabeth, had their first child, whom they named Samuel Finley Breese Morse. Jedidiah had gained an international reputation as the author of geography texts such as *The American Universal Geography* in 1793, and he had occasionally rubbed elbows with the likes of Benjamin Franklin and George Washington. His fervent interest in politics and his religious bias against Catholics were passions that he would pass down to his son.[4]

At eight years old Samuel Morse attended Phillips Academy in Andover, Massachusetts, and later attended Yale University. At Yale, Professor Benjamin Silliman's lectures introduced Morse to the emerging science of electricity. While drawn to science, Morse was drawn even more strongly to painting. He lived in Europe for three years to study under Washington Allston, a widely known American painter in the same league with earlier famous American painters including Benjamin West and John Singleton Copley.

Morse returned to the United States to earn money to continue his education and career as an artist. While back home, Morse attempted to sell ten paintings, including two of his large historical paintings, *Dying Hercules* and *Judgment of Jupiter,* but after public displays in Philadelphia and Boston, neither painting sold. In 1817 Morse had his "first commercial success" as a painter when he went to Charleston, South Carolina, and was commissioned by the city to paint President Monroe. In 1821 he made a bold attempt at a major painting featuring the interior of the US House of Representatives. (Morse's painting, *The House of Representatives,* was recently moved to the National Gallery of Art in Washington, DC.) Morse then moved to New York and was chosen by the city to paint the hugely popular General Lafayette, who was touring America at the time. More high-profile portraits followed.[5]

And yet sustained commercial success eluded him, and at forty-one Morse accepted a post as a professor of painting and sculpture at the University of the City of New York (later New York University). Morse the painter then turned politician. In 1834 alone he wrote twelve articles on the "Catholic peril" for the *Observer*, his brother's newspaper. In 1835 he wrote, "This Jesuitical artifice of tyrants . . . *if not heeded will surely be the ruin of Democracy.*" Jesuits were working "upon the passions of the American community, managing in various ways to gain control" of the country in a "conspiracy" to overthrow the government. The only thing worse to Morse than Catholics were immigrants, and he went so far as to run for mayor of New York City in 1836 under the banner of the Native American Democratic Association. Needless to say, this was not a group that supported the rights of actual Native Americans. Morse was not alone in his anti-immigrant nativism, especially in cities in the Northeast, but he was set apart by his very public aggressive views. Thankfully, he failed at politics and turned instead to technology.[6]

MORSE TURNS TO INVENTION

In 1837 two Frenchmen came to the United States to demonstrate what they called the telegraph. Actually, the Frenchmen demonstrated what is best described as an optical or visual telegraph called a semaphore. They used a device that looked like a football goal; the arms of the goalpost could be tilted and bent, with each position indicating a letter and number detailed in a codebook. At the time, Morse had a crude but working version of his telegraph. The key difference from other attempts was that Morse's telegraph was based on electricity and magnetism, rather than on physical symbols, as with the French invention.[7]

What drove Morse to technology? Jilted by his first love—painting—perhaps Morse was turning to his second love at Yale—science, particularly electricity. And then there was his chance encounter in 1832, on his return voyage from Europe on the ship *Sully*. One evening on the *Sully* the conversation turned to Ampère's work on electromagnets. One gentleman asked if the power of an electric current diminished over long distances, and Dr. Charles T. Jackson, a physician from Boston, responded that the answer was no, because Benjamin Franklin had shown electricity could be transmitted for miles. Jackson went on to say that his research in Europe had confirmed this. Morse responded that he believed there was "no reason why intelligence might not be instantaneously transmitted by electricity to any distance." He relished the thought that this idea was original to him. However, if he had read more widely on the topic he would have realized the concept was already in the public sphere. Morse pursued the idea in his notebook as the trip on the *Sully* continued, writing out his first rudimentary code for possible messages.[8]

After the voyage, Morse was low on funds and needed financial help from his brothers in New York. While teaching there, Morse continued his work on the telegraph and in 1837, soon after seeing the Frenchmen demonstrate their semaphore, Morse sought notice from the US Patent Office for the invention of the "American Electro-Magnetic Telegraph." He submitted his actual patent application in April 1838. In that same year, on his way to pursue funds from Congress, Morse demonstrated his telegraph at the Franklin Institute in Philadelphia. The first scientific assessment of his invention garnered positive reactions. Mentioning competing devices in Europe, the Franklin Institute reported Morse's plan was "entirely different from any of those devised by other individuals."[9]

THE BUSINESS AND POLITICS OF THE TELEGRAPH

Morse began to enlist partners for his business venture to commercialize his telegraph. He started with colleagues at New York University, including Professor Leonard D. Gale. Alfred Vail, a former student of the university, also became an active player in the venture. Morse also turned to his brothers and to Vail's family for funding. After a high-profile demonstration before Congress, he asked his listeners for $26,000 to conduct a fifty-mile test and suggested that the federal government take ownership of the telegraph. He proposed that the government keep the rights to the technology but allow private firms to build lines from point to point. He saw this arrangement—what today would be called a public-private partnership—as the best of both worlds.[10]

Congressional committee chair Francis O. J. Smith oversaw Morse's request. Ignoring concern about a conflict of interest, Smith took leave from Congress (he did not resign) and became one of Morse's partners. Morse had three partners in his venture, and each gained a share of the patent rights for his telegraph. Smith secured four-sixteenths of the US patent rights and five-sixteenths of the foreign patent rights; Vail secured two-sixteenths of both the US patents and foreign rights; Gale got one-sixteenth of both. Morse was left with nine-sixteenths of the rights in the United States and eight-sixteenths in Europe.[11]

Morse traveled to Europe once again, but this time it was to secure patents for his invention rather than to paint. A British judge rejected his patent request and told Morse to be satisfied with his US patents. It seemed clear that part of the reason for the judge's denial of a patent for Morse was that there were British competitors. While he was rejected by the British, the French granted Morse a patent under the condition that he sell the telegraph only to the French government.[12]

Morse returned home from Europe to be confronted with a host of problems. First, he had no money. Second, Jackson, who had been his fellow passenger on the *Sully*, continued his claim that the idea of the telegraph was his own. In addition, New York University suffered a declining student base, putting Morse's day job in jeopardy. And Congress still had not approved funding for his telegraph. Finally, in 1843, Congress once again considered the issue of funds for Morse's experimental line. By a narrow vote—eighty-nine to eighty-three—the House passed the funding proposal for $30,000. In the Senate, the funding bill passed unopposed, and the bill was signed by President John Tyler.[13]

On May 24, 1844, Morse completed his forty-mile line from Baltimore to Washington. It terminated at the US Supreme Court chamber in the Capitol. Morse had asked the US patent commissioner's daughter to come up with an appropriate first message. Annie Ellsworth, at her mother's suggestion, chose from Numbers 23:23. The first message over the first commercial telegraph line in the United States was "What hath God wrought."[14]

The carefully orchestrated experiment ultimately won Morse's line the publicity it needed. The Democratic Party held its convention in Baltimore, and Vail used Morse's line for a blow-by-blow account of the deadlocked voting. At the start, Martin Van Buren was in the lead, but on the eighth ballot James K. Polk entered the race. Polk won the nomination. On the convention floor, a delegate from New York reported that the "telegraph has announced the nomination of Mr. Polk to our friends in Washington, and I am happy to inform the Convention that his nomination is well received." He asked for the convention's indulgence to read the message sent from Washington, which read: "Three cheers for James K. Polk, and three cheers for the telegraph."[15]

Interestingly, while Morse was very political, he made sure to keep his powerful technology neutral. Before the 1844 election Morse wrote in a letter to Vail, "As there is great interest by the citizens generally of both political parties in the results of the various elections occurring . . . you will be especially careful not to give a partisan character to any information you may transmit."[16]

THE GREAT PATENT WAR

The patent war over the telegraph began when Henry O'Reilly, a competing investor, attempted to expand the telegraph system west using equipment designed by Royal Earl House, a scientist and inventor from Vermont. Morse believed that House's design infringed on his patent, and, in 1848, a judge in Frankfort, Kentucky, ruled in Morse's favor. O'Reilly appealed the Kentucky decision to the US Supreme Court in the case *O'Reilly v. Morse.* Chief Justice Roger B. Taney wrote the opinion for the court, in which he asked three questions. First, was Morse the "first and original inventor of the Electro-Magnetic Telegraphs described in his two reissued patents of 1848?" Second, if he was, were his patents "issued conformably to the acts of Congress?" Third, if Morse's patent was conforming, was the "telegraph of O'Reilly . . . substantially different from that of Professor Morse" so that there was no infringement on the Morse patent by O'Reilly?[17]

Suffice it to say, *O'Reilly v. Morse* was an important case in confirming that Morse should be credited with inventing the telegraph, but it was only one of many. By one count there were sixty-two opponents of Morse in litigation. In one case, the press, only half-joking, warned a competing inventor in Europe not to bring his children to America for fear that Morse would claim them as his own inventions. Indeed, the legal attacks continued for the remainder of Morse's life. In 1872, the year he died, Morse received a draft chapter on the telegraph from a book titled *Great Industries of the United States*. The author argued that Morse's claim to be the inventor of the telegraph was a "pious fraud." The book went on to say that Morse could not have produced the so-called Morse telegraph because "he lacked the requisite scientific knowledge, mechanical skill, and entrepreneurial ability." These traits "had been supplied by its actual inventors: by Joseph Henry, 'the legitimate father of the American electro-magnetic telegraph'; by Alfred Vail, 'the brains of the mechanical portion'; and F. O. J. Smith, who 'made it a commercial success.'"[18]

THE TRANSATLANTIC TELEGRAPH CABLE

One remaining challenge for the telegraph was to link North America and Europe with a transatlantic cable. Laying a cable under the rough waters of the Atlantic and making it work posed special challenges. John Brett first set out to link England and France with a cable under the Channel, and in 1852 the first message was sent from London to Paris. In 1853 England was connected to Ireland and later across the North Sea to Germany, Russia, and Holland, but the transatlantic cable was a feat waiting to be attempted. In 1854 English inventor Frederick Gisborne visited American businessman Cyrus Field, who had made his fortune in the paper trade and retired at thirty-three. Gisborne convinced Field that he should invest his fortune in the telegraph. Intrigued by Gisborne's proposition, Field learned that the raised seabed between Newfoundland and Ireland could potentially serve as a platform for the telegraph cable. Field turned to Brett, who had laid the cable under the English Channel, and together they formed the Atlantic Telegraph Company and secured backing from the US and British governments. Field then hired English surgeon and electrician Dr. Edward Orange Wildman Whitehouse as his science adviser. It was a decision he would come to regret.[19]

In July 1857 Field made his first attempt to lay the transatlantic cable. Again pursuing a public-private partnership, Field won the use of two ships, America's USS *Niagara* and the British HMS *Agamemnon*. About 350 miles out from the Irish coast, the cable snapped and Field met his first failure. By June of 1858 Field was ready to try again. At last, after suffering three broken cables, the 2,050 miles of cable were completely laid on August 5, 1858. North America and Europe were connected for the very first time. The celebrations were triumphant, as was the hyperbole. It was said that the "cable had reunited the British and American peoples" and had "half undone the Declaration of 1776." The first message transmitted along the cable, which took a week to send successfully, said, "Europe and America are united by telegraphic communication. Glory to God in the highest, on earth peace, goodwill to men." The triumph was premature. It took more than sixteen hours to transmit another message sent from Queen Victoria to President Buchanan. Within one month the cable stopped working altogether.[20]

A committee was formed to determine what went wrong. The star member was Professor William Thomson, one of Maxwell's greatest influences and a rival of Field's science adviser, Whitehouse. Thomson concluded that much of what Whitehouse recommended had been wrong. Whitehouse had suggested that a small-diameter cable form the core of the whole cable, but Thomson found that the core, which contained the conducting wires, was too small to function consistently. Whitehouse had recommended high voltage, but Thomson found that the high voltage destroyed the cable's protective insulation. Thomson also suggested that the cable would benefit from a better receiver, and recommended his own mirror galvanometer. Drawing on Thomson's diligent science and technological expertise, Field embarked on his next attempt on June 24, 1865. On August 2, two-thirds of the way across the Atlantic Ocean, the cable broke and was lost. Field tried again and finally succeeded in July 1866. Luckily, the cable lost on the previous attempt was recovered; in the end, therefore, two Atlantic cables were ready for service—albeit after nine years of trying.[21]

Thomson was knighted by the British for his scientific contribution, while Field was given a special medal by the US Congress and heralded as the "Columbus of our time" by John Bright, a high-profile English politician, orator, and free-trade advocate. On top of the awards and honors, the transatlantic cable was profitable and Field was able to pay off his debt in a single year. Public-private partnerships had a key role in the commercialization of electric technology. Notable, too, is the fact that failure was a defining part of the journey to success with this first-of-a-kind technology.[22]

THE VICTORIAN INTERNET

As Tom Standage puts it in his book *The Victorian Internet* (1998), "In the nineteenth century there were no televisions, airplanes, computers, or spacecraft; nor were there antibiotics, credit cards, microwave ovens, compact discs, or mobile phones. There was, however, an Internet." The Victorian internet was a "patchwork of telegraph networks, submarine cables, pneumatic tube systems, and messengers combined to deliver messages within hours over a vast area of the globe." As with today's new communications technologies, the primary advantage of the Victorian internet was that it cut the time needed for communication. When Morse installed the first link in 1844 with his Baltimore–Washington line, it took over two months to send and respond to a message from London to Bombay. Thirty years later, it took minutes. Soon, person-to-person communication wasn't its only use. Some used the telegraph to track the stock market and to buy or sell based on the information they received; others used it to bet on horse races. The Victorian internet was used for fun and romance; telegraph operators even played checkers over the wires. The 1879 novel *Wired Love* by Ella Cheever Thayer was based on the notion of online flirting and dating, and couples could get married online.[23]

The telegraph disrupted the old ways of the news business. Before the telegraph, readers accepted that anything they read was likely to be days or weeks old. With the advent of the telegraph, people experienced news in real time. As is true today, there was much debate over whether the news business, as it was known, would be enhanced or rendered obsolete by this communications revolution. With emerging newswire services, like the *Associated Press* and *Reuters*, the telegraph quickly became essential for the news business.[24]

Issues of privacy, policy, and law also arose. Some wanted messages to be private, so they encrypted them, and in 1875 regulations were proposed to limit the coded messages. Legal questions arose. For example, if a loss was incurred on a stock transaction because a telegraph message was in error, who was liable? One person sued Western Union for such a stock loss and the case went to the Supreme Court, which ruled that the plaintiff was only due a refund of the price he paid to send the original telegram: $1.15 versus the $20,000 he allegedly lost.[25]

AN OUTSIZED ECONOMIC IMPACT

The telegraph had an outsized economic impact, as do most innovations in the science and technology of electricity. Because it introduced real-time communication, it enabled the creation of bigger markets, and thus bigger businesses. The telegraph itself became a global business, enveloping not only the United States but Europe and Asia as well. In about twenty years—from 1846 to 1866—the US telegraph system grew from a few hundred miles of lines to more than one hundred thousand miles, and from thousands of dollars of investment to $40 million. Its impact was outsized in the sense that it could not be measured by the amount of money invested on the poles and wires and stations or on the amount of money spent each year on sending messages. Rather, its effects on the economy could be measured only by looking at what it enabled other businesses to do and how it empowered businesses to grow.[26]

Before the telegraph, information traveled no faster than people could get from one place to another. Travel time in the United States was shortest in the Northeast because of the network of rail lines; rail time was about a half day from New York to Boston. Outside the Northeast, the time needed to deliver a person or a message was much longer, two weeks from New York to New Orleans, months to California. Within seven years after Morse's Baltimore–Washington line opened in 1844, however, there was real-time communication between the East and stations on the Mississippi River. In a head-to-head price comparison between the US Postal Service and the telegraph, the Postal Service might have looked like a better deal. Before 1845 a one-page letter cost up to twenty-five cents to send; in 1851 the price fell to just three cents an ounce. For the telegraph, a ten-word message cost $2.05 to send; by the 1870s its price dropped to fifty cents, but it was still much higher than a one-page letter delivered by mail. In the end, because for many businesses time often mattered more than savings, the telegraph's high prices did not diminish its growth.[27]

The telegraph's creation of real-time communication enabled the growth of businesses and markets that ultimately benefited the nation's economy. A big market was better for buyers since they could shop for better prices and better-quality products over a broader geographic range. A big market was also good for sellers, since they could increase sales and thereby increase the scale of their enterprises, which enabled them to use newer, larger-scale, lower-cost technology to produce their products. Just as the marketplace on Amazon.com has transformed the way we shop today, the telegraph cut the costs of communication for the buyer to find the right supplier and the supplier to find and to serve a larger number of customers. This also encouraged the rise of big firms as companies became more integrated; they took on more of the chain of activities needed to produce and deliver a final product.[28]

The telegraph helped to create bigger markets and bigger firms that reached into nearly every aspect of ordinary life. It enabled a much bigger and more transparent market for stocks and other financial instruments. The meatpacking business in the Midwest was able to serve the East with modern centralized slaughterhouses because the telegraph allowed them to sell to a bigger market as well as to integrate delivery by refrigerated railcars.[29]

The telegraph and rail businesses developed a symbiotic relationship. To ensure access to the best rights-of-way, the telegraph companies often ran their lines along the railroad rights-of-way. Businesses, meanwhile, used the railroads to deliver products to big geographic markets recently created by the telegraph. America's rise as an economic powerhouse of the nineteenth century was fueled by the dual developments of rail and telegraph: they enabled the collapse of space and time in a nation that was still relatively new.[30]

A ROLE FOR GOVERNMENT, INEVITABLY

From the start, Morse wanted the government to buy and control the telegraph. But Congress did not want the government to own the patents or assume the responsibility of building the lines; instead, private ventures built an extended and overlapping network of telegraph lines. The telegraph was never subsidized or politicized to the extent the railroads in America were. State governments, however, did get involved, and their role changed over time. At first the states promoted and enabled private-sector investment in the telegraph. The states primarily helped ensure rights-of-way and set penalties for those who damaged telegraph lines. Then the states began to protect the public from bad performance by the telegraph companies. Protection included a requirement that the messages from competing lines had to be accepted, a certain priority of messages had to be followed, and the content of messages was automatically considered confidential.[31]

In an effort to end the chaos of competing telegraph lines in the 1850s, government also facilitated the consolidation of the private telegraph companies. Industry attempts at self-regulation had not worked. In his book *Wiring a Continent* (1972), Robert Luther Thompson writes of the telegraph industry's "irresistible trend toward monopoly." Amos Kendall, whom Morse hired to run the business side of the telegraph, had hoped for better organization of the telegraph industry through individual management of lines by the Morse patent team; however, quarrels among the team undermined such structured management. Competition from other technologies contributed to the fragility of the industry, and, as competition increased, profits vanished. Several other attempts to coordinate the telegraph industry through self-governance also failed. O'Reilly tried a democratic council, while Kendall's American Telegraph Confederation also failed. In 1857 a last-ditch effort resulted in the Six-Party Contract that assigned areas to competitors and set rules for their interaction. Out of the Six-Party Contract grew the North American Telegraph Association, which, according to Thompson, became a major force in the industry. In the end, however, it was consolidation that allowed Western Union to transform into the first truly nationwide industrial monopoly in 1866 after acquiring the United States and the American Telegraph Companies.[32]

THE TELEGRAPH'S LEGACY, AND MORSE'S

The telegraph was the first commercial application of the new science of electricity and electromagnetism. It did no more than send a stuttering on-and-off flow of electricity through a wire that, when the electricity was on, magnetized a block of iron. This simple device, however, revealed the commercial importance of electricity and revolutionized communication. Electricity had an essential role in the birth of modern telecommunications. Eventually, the electricity supply and the telecommunications businesses would go their separate ways, but they remain twins born of the same great science. As the narrative unfolds it is worth asking why one twin—the telecommunications business—is today a symbol of technological advance while the other—the electricity supply business—is generally seen as stodgy, with little technological innovation achieved.

The outsized economic impact of the telegraph is difficult to overstate; it enabled the globalization of business. Communication technology was and is an essential parallel development to industrialization. Together they allowed the United States to begin to exploit one of its greatest assets—its size. This same outsized economic impact, although different in nature, would be seen again with new electricity technologies after the telegraph.

The fact that the commercialization of the telegraph was messy, contentious, and litigious is another important part of its legacy. The great patent war makes this plain. The judgment of a court—most likely the judgment of one judge—was a potent force in deciding, in effect, the paternity of this new technology. Patents are a powerful means for government to affect the nature and pace of technological development, and the question of whether patents promote or deter innovation is a hot topic even today. As we will discover later on, the range of issues decided by the courts broadens over time to include all issues of the modern electricity business—a development that unsettles many people.

The telegraph's evolution also set a model for the relationship between government and private corporations. The telegraph got a jump-start from the federal government's funding for Morse's Baltimore–Washington line. The private sector then built the system with unbridled enthusiasm—and little interest in coordinating efforts. State governments first enabled the private investment, then helped mitigate abuses. Private sector control over the telegraph had mixed results, as evidenced by Western Union emerging as a monopoly. In the end, it came full circle with a new technology—the telephone—protecting the consumer by offering a better, less expensive service to replace the telegraph monopoly. The lesson taken from the telephone eclipsing the telegraph is of great importance: the best government oversight might be simply keeping the door open to new competitors with new technologies.

Morse's personal legacy is complicated. He died on April 2, 1872, at the age of eighty. Flags across America were flown at half-mast, telegraphs were covered in black, and most obituaries said that he was a great man, though some countered that view by citing his defense of slavery. Among other positions, Morse had actively campaigned with a group of prominent men hoping to revoke the Emancipation Proclamation. Great men are not great, or right, all the time, and for some the bad surely grows to outweigh the good.[33]

A poem in tribute to Morse summed up his most lasting achievement: "And Science proclaimed from shore to shore, / That Time and Space ruled man no more." Ironically, this once-ambitious painter was memorialized in grandiose paintings done of him by others. A telling example is a fresco painted by Constantino Brumidi in 1865, depicting George Washington "rising to the heavens in glory." Around him are six themes: the one for science has the goddess Minerva instructing Benjamin Franklin, Robert Fulton, and Samuel F. B. Morse.[34]

CHAPTER 5

Thomas Edison's Light

> It has sometimes been said that we live in an industrial
> age. It might better be said we live in the age of Edison.
> —Henry Ford (1931)[1]

With his 1,093 US patents, Thomas Alva Edison remains one of the most prolific
and celebrated inventors in American history. While few of these patents
single-handedly brought about revolutionary change, many marked significant advancement for specific technologies, and the breadth of the
technologies to which he contributed is remarkable. Edison began with
inventions that advanced telegraph technology, and despite Alexander
Graham Bell beating him to the punch, Edison made important contributions to the telephone. And then Edison went even further: he invented
the phonograph and motion pictures as well.[2]

Not only was Edison a prolific inventor, he also redefined the process
of invention. Edison, the symbol of the solitary inventor, was actually the
first to create a corporate research laboratory in which he and his team
became adept at drawing out the commercial value of the science of the
day. As Edison scholars Robert Friedel and Paul Israel note, "In Edison

we find the transition from the common, ingenious invention that seemed to move much of the world forward in the nineteenth century to the specialized, scientific technology that was to be a dominating social and economic force in the twentieth."[3]

Edison actually invented the entire electricity business as it is known today, from top to bottom. He started with painstaking research to create an electric lightbulb, meant to replace the gas-fired lights of his day; a key hurdle was finding a material for a filament that would make electric lighting commercially feasible. He also had to invent the technology, or at least improve existing technology, for the rest of the business, such as the wall switches, power lines that snaked under city streets to deliver electricity to the lights, and dynamos that produced the electricity to initially jostle the electrons in those lines and illuminate the lightbulbs.

Edison was no stranger to hard work or to hard times. His work ethic is legendary; he often worked well into the night and demanded the same from his colleagues. He put his own money on the line, and on occasion he literally bet his home on an invention. But Edison also needed and received money to fund his research from some of the most famous names on Wall Street: Cornelius Vanderbilt and J. Pierpont Morgan were both investors. In the end, however, it was the moneymen who erased the Edison name from his own company, Edison General Electric Company, and changed it to what is known today as General Electric, or simply GE.

However, Edison could not be erased from American mythology. It is no accident that drawing a lightbulb above a person's head symbolizes a great idea. Not only was the lightbulb one of America's brightest ideas; Edison was one of America's brightest lights.

INCUBATION

A large part of Edison's legend grew from what we know about how hard he had to work, how much he had to risk, and how sharp his vision had to be to achieve success. Edison was born February 11, 1847, in Milan, Ohio. In 1854 his family moved from Milan to Port Huron, Michigan.

The family left Milan when railroads disrupted the success of canals, including the canal that connected Milan to Lake Erie and allowed the town to prosper. Edison received little formal education in Milan because he was often sick and his family didn't have much money; their financial problems continued in Port Huron. Edison's mother became the most influential teacher in his life, introducing him to books including Edward Gibbon's *History of the Decline and Fall of the Roman Empire*.[4]

By age twelve Edison had his first job, selling snacks and newspapers in the aisles of the Grand Trunk Railway. He was already an avid experimenter and built his own mini-telegraph. Later he got interested in chemical experiments, and, in one such experiment, he broke a bottle of phosphorus and started a fire on the train. The conductor on the train is said to have "boxed Edison's ears," and some biographers trace his lifelong hearing difficulties to this incident. But good fortune also visited Edison while he worked on these trains. One day Edison saved a young boy's life when he pulled him from the path of an oncoming train. The boy's father, James Mackenzie, rewarded Edison by offering to teach him how to be a telegraph operator.[5]

With this practical education in hand, Edison became a traveling telegraph operator for five years, from 1863 to 1868. It was in this role that Edison produced what some call his first invention. As a telegraph operator, Edison listened to a "sounder," which made the dots and dashes of Morse code audible. A skilled operator wrote out what he or she heard from the sounder at a pace of about forty words per minute. Edison wanted to slow down the sounder, and to do so he drew from an older technology in which a Morse register would punch the dots and dashes on paper. Edison linked this Morse register to a sounder so that the audible dots and dashes came out at a slower pace of about twenty-five or thirty words per minute. The invention was called the Morse repeater, and it slowed down reception and transcription to give time for longer news reports. It did not help with the intense messaging of breaking news events, so it suffered an early demise. Still, even this early failure reveals the reasons for Edison's success: a skill with and interest in advancing practical technology for cutting-edge industries.[6]

In 1867 Edison moved to Cincinnati and, at twenty, started to get more serious about his life as an inventor. There were a small number of cities that served as laboratories for new telegraph technology, and Cincinnati was one of them. It was there that Edison began his constant study of telegraph technology, reading about the science behind electricity, including Faraday's *Experimental Researches in Electricity*. With a life of invention in mind, Edison focused on two devices to improve the telegraph: relays and repeaters. A relay directed an incoming telegraph signal to the local sounder. A repeater was like a booster shot for a telegraph signal—it strengthened the signal as it went on to the next station. Edison also worked on duplex technology, which would allow two messages to be sent simultaneously over a single telegraph line.[7]

In 1868 Edison moved to Boston. Local investors there financed two of his new inventions outside of the telegraph business: his electric vote recorder and his stock printer. In 1869 Edison filed his first patent application for his vote recorder. It was not a commercial success, since legislators did not want to speed up the voting process; a slower pace gave them time to win over votes from their rivals. Edison did have some success with his gold and stock quotation service in Boston. However, his real success came after he moved to New York in 1869.

A FULL-TIME INVENTOR

As is true today, New York financial houses needed information as fast as possible, and two of the companies competing in this field were Edward Calahan's Gold and Stock Telegraph Company and Samuel Laws' Gold and Stock Reporting Telegraph Company. Edison found his way to Laws' company first, where he met with Franklin Pope, who knew of Edison's stock printer in Boston. While Pope could not offer Edison a job, he did allow Edison to use the company's shop for experiments and to sleep on a cot in the basement. As luck and skill would have it, Edison proved himself on one especially high-pressure day. In the early stages of Jay Gould's attempt to corner the market for gold, Laws' machinery suddenly stopped working, and his clients panicked at the loss of information at such a critical time. Pope and Laws did not know what to do, but Edison calmly bent over the broken machine, found the problem, fixed it, and thereby earned a job that day.[8]

After the sale of Laws' company, Edison and Pope began the first electrical engineering firm in America. Later, on his own, Edison found success selling their inventions to bigger companies like Western Union, and this created the financial foundation for the laboratory space that would redefine the path to innovation and the electricity business to come. Edison and his partner found lab space in Newark, New Jersey, and formed the Newark Telegraph Works. There, Edison recruited a team of men who would become friends and fellow inventors for years to come. By 1876 Edison and his team had secured more than two hundred patents. Edison had risen from near failure when he first arrived in New York to great renown as an inventor and considerable financial success, though his rise was not without risk or controversy. During his career, Edison once had to give up his house to save the business, and he occasionally flirted with conflicts of interest. And in one instance, although Western Union had sponsored the development of a technology, Edison sold the rights to Gould. Western Union sued, painting Edison as a traitor, and eventually won the suit.[9]

THE WIZARD OF MENLO PARK

Edison once said, "My one ambition is to be able to work without regard to the expense . . . what I want is a perfect workshop." In 1876 he took a big step toward the perfect workshop by creating his laboratory at Menlo Park, New Jersey. Menlo Park forever changed the invention process by becoming the first "invention factory."[10]

Western Union was key to funding Menlo Park, as it hired Edison to invent a "speaking telegraph" to compete with the telephone Alexander Graham Bell patented. In 1877 Edison filed a patent for the speaking telegraph transmitter, and Western Union formed a subsidiary called the American Speaking Telegraph Company. How did an amateur and outsider like Bell beat the professional inventor Edison? Biographer

Paul Israel suggests that Bell won the race because he was an outsider; he brought a unique perspective to the field. Bell's father had developed "Visible Speech" for the deaf, and later Bell created an "electromechanical ear called the phonautograph to allow the deaf to see speech patterns." Israel emphasizes that Bell was aiming to do something brand-new in transmitting speech while Edison was more narrowly trying to improve the existing technology of the telegraph. Edison did, however, make an important contribution to telephone technology. Bell's receiver was better than Edison's, but Edison's transmitter was better than Bell's.[11]

In 1877 Edison pushed forward with experimentation that would lead him to inventing the phonograph. In the wake of news reports, an Edison assistant sent a letter and sketches of the invention, called the phonograph, to *Scientific American*. The editor then published the letter in the magazine, prefacing it with a note about how startling it was that speech had become "immortal." Edison then asked his engineers to build the device. Edison shouted out the nursery rhyme "Mary Had a Little Lamb" into the recorder, and then startled his colleagues by playing it back to them from his new tinfoil device. It was the invention of the phonograph, not the electric light, that first made Edison famous and won him his oft-repeated nickname: the Wizard of Menlo Park. But Edison soon put his first famous invention aside for ten years to turn to the challenge of inventing the electric lightbulb—as well as the electricity industry as a whole.[12]

INVENTING AN INDUSTRY

Dead Aim on the Competing Technology

At thirty-one, Edison was already a widely known, accomplished inventor. Exhausted from his work on other inventions, he jumped at the chance for a somewhat exotic vacation: he would travel to Rawlins, Wyoming, to observe the solar eclipse on July 29, 1878. One of his fellow travelers,

Professor George Barker of the University of Pennsylvania, suggested that Edison meet with William Wallace in Ansonia, Connecticut, who had been experimenting with electricity and had developed his first dynamo, or electric generator, in 1874. Later that year, Edison met with Wallace to witness firsthand the Wallace-Farmer dynamo and the lights it kept illuminated. It was at this meeting with Wallace that Edison got excited about electricity and decided to take on the challenge of developing an electric light. Edison said upon his return that, "[in Wallace's shop] I saw for the first time everything in practical operation. It was all before me. I saw the thing had not gone so far but that I had a chance. I saw that what had been done had never been made practically useful. The intense light had not been sub-divided so that it could be brought into private houses."[13]

The established competitor to electric lighting was the gaslight, and throughout his work Edison took dead aim at the goal of beating the gaslight on price and performance. Electric lighting had stalled with the arc light, developed first by Faraday's mentor, Humphry Davy. It was called an arc light because when two pieces of carbon are held with a short gap between them and high-voltage electricity is put through both, a dazzling arc of light is created across the gap. In 1880 a part of New York's Broadway was illuminated with arc lights and that is how the street earned its nickname as the Great White Way.[14]

The problem with the arc light was that it was much too bright. Using a common measure of the day—candlepower—an arc light emitted thousands of candlepower while the gaslight emitted around ten to twenty candlepower. In the context of the arc light, there was talk of *subdividing the light* or splitting the arc light into smaller, less glaring pockets of light. This would require a new approach: incandescent lighting.[15]

With incandescent light, a material is heated until it glows, which provides the required light. The primary technical concern for Edison was that most materials melt before they can reach the temperature at which incandescence would be achieved. Edison had to find a thin, light material or refine the process to provide light. Early on, Edison focused on creating a "regulator" for his lamp. A regulator would reduce electric current when the temperature approached the melting point of the incandescent material being used. Edison also had a primary commercial concern. It was thought that a high electric current would be necessary to light Edison's lights and a great deal of expensive copper would be needed to transmit that high current through copper wires that would run from the electric generator to lights in homes. To effectively commercialize his invention, it would have to make sense financially.[16]

To address these and other concerns, Edison needed money to fund research. He turned to Grosvenor Lowrey, his attorney, to arrange financing. Lowrey attracted two Western Union officials, Tracy Edson and Hamilton Twombly, as key investors, plus various members of the Vanderbilt family, which was also a major investor in both the telegraph and gas utilities. The deal involved the formation of a stock company with $300,000 of funding. Half of the funding came from Edison's patents and the other half was in cash from the investors. It is interesting to note that Edison made much of the fact that the Vanderbilt family "is the largest gas stock owner in America." By investing in Edison's company, the Vanderbilts were, in effect, competing against their own companies, since Edison was out to beat the gaslight. Here is evidence that savvy private investors often have no emotional or political attachment to inventions of the past; instead, they are ready to fund the transition to new, better ways. Friedel and Israel state, "The backing given [Edison] in 1878 and 1879, however, marked a new stage in the relationship between money and invention in America—a glimpse of the era in which giant corporations would routinely expect technical expertise, backed by science and laboratory resources, to turn out newer and better products for profit."[17]

Edison's efforts marked a new stage in history, but on the most essential level they marked a new stage of product development for the electricity business. Regarding the latter, it is useful to think in terms of at least four stages of life for any product: research, development, demonstration, and commercialization. The pure scientific research had long ago been done by Franklin, Faraday, and Maxwell. Edison was well into product development with, for example, his testing of filaments. He would soon step into demonstration with his Pearl Street Station and then into full commercialization with the proliferation of the Edison companies.

Name Dropping (Terminology)

An electric light is one of three parts of an electric circuit; it is itself the user of electricity. The second part is the electric power plant, which can be far from the electric light; it is the source of electricity. The third part is the pathway from the power plant to the light—typically metal wires that run along the highway to a home or business, including the copper wires in the walls of a building. When the light switch on the wall is on, the electricity flows in a circuit and the lamp lights. If a problem is detected, a circuit breaker intervenes to stop that flow.

To measure the electricity in a circuit, George Mason University physics professor James Trefil and earth science professor Robert M. Hazen offer an analogy between electrons that flow through a wire and water that flows through a pipe. For water, important measurements include the amount of water flowing and the pressure in the pipe. As to measuring the current of electrons "flowing" through wires, the metric is called the ampere—or *amp*, as we usually say—named after French physicist André-Marie Ampère. The pressure behind the flow of electrons, meanwhile, is termed the voltage—or *volts*—named after Italian scientist Alessandro Volta. A third metric, the amount of power delivered to the electric lamp (or any other electrical device, such as a motor), is defined in watts, named after James Watt. The formula goes: power in watts equals current in amps times voltage in volts. So, more amps or more volts result in more power. Finally, the term *resistance* measures how hard it is to get electrons though wires. High resistance means that more of the electrical energy shows up as heat. This is why low-resistance metals like copper are used to transmit electricity over long distances, so less electricity is lost as heat. High-resistance materials are used in devices like toasters that have a good use for the heat. Resistance is measured in ohms, named after the German physicist Georg Simon Ohm.[18]

Using Science

Edison's investors wanted to know the status of other efforts to produce electric lighting, and, fortuitously, Edison hired Francis Upton to find out. Upton would become one of Edison's closest collaborators, and would be one of the few on his team with training and experience in pure science—an invaluable asset. In the end, Edison built original versions of just about everything needed to produce electricity (the dynamos, or generators), to deliver it (the underground copper conductors), and to use it (the electric lightbulb itself). One cannot overstate the extent to which Edison and his team created the electricity business from scratch, and their success was based on a unique mix of skills in science, technology, manufacturing, and finance.

A great example is how Edison and Upton discovered that they could reduce the amount and cost of copper wiring if they used a high-resistance material for the filament inside their lightbulb. They made this discovery by applying science of the day—a combination of Joule's and Ohm's laws. Their thinking might best be explained with three equations. Start with the equation that electric power equals current times voltage. Then add the equation that voltage equals current times resistance. Finally insert the second equation into the first, showing that power equals current squared times resistance. So to get the same power Edison could lower current—which meant less copper was needed—and raise resistance. This discovery launched Edison's worldwide search for the best high-resistance material for his filament.[19]

At the start Edison preferred platinum; a long platinum wire sat tightly spiraled in the center of a sealed glass globe, inside of which was a vacuum. The light looked very much like the lightbulbs used today. Edison's platinum light failed, however, and he had to renew his search for the right material for the filament in his electric lightbulb; platinum was expensive and it had a low resistance. In October 1879, he and his team made an important breakthrough by finding that a carbon-based filament worked.[20] Edison filed a patent for his carbon-based light on November 4, 1879.[21]

On December 21, 1879, an article in the *New York Herald* titled "Edison's Light: The Great Inventor's Triumph in Electric Illumination" created great anticipation in readers' minds. In the article, Edison announced his first public demonstration for New Year's Eve, 1879—ten days later. Marshall Fox, the *Herald* reporter, described that demonstration vividly: "Edison's laboratory was tonight thrown open to the general public for the inspection of his electric light." He reported that hundreds of people came despite the inclement weather. Edison's offices were "brilliantly illuminated," as was a path to the train depot. Edison and his team explained their new system and ran it through a "variety of tests."[22]

Before the event there were skeptics, and some scientists called the demonstration a fake. Others saw it as quite real; soon thereafter, the stocks of gaslight companies nosedived while the value of Edison's company stock skyrocketed. To biographers Friedel and Israel, however, the most important reaction was that of the public, especially the throngs who came to Menlo Park that New Year's Eve. Instead of evoking fear or distrust, Edison's electric system was a source of hope. Specifically, these Edison scholars conclude that "this attitude toward the powers of science and technology is one of the nineteenth century's most important legacies, and no single instance exemplifies it better than the enthusiasm with which the crowds ushered in the new decade at Menlo Park."[23]

BUILDING THE PEARL STREET STATION

Edison's New Year's Eve demonstration at Menlo Park took his electric lighting system up through two more of the stages in the research, development, demonstration, and commercialization process. Still, commercialization remained ahead of him. Almost three years after the Menlo Park demonstration, on September 4, 1882, Edison began to operate his commercial-scale power plant at 257 Pearl Street in New York City and illuminated lights in a one-quarter-square-mile radius in the Wall Street area, including J. P. Morgan's offices. It is still one of the best-remembered commercialization efforts in American history.[24]

The Pearl Street station, however, was not the inaugural commercial project for Edison's lighting system. A man by the name of Henry Villard saw the Menlo Park demonstration and was quite impressed. Villard was the head of the Oregon Railway and Navigation Company, and he asked Edison to install an electric lighting system on his company's new steamship, the *Columbia*, which was being built in Pennsylvania. Edison agreed and the system was installed in the spring of 1880, delighting passengers from the steamer's first voyage.[25]

In addition, Edison and his partners achieved fame in Europe. In 1881 Edison's electrical system was exhibited at the Paris International Electrical Exposition. Reviewers at the time reported the Edison system was prominent because he, unlike other inventors in the same field, had figured out all the details and developed a completely functioning electrical lighting system. Edison's system also was unveiled at a similar exhibition near London in 1882. The "Wizard" went even further there when he opened his full-scale lighting system in the center of London, at 57 Holborn Viaduct, with a generator, underground wires to deliver the electricity, and lamps. By April 1882 the Holborn Viaduct project consistently kept 938 lamps alight, and it served as something of a dress rehearsal for the Pearl Street station, which opened five months later.[26]

Back in America, Edison turned his full attention to the Pearl Street project by focusing on the manufacture of all the necessary equipment. This did not mean, however, that further improvements on the technology ceased; Edison and his team filed for 256 patents in the period from 1880 to 1882. The effort to demonstrate Edison's electrical system through the Pearl Street project was impressive in its detail and depth. For example, Edison and his team persisted in seeking the best material for the filament in his electric lightbulb. In 1880 they finally settled on bamboo, but not just any bamboo—their preference was for Japanese bamboo, which outlasted the previously popular Calcutta bamboo. In addition, whatever choices were made on materials had to also work in a manufacturing setting. There was some difficulty manufacturing the bamboo filaments, but the Edison team found that the problem could be addressed with a subtle change in the carbonization process by holding the bamboo "edgeways" instead of "flatways." The source of Edison's competitive advantage was well illustrated with the bamboo filament because he had the resources to search worldwide for the best material, and he recruited people who had the patience and the perceptiveness to find the subtle differences that solved problems. Today, filaments are made of tungsten, a rare metallic element with a high melting point.[27]

No less impressive was the array of new companies Edison established. With the Pearl Street project, Edison created a corporate structure that assigned key responsibilities among his team and key risks among investors. The parent corporation was formed in 1878 as The Edison Electric Light Company; it was used to finance the invention of the full lighting system. The Edison Electric Illuminating Company of New York was formed in 1880 to build the Pearl Street project, and Edison Machine Works was formed in 1881 to build the "Jumbo" dynamos for the Pearl Street station and elsewhere. The Electric Tube Company was formed in 1881 to produce the copper conductors for the underground delivery of electricity from the generator to the electric lamps. The Edison Lamp Works was formed in 1880 to manufacture the electric lamps themselves, and Bergmann and Company was formed in 1881 to manufacture accessories, such as lamp sockets and switches.[28]

With all these advances in technology and the development of a sophisticated corporate structure, the tangible result—the Pearl Street power plant—might seem slightly anticlimactic. Edison bought the buildings at 255 and 257 Pearl Street for $65,000. Coal was stored in the basement, and a conveyor transported the coal to the first floor, where it was burned in four Babcock & Wilcox boilers to make steam. The Jumbo generators were on the next floor up, each measuring 168 by 105 inches and weighing thirty tons. Each had the capability to produce about 51,500 watts and could light seven hundred class A lamps. The next floor housed a set of electric lamps for testing and to measure performance.[29]

While the Pearl Street building itself was relatively unremarkable, what went on inside the building must be recognized as a triumph for Edison and for mankind. Within that building an amazing transformation of energy occurred, as burning coal unlocked the chemical energy that had been forming in the coal for thousands of years and transformed it to steam in the Babcock & Wilcox boilers. Inside the Porter Allen engines, the steam was transformed to mechanical energy, and, finally, that mechanical energy pushed and pulled the magnet inside the generator to induce the flow of electricity in a coil of wire—just as Faraday proved it would. From the Pearl Street power plant, electricity flowed through over eighty thousand feet of underground wire conductors that Edison's team had laid. Those wires were laid to compete for the business of about 1,500 customers in the Pearl Street District, who used a total of about twenty thousand gas lamps. With the electric lights aglow on September 4, 1882, Edison's reply to a reporter neatly summed up what he had done: "I have accomplished all I promised."[30]

PUSHED OUT

Indeed, Edison had done all that he promised and, in the process, invented an entire industry. Not long after the Pearl Street triumph, however, Edison suffered a defeat that pushed him out of the industry he invented. He had a preview of what was to come when he asked Villard to sort out a few problems with his European enterprises. Under Villard's plan in 1887, the name of Edison's Company in Germany, the German Edison Company, was changed to AEG or Allgemeine Elektricitäts-Gesellschaft—in English, the General Electric Company. The Edison name had disappeared from the German enterprise. Villard returned to the United States with the purpose of negotiating with Drexel, Morgan and Company on behalf of German banks looking for investments in American industries, including the electricity business. In 1889 Villard helped reorganize all the Edison Companies into a new company called Edison General Electric Company. Over half of the money to consolidate the Edison companies came from German banks and interests. Villard became the head of the new Edison General Electric Company, which combined all the companies that helped create the Pearl Street project. Edison apparently signed off on this because it would allow him to get back to inventing. Ultimately, his trust was misplaced.[31]

At the same time that these financial deals were being made, the electricity industry was embroiled in what was termed the "Battle of the Currents." Edison backed a system based on direct-current (DC) power, while others, most notably George Westinghouse and the Thomson-Houston firm, backed a system based on alternating-current (AC) power. Edison's DC power eventually lost the battle and this loss was partly the reason Edison lost his company. A merger was eventually proposed between Edison General Electric and Thomson-Houston. In 1892 Morgan and Vanderbilt interests took over the negotiations on the merger and closed the deal; the new company was simply named General Electric. At age forty-five, Edison was no longer a key businessman in the industry he had created, and his name had been erased from the company he built. While it was the end for Edison, it was the beginning for General Electric, a company still central to the science and technology of the electricity business today.[32]

EDISON'S LEGACY

Thomas Edison invented the electricity business from top to bottom—from lightbulb to dynamo. The business needed to prove that it could employ the science of electricity and magnetism on a commercial scale, and that is precisely what Edison did at the Pearl Street Station and with several other projects thereafter. He succeeded because he was a practical genius who could travel in three worlds: science and technology, manufacturing and business, and, at least for a time, high finance. Edison connected the dots within and between these three contexts. The effort he expended, the bet-it-all risks he took, and the clarity of and commitment to his vision set a high standard for the entrepreneurs of any age.

Despite his genius, Edison's judgment was not infallible. At a crucial fork in the road he chose the wrong way—direct current instead of alternating current. That turn does not diminish his legacy, however, because being proved wrong is not the same as being proved useless. In addition, despite his reputation and savvy, he was not invincible against attacks from financiers. Being pushed out of his own company must have felt catastrophic, but it is often a consequence of using private investment, especially in a time of concentrated wealth. It is to Edison's credit (and to his investors') that funds were directed toward his laboratory to launch a new industry.

Edison's legacy, of course, goes well beyond the electricity business. He was among the most important inventors in American history. He showed a versatility and can-do attitude that were essential to invention in his time, or any time for that matter. He must still be an inspiration to any individual with an idea. Yet that inspiration can be felt beyond the individual because Edison fostered research and development that climbed to a new level, creating a template for corporate labs moving forward. He essentially created corporate R&D, and his model is still used to this day.

In 1928, to honor his mentor and friend and celebrate the fiftieth anniversary of the incandescent bulb, automaker Henry Ford began the construction of a museum in Dearborn, Michigan. It would contain a painstaking recreation of Edison's lab from Menlo Park. A report from the *Detroit Free Press* captured the moment by reporting that at his own commemoration Edison "sat there, silent, his arms folded, an indescribably lonely figure, lonely in the loneliness of genius, of one who, somehow, has passed the others, who no longer has equals to share his world, his thoughts, his feelings."[33]

On October 18, 1931, at the age of eighty-four, Edison died at his home in Glenmont, New Jersey. On the day of his memorial service, October 21, 1931, at about 10:00 PM eastern time, all but essential lights were turned off for one minute across America, including those on Broadway and the Statue of Liberty. *Time* magazine quoted Ford, who observed, "It has sometimes been said that we live in an industrial age. It might better be said we live in the age of Edison."[34]

CHAPTER 6

⚡ ⚡

George Westinghouse's and Nikola Tesla's Alternative

Mr. Edison does not hesitate to say: "My personal desire
would be to prohibit entirely the use of alternating currents."
—George Westinghouse Jr. (1889)[1]

Thomas Edison must be given his due: he invented the electric lightbulb and
commercialized the first full system for electric lighting. He was convinced, however, that electricity should be delivered with a direct-current
(DC) power system, which meant that power would be delivered economically only within a one-mile radius or so of the centralized dynamos he
wanted to build. To Edison a direct-current system was safe, allowing the
use of low-voltage electricity; but the limited reach of his DC power plants
meant the plants would be small and their costs would be high.

George Westinghouse had an alternative to Edison's DC system. He
believed in an alternating-current (AC) power system, which could deliver

power economically over hundreds of miles. This ability to transmit electricity over great distances meant that power could be generated from remote sources, such as the great hydroelectric dam at Niagara Falls, and power plants could be larger since they would serve a much broader geographic region. Edison chose direct current because he believed the high-voltage power of the Westinghouse system was dangerous.

Since alternating-current systems dominate today, it is fair to say that Westinghouse, not Edison, had the last word on the technological foundation for modern electricity systems around the world. In addition, Westinghouse established the economic foundation for the electricity business by demonstrating that bigger was better because economies of scale meant lower costs. Economies of scale would become a critical concept in the modern electricity business with profound consequences. Westinghouse demonstrated the superiority of alternating current in high-profile events such as the 1893 World's Fair in Chicago and the hydropower developed at Niagara Falls.

Westinghouse was already a successful inventor and businessman when he came to the electricity business; he had invented the air brake for the railroad industry, which dramatically improved safety. Westinghouse clearly had the intelligence to achieve great things, and he even matched Edison in his people skills; he, too, developed a team of loyal, devoted colleagues. He was equal to Edison in vision as well—he envisioned the full AC system that he planned to establish, just as Edison had done with the DC system. Westinghouse, however, was not a "wizard"; he did not develop surprising new technology from scratch, as Edison seemed magically able to do, at least in the eyes of the public. Westinghouse often, and without hesitation, built upon the ideas of others. There is no more important example of this than his reliance on Nikola Tesla, the brilliant, quirky engineer who gave Westinghouse the science for the AC system and the essential crown jewel of that system: the AC electric motor.

Although Westinghouse and Tesla won the Battle of the Currents, it is Edison who gets credit when the electric lightbulb and the invention of electric systems come to mind. Westinghouse never experienced the name recognition that Edison enjoyed. Mainstream history often neglects

Tesla as well, although he made crucial contributions to creating today's electricity system. In this chapter we'll turn to these two lesser-known but equally intriguing—and in some ways more successful—figures.

TESLA'S PATH TO DISCOVERY

Nikola Tesla was born in 1856 in what is now part of Croatia. According to the family legend, a powerful lightning storm marked the night of his birth. The midwife thought this a bad omen, declaring that he would be a child of darkness. His mother, as would any mother, disagreed—he would be a child of light. It might be most fitting to call Tesla a child of lightning. He surely kept faith with the science of Franklin that proved lightning was an electrical phenomenon; Tesla himself would eventually make his own artificial lightning in spectacular experiments.[2]

His father, Milutin Tesla, was a priest in the Serbian Orthodox Church, and his mother, Georgina Mandić, was bright but unschooled. Nikola benefited from his mother's inventiveness and his father's daily testing to strengthen his memory and reason. His father expected him to follow in his footsteps as a cleric, but Tesla's passion was in engineering. After he survived an attack of cholera, his father relented and let him pursue his studies in engineering at the Austrian Polytechnic School in Graz.

In 1884, at the age of twenty-eight, Nikola Tesla arrived in America and, having worked for Edison's company in Paris, he was eager to continue working for the Wizard. Tesla was tall, slender, and elegantly dressed. Edison was "rumpled, weary," and in need of some help. He had, at that time, built isolated electric systems for mansions belonging to the wealthy, including J. P. Morgan and Cornelius Vanderbilt, and, as with all new technologies, there were problems. Morgan employed a full-time engineer at his house to ensure he had light in the evenings, yet short circuits still set the drapes afire, and the lights frequently flickered. Not unlike neighbors of electric power plants today, Morgan's neighbors complained of the noise, vibrations, and soot from the coal used to generate electricity. Mrs. Vanderbilt had even called to demand that the Edison system be removed from her house after it had caused a fire that she had to extinguish herself. As if that was not enough, a major ship using the Edison electric system—the SS *Oregon*—had been delayed in port in need of electrical repairs. Edison sent Tesla out to the ship to do the repairs, which he successfully completed. This won Tesla a job as a part of the Edison team.[3]

Tesla did not stay with the Edison team for long. He offered to make Edison's direct-current electric generators more efficient, and Edison said he would give Tesla $50,000 if he could. Tesla succeeded after months of hard work on his own, but when he went to Edison for the $50,000, Edison backtracked and said he'd meant the offer as a joke. Tesla quit Edison's company the same day. But he needed money, and for a time he resorted to literally digging ditches; he could not have relished the work, since they were intended to bury the transmission lines for Edison's direct-current systems. News got around, however, that the talented Tesla had ideas.[4]

Investors approached Tesla to develop an improved arc lighting system. In 1885 he formed the Tesla Electric Light & Manufacturing Company with headquarters in Rahway, New Jersey. It was not focused on the alternating-current system he wanted to work on, but it was work nonetheless. Tesla succeeded in improving the arc light system, but soon was pushed out of the company and struck out on his own.

TESLA'S AC POWER SYSTEM

To understand Tesla's major contribution, it is important to define the difference between direct-current and alternating-current power. Electricity is most easily thought of as a flow of electrons. With direct current, the flow of electrons goes in one direction, but with alternating current, the direction of the flow alternates—it moves first in one direction, and then in the reverse direction. In an article written in 1890, George Westinghouse explained why the reversal occurred, writing that when "a coil of wire with united ends is caused to approach a magnet," then "a pulsation of electricity" is produced in the wire "flowing in a definite direction." Westinghouse went on to explain that "when the coil is withdrawn from the magnet" another "pulsation" is created in the wire "but this time in the opposite direction." This was an exact match to Faraday's experiments, which showed the induced electric current occurring only when the wire moved back and forth within a magnetic field.[5]

For electric motors, this alternating current is exactly what works best. Why is that? As noted earlier, the purpose of an electric motor is to take electrical power and convert it into mechanical power so that work can be done. Writer David Bodanis offers a clever way to think about how an alternating-current electric motor works. He imagines "a clock face with only a single long minute hand, pointing straight up to twelve o'clock." He then inserts an electromagnet on the face of the clock at the three o'clock mark and does the same at nine o'clock. He turns on the electromagnet at the three o'clock position, which attracts the minute hand. Then, just before the minute hand reaches three o'clock, he shuts the electromagnet off and turns on the electromagnet at the nine o'clock position. The minute hand spins with attraction to nine o'clock. If that on-off sequence is repeated, the minute hand whirls around clockwise, and that movement can do all sorts of work. If the magnets are strong enough, the motor could "drag a ton or more of elevator straight up a shaft in a tall building," or spin the "wheels of electric street cars," in Bodanis' words.[6]

If motors work best with alternating current, how did direct-current motors work? Direct-current motors worked by mechanically changing the direction of the current flow with a device called a commutator. One of Tesla's professors at Graz scoffed at him for suggesting that one day he would eliminate the commutator and make the electric motor much more efficient, but that is exactly what Tesla did.[7]

The impact of Tesla's discovery and invention did not stop at the electric motor; it traveled back through all the elements of an alternating-current electrical system. Most important, alternating-current power could be transmitted over long distances because the voltages of alternating-current power could easily be changed from higher to lower as needed—that is, the voltages could be *transformed*. Transformers change the voltage of electricity passing through the power lines, for example, as they enter a home. Transformers lower the voltage from a high level, which facilitates the transmission of electricity, down to a low one, which makes electricity safe for home use.[8]

Tesla wasn't careless about the business side of his scientific work. In 1887 he filed for seven patents for his polyphase AC motor and transmission system. The following year, Tesla, then thirty-one years old, gave a speech at Columbia University before the American Institute of Electrical Engineers titled "A New System of Alternate Current Motors and Transformers." George Westinghouse, the successful industrialist from Pittsburgh, was not in the audience for Tesla's speech, but he arranged a visit to Tesla's lab. Impressed by what he saw, he paid $5,000 in cash, $55,000 in Westinghouse Corporation stock, and a promise of royalties equal to $2.50 per horsepower of electrical capacity sold.[9]

WESTINGHOUSE'S PATH TO SUCCESS

Westinghouse was born in upstate New York on October 6, 1846. His father, George Westinghouse Sr., owned a machine shop that designed and made farm equipment; he also was something of an inventor, having seven patents to his name. It was in his father's machine shop that young George Jr. got his first hands-on experience with machines, including the all-important steam engine. By the time he was twelve, Westinghouse knew how to assemble a steam engine and developed a great interest in building a rotary version. The steam engines of his time were reciprocating engines, in which the steam pushed pistons up and down to facilitate mechanical work. This vertical motion had to be converted with rods and shafts into the rotary power needed for manufacturing, and this conversion meant that a lot of the steam power was wasted. In 1865, at age nineteen, Westinghouse won his first patent for a rotary engine that would help avoid wasted energy. It was the first of many; he went on to win 361 patents of his own and his companies held more than three thousand patents in total.[10]

The Air Brake

The growth of railroads empowered the American economy in the second half of the nineteenth century, as it set off explosive growth in coal, steel, machinery, and mining. It was in the railroad business that the young Westinghouse found his first technological and commercial successes. Railroad accidents were a major problem, both in terms of lost lives and lost revenue. It was time-consuming to right trains that had been derailed, meaning a tedious delay for travelers and businesses. Westinghouse's first invention for the railroad industry was the "rerailing frog," which efficiently put derailed trains back on the tracks. To manufacture his rerailing frogs, Westinghouse needed a steel foundry and he chose Pittsburgh's Cook & Anderson.[11]

The Civil War had transformed Pittsburgh into America's leading center of steel production. It was a powerhouse: 45 percent of the nation's iron was produced there, and 50 percent of its glass. This is why leaders in business, such as Andrew Carnegie, Benjamin Jones, and James Laughlin set up their businesses there. Pittsburgh thrived on industry, growth, and innovation.[12]

Westinghouse's greatest contribution to the railroad industry was not, however, in rerailing cars after accidents. Rather, it was in preventing train wrecks in the first place with his new braking system and new signaling and switching devices. Westinghouse first got the idea for his air brake when he was delayed by a train wreck near Troy, New York. At the time, braking power was applied physically to the individual train cars by brakemen every few cars. The core of the new air brake idea was that the train engineer alone could apply brakes with enough power to distribute to every train car. At first, steam was considered for the job since it was already being produced to power the train by burning coal in boilers. Steam could potentially be distributed to brake cylinders on each car through pipe, but the problem was that steam condensed and weakened as it traveled through the pipe.[13]

Westinghouse decided to try an alternative to steam after reading a European magazine article about a mine project where the builders had switched from steam to compressed air. The compression sped up the drilling significantly for the Mont Cenis Tunnel connecting the railroads of Italy and France. From this, Westinghouse saw the application to railroad brakes. In 1868 Westinghouse filed a patent for a compressed-air version of a brake called the "straight-air brake"; steam was used to compress air. The locomotive would then have a reservoir of compressed air, and when braking was required, the engineer would release the compressed air to each of the cars through a flexible pipe. When released, the compressed air would push the brake shoes against the train wheels and stop the train. His concept was met with skepticism from investors, but the brake did what it was supposed to do; it cut the average stopping distance from 1,600 feet to just 500 feet.[14]

In 1869 Westinghouse formed Westinghouse Air Brake Company with capital of $500,000.[15] He struggled to keep up with orders for his new air brake, so he made important efforts to standardize the manufacturing process. By 1881 the air brake dominated the passenger railroad brake business worldwide, and Westinghouse had manufacturing facilities in England, France, Germany, Belgium, and Russia. This made him one of the nation's largest international manufacturers, as well as a multimillionaire. More to the point here, it gave him important experience in mass-producing a technology that was in high demand. That experience would serve him well when he jumped into the electricity business.[16]

The air brake was used widely for passenger trains, but it took longer to catch on with freight trains. The obstacle was that freight trains, unlike passenger trains, frequently had to switch cars from one railroad company to another, so air brakes could be used only if all companies used them or if railroad equipment became standardized. In 1886 the railroad industry called for trials of competing brake systems, pitting the air brake against a competing technology, the vacuum brake. By 1889 the air brake became the standard recommendation by the Master Car Builders Association, and 90 percent of all cars had Westinghouse brakes by 1920. Westinghouse's success was founded on technological innovation and his

Garrison Alley research facility in Pittsburgh was a vibrant, productive place. Impressively, Westinghouse had achieved success without investment from the big banks.[17]

WESTINGHOUSE'S ELECTRICITY BUSINESS

Westinghouse made his move into the still-young electricity business in 1885. That year, Guido Pantaleoni, a Westinghouse employee who was also the son of a Westinghouse family friend, traveled to Italy and met with two men, Lucien Gaulard and John Dixon Gibbs. Gaulard had installed an alternating-current distribution system, and the two men held the relevant patents. Westinghouse became especially interested in the transformer they had used, and, in 1885, Westinghouse asked Pantaleoni to secure the rights to the technologies for the US market on Westinghouse's behalf. Biographer Quentin Skrabec believes that Westinghouse's interest in electricity evolved from his use of electric power in his railroad signals and switches. Westinghouse built his new electric company using his Garrison Alley lab, which rivaled Edison's Menlo Park facility.[18]

The Westinghouse Electric Company was formed in 1886 with $1 million in capital plus twenty-seven patents as assets, including the US rights to the Gaulard and Gibbs technology. Westinghouse put two of his most enterprising people in charge of the development of alternating-current power, Albert Schmid and Oliver B. Shallenberger. He also began a three-part research endeavor covering the alternating-current generator, alternating-current incandescent bulb, and alternating-current motor. He had already built three very successful businesses without the big banks, and he had substantial cash flow from his successes—Air Brake, Union Switch and Signal,[19] and Philadelphia Company (a natural gas supplier). Unfortunately, the electricity business demanded more investment capital than Westinghouse could muster on his own. He had to rely on the banks, which eventually led to financial crisis.[20]

Dirty Tricks in the Battle of the Currents

With the success of Westinghouse's alternating-current power system, the Battle of the Currents heated up. It was in this context that an odd sideshow emerged. In 1888 Harold Brown, a consulting electrician, wrote a letter to the *New York Evening Post* arguing that laws should be adopted to block the use of high-voltage alternating-current power because it was dangerous. A proposal to limit voltages was brought before the New York Board of Electrical Control, and Brown approached none other than Thomas Edison to help prove that alternating-current power was dangerous. Edison agreed to help Brown and assigned his colleague Arthur E. Kennelly to the task. Later that year, at the Columbia College School of Mines, Brown put on a spectacle by executing a seventy-six-pound dog with alternating-current power. In later demonstrations, he did more of the same. In addition, around this time, New York was looking for an alternative to hanging as a form of capital punishment; in light of Brown's spectacles, electrocution was proposed. Edison spoke in favor of this idea, and in the end the commission agreed to recommend the use of alternating-current power for electrocution, but left the logistical details to a panel of experts.[21]

When electrocution became the legal means for capital punishment in New York, Brown eagerly—and successfully—sought help from the Edison Electric Light Company and the Thomson-Houston Electric Company to have Westinghouse generators shipped to Auburn Prison to be used for executions. A man named William Kemmler was sentenced to electrocution there, but his legal counsel challenged his sentence, arguing that electrocution was cruel and unusual punishment. Edison testified that electrocution with one thousand volts of alternating-current power would bring a quick and painless death. The judge ruled in favor of Edison's testimony, and Kemmler was executed in August 1890. However, the process was neither quick nor painless. A *New York Times* headline read, "Far Worse than Hanging: Kemmler's Death Proves an Awful Spectacle." The first seventeen-second application of voltage did not kill Kemmler and the second sixty-second application had gruesome results, bursting blood vessels and singeing his body.[22]

Brown and Edison's efforts did nothing to stop the victory of alternating-current power. They did tarnish Edison's image, however, because he tried to impede the full commercialization of the alternating-current system, the eventual winner in the Battle of the Currents. Westinghouse's reputation and character were enhanced by the way he handled Edison's dirty tricks. In an article titled "A Reply to Mr. Edison," published in the *North American Review* in 1889, Westinghouse calmly put forth substantive proof of why alternating-current systems were safe. Westinghouse's transformers were at the heart of his arguments on safety. In essence, he said the lines that delivered electricity through the streets—the *street mains*—were completely separate from the lines that delivered electricity into a home, the *house mains*. Westinghouse wrote that "one of the most beautiful features" of an alternating-current system is the fact that the electricity sent through a "coil of thin wire" induces "an equivalent amount of electrical energy" in a "neighboring coil of thicker wire"—just as Faraday had proved. And those separate wires could be designed to have fundamentally different voltages.[23]

Success at the 1893 Chicago World's Fair

Although Westinghouse had secured the rights to Tesla's patents and worked diligently on the alternating-current system, he had no major commercial success for five years. Some of the delay was caused by technical choices the Westinghouse team made. At last, Westinghouse won his first major victory in the Battle of the Currents at the 1893 Chicago World's Fair, or the Columbian Exposition. Electricity took center stage at the fair, and it was Westinghouse's alternating-current power that made the show. After an aggressive competition with General Electric, Westinghouse won the lighting contract for the fair itself. Westinghouse was able to come in at about half the General Electric bid of $1 million because his alternating-current system used less copper wire, and a single system, with transformers, served both motors and lighting. In retaliation, General Electric won a court order that blocked Westinghouse from using Edison lamps of any type, so Westinghouse was forced to use the inferior "stopper lamp." This was a lamp designed and manufactured by Westinghouse to avoid infringing on the Edison incandescent bulb. It used inert nitrogen gas within the bulb rather than making a vacuum as Edison did. The stopper lamp had a shorter lifespan, but it was the quick fix needed to meet the demands of the World's Fair.[24]

Victory at Niagara Falls

The alternating-current system triumphed in an even more compelling way by winning the right to capture the power of Niagara Falls. There had been several failed plans to accomplish this over the years. In 1889, however, a group of bankers formed the Cataract Construction Company and finally garnered enough interest from investors including Morgan, Vanderbilt, and Baron Rothschild.[25]

To reach a sufficiently large market, the power had to be transported from the falls to Buffalo, New York. However, there was no consensus on how to move it over the twenty miles between the two. The options for what form the power should take included water under pressure, compressed air, and electricity. It soon became clear that the scientists and engineers favored the electricity option, but the choice between alternating current and direct current still hadn't been made.[26]

The Cataract Construction Company invited bids from companies around the world. By March of 1893 both Westinghouse Electric and the newly formed General Electric (the result of the merger of Edison General Electric and Thomson-Houston) submitted proposals for an alternating-current system. Initially, both bids were rejected because Cataract decided to design its own generators. In the end, Westinghouse modified the design of the Cataract company turbines before manufacturing them, and electric power was flowing in Niagara Falls by 1895 and to Buffalo by 1896.[27]

With the engineering and financial success at Niagara Falls and the Chicago World's Fair, Tesla and Westinghouse had decisively won the Battle of the Currents with their alternating-current system.[28]

TESLA'S WIRELESS WONDERS

After the victory at Niagara Falls, Tesla was famous. His contribution to alternating-current power was widely celebrated, and yet all was not well. In March 1895, Tesla's lab and all his records had been lost in a fire, and while Niagara Falls made him famous, it did not make him rich. Tesla's deal with Westinghouse called for a royalty of $2.50 per horsepower for the alternating-current systems actually developed with Tesla's design; he should have profited handsomely. But at the time, Westinghouse was opposing Morgan's effort to form an electricity trust, and his finances were on shaky ground. Westinghouse told Tesla he could not afford to pay the royalty and asked him to waive the contractual requirement. Purely out of gratitude for Westinghouse's faith in him from the beginning, Tesla nullified the royalty requirement. It was a decision that ultimately left Tesla struggling financially for the rest of his life.[29]

Tesla was eager for a new lab and funding; he wanted to get back to his work on wireless transmission. In 1895 he found additional incentive to do so because of advances in wireless communication. Guglielmo Marconi had built a wireless system that could send and receive Morse code over a distance of one and a half miles, but Tesla was convinced he was ahead of Marconi in creating what would be, in essence, radio transmission. In 1893 Tesla had demonstrated his own system, which had all the elements of modern radio technology, and he received his first patent for it in 1900. Marconi, meanwhile, filed his first patent in England in 1896 and another in the United States in 1900. For three years the US Patent Office refused to grant Marconi a patent, arguing that Tesla's patent preempted it, but in 1904, the Patent Office reversed its decision and granted Marconi's request for reasons never made clear. In 1909 Marconi won the Nobel Prize, and an embittered Tesla sued for patent infringement in 1915. He had no funds to pursue the case, but in 1943, many years after Tesla's death, the US Supreme Court named Tesla the primary inventor of the radio. Why did the court take up the case at this time? A popular explanation is that the Marconi Company sued the US government for using its patents in World War I. By restoring Tesla's patent, the court avoided the case.[30]

Tesla's radio work is germane to the history of electricity because he believed that, just as he had invented wireless transmission of radio waves, he could invent a mechanism for the wireless transmission of electrical energy. It was this belief that led to two amazing experiments; one in Colorado Springs, Colorado, the other in Long Island, New York. Tesla attracted $30,000 of funding from Colonel John Jacob Astor, owner of the Waldorf Astoria in New York, and his friend Leonard E. Curtis helped secure land and power for this work in Colorado. Tesla set up an experimental station with a few assistants in Colorado Springs in May 1899. There, he built an ordinary-looking, barnlike building with an extraordinary eighty-foot wooden tower resembling an old-fashioned oil derrick protruding out of the roof. Attached to the tower was a skinny 142-foot-tall mast with a copper ball on the end.[31]

Inside the barnlike building lay a large Tesla coil. In 1891 Tesla had patented his coil, which was a type of transformer that could take electricity at an ordinary frequency (sixty cycles) and increase it to frequencies of hundreds of thousands of cycles. The Tesla coil could also increase voltage to extremely high levels, so it was central to Tesla's research on high-frequency, high-voltage electric currents. This research set the pattern for his ultimately futile attempts to achieve wireless transmission of electrical energy.[32]

Tesla predicted that, far beyond transmitting wireless telegraphic messages, man could "transmit power in unlimited amounts to any terrestrial distance and almost without any loss." Tesla's biographers Margaret Cheney and Robert Uth point to one report that Tesla successfully tested his views by lighting two hundred incandescent bulbs from a distance of twenty-six miles from his lab. They conclude, however, that there is no evidence of such success and, moreover, no one recreated the alleged feat.

The same way natural lightning can cause a blackout, Tesla's first man-made lightning caused the entire city of Colorado Springs to lose power when it burned out the local generator. Tesla spent nine months on his experiments in Colorado Springs, and though there is little evidence he actually transmitted electricity wirelessly there, he seemed pleased with what he learned. Before his return to New York in 1900 he said, "The practicability of the system is thoroughly demonstrated."[33]

Tesla's friends Robert and Katharine Johnson were intrigued by his tales of the Colorado Springs experiments. Robert asked Tesla to write an article for *Century* magazine, which Tesla titled "The Problem of Increasing Human Energy." In it he painted a futuristic view of the importance of electric power; his ideas on energy ranged from capturing and storing the energy from lightning strikes to solar power. Tesla's article caught J. P. Morgan's attention, and, with the promise of major improvements in communication speed, Morgan funded Tesla with $150,000 to build a power plant and tower. Tesla used Wardenclyffe, a two-hundred-acre site near Shoreham on Long Island. Construction started in 1901 on an enormous tower that extended 187 feet into the sky with a fifty-five-ton ball made of steel at the top.[34]

Unfortunately, Morgan withdrew his support as suddenly as he had granted it. When Marconi successfully established wireless communication across the Atlantic in 1901, Morgan became skeptical of Tesla's focus on the great tower he was building with Morgan's money. When Tesla inevitably ran out of funds, he turned to Morgan in 1902 for more, and Morgan simply said no. By 1905, beset by financial problems, Tesla had to stop work, and with $20,000 in unpaid hotel bills, he turned over the deed for Wardenclyffe to the Waldorf Astoria.[35]

In 1916 Tesla was almost sixty years old and so broke that he had to file for bankruptcy. With Tesla's financial troubles as a contributing factor, the American Institute of Electrical Engineers awarded him the Edison Medal. Tesla was furious when first approached about the award: the fact that it was the *Edison* Medal must have felt like a provocation. Hamstrung by his financial difficulties, Tesla ultimately agreed to accept the award, and the ceremony took place in 1917.[36]

In 1931 a seventy-fifth birthday celebration was arranged for Tesla, and *Time* magazine put on its cover a reproduction of the *Blue Portrait* of Tesla. The portrait was painted in 1916 by Princess Lwoff-Parlaghy, a noted portrait painter and short-term wife to a Russian prince. It was painted under artificial light suggested by Tesla; the light from a strong incandescent bulb was filtered through blue glass.[37]

In 1937, on his way to feed the pigeons at Bryant Park, Tesla was struck by a taxi; refusing help, he made his way back to his hotel alone. Former colleagues at Westinghouse's company, worried about Tesla's finances, arranged a $125 monthly stipend from the company. Still, in 1942, a friend was alarmed to find him surviving on warm milk and crackers. On January 7, 1943, Tesla died in his sleep in his room at the New Yorker Hotel.

PUSHED OUT

Westinghouse continued to make significant contributions to lowering the cost of electricity generation with advances in the steam turbine as a substitute for steam engines. The steam turbine was first developed by Sir Charles Parsons in England in 1884. Westinghouse licensed the Parsons technology for manufacturing in the United States, and he became the first to manufacture steam turbines in America.

Westinghouse, as already noted, generally tried to avoid the big banks when financing his enterprises. Instead, he sought loans and equity investment from a relatively small circle of friends and associates. He also invested his own money heavily, if not exclusively, in his own endeavors. In 1907 a nationwide financial crisis led to his small circle of investors requesting repayment for many of the loans. As a result, three of his companies went into bankruptcy or receivership: the Westinghouse Electric Company, the Machine Company, and the Security Investment Company. Two of his other companies had no debt and survived the financial crisis: Air Brake and Union Switch and Signal.

Westinghouse Electric had borrowed a great deal of money from banks and, by 1908, needed a reorganization plan in which the debt from Westinghouse Electric was converted to stock ownership. This increase in outside stock ownership brought a new board of directors. While Westinghouse was retained as the president of Westinghouse Electric, his power was drastically limited. By 1911 Westinghouse was pushed out of the business altogether, his ouster a striking parallel to Edison's fall.[38]

THE WESTINGHOUSE-TESLA LEGACY

As in Edison's case, Westinghouse's and Tesla's contributions to the commercialization of the electricity business cannot be overstated. Together they built a commercial-scale alternating-current power system from end to end, just as Edison had done with the direct-current system. Notably, and also like Edison, they started to create their system by perfecting a technology for the use of electricity. Edison started with the lightbulb, while Westinghouse and Tesla started with the electric motor. This primary focus on inventions that transformed daily life and livelihoods would be lost as electric utilities made a shift to focus on the generation of electricity.

Who made the more important contribution to the history of electricity: Edison or Westinghouse and Tesla? It is crucial to remember that the last step in a journey is no more important than the first. Yes, Tesla and Westinghouse were right in taking the alternating-current fork in the road, but Edison's choice of the direct-current path was an essential point of comparison for proving the alternating-current path was better. The electricity business would not have developed at the same pace had they not both made important contributions. Indeed, as is often the case, the competition between these two efforts prodded them to move faster and take more risks than they might have otherwise.

Westinghouse was arguably better at recognizing and winning over talent than Edison; their respective treatments of Nikola Tesla are a case in point. Moreover, the dedication plaque on the Westinghouse Memorial in Schenley Park in Pittsburgh reads, "This memorial unveiled October 6, 1930, in honor of George Westinghouse is an enduring testimonial to the esteem, affection and loyalty of 60,000 employees of the great industrial organizations of which he was the founder."[39]

In a survey published in 2003, fifty-eight professors of history and management were asked to name "the greatest entrepreneurs and businesspeople in American history." The top five were Henry Ford, Bill Gates, J. D. Rockefeller, Andrew Carnegie, and Thomas Edison. George Westinghouse was ranked twentieth, which is not bad, but it is too far behind Edison. Westinghouse deserves better for what he accomplished and how he did it. He achieved enormous impact in two new, big industries of his time: rail transport and electricity.[40]

Importantly, Westinghouse won the Battle of the Currents by betting on Tesla's science; Westinghouse then merged Tesla's scientific genius with his own technological and business know-how. His technology, with its large scale and wide reach, powered the economics of the electricity business for many decades to come. As we'll see in the next chapter, Westinghouse's technology was the means by which Samuel Insull captured economies of scale and scope to electrify more homes and businesses than ever before.

CHAPTER 7

Samuel Insull's Electric City, Chicago

. . . the Ishmael or Insull whose hand is against every man's.
—Governor Franklin D. Roosevelt (1932)[1]

Edison, Westinghouse, and Tesla created the electricity industry **by commer**-cializing the full range of technologies, from power plants to lightbulbs to electric motors. Samuel Insull created the electricity *business* by establishing the practices that made electricity available and affordable to all.

Insull immigrated to America from England in 1881, at the age of twenty-one, expressly to serve as the private secretary for Thomas Edison. He became a trusted member of the Edison team, bringing rationality and sound financial practices to Edison's enterprise. Although Edison himself was eventually pushed out of the company that bore his name, Edison General Electric, Insull was asked to stay on in a top position at the newly created General Electric Company. Insull rejected the offer, however, and in 1892 he chose instead to move to Chicago to become president of a much smaller business called Chicago Edison. Insull believed the future of the industry was in selling electricity through electric utilities like Chicago Edison, which would generate and deliver electricity to homes and businesses, rather than in companies like General Electric that manufactured equipment. Chicago was fertile ground for Insull to test his ideas about the potential importance of electricity to America because it was one of the largest and most important commercial cities. Insull soon became one of its most prominent citizens.

From his new base in Chicago, Insull began to identify and implement the business and regulatory practices that would dramatically cut the cost and, consequently, the price of electricity. He was eager to gain new electricity customers because the more customers he had to share the fixed cost of generating and delivering power, the lower the average price of electricity per customer would be. These were the economies of scale and scope that Insull and others would use to justify the concept of a natural monopoly for local electric utilities.

Insull wanted and achieved a monopoly on electricity sales in Chicago (and elsewhere), but unlike the profit-hungry businessperson who would use a monopoly to raise prices, Insull used his monopoly to lower them. This is where his influence on regulatory practices comes in: Insull did not want an unbridled monopoly, and he actually requested that electricity prices be regulated by state government.

From his modest beginnings at Chicago Edison, Insull enjoyed a meteoric rise into Chicago's electricity business and then nationwide. However, it is his spectacular fall for which he is most remembered. Insull's financial downfall was brought on by a combination of his own strategic mistakes and the hard line taken by Chicago and New York banks during the Great Depression. On top of his financial ruin, politicians big and small contributed to Insull's political and personal losses. The collapse of his investment holding companies in the early years of the Great Depression made Insull a target for politicians looking for someone to blame for the nation's economic collapse. Among the most notable politicians was then Governor Franklin Roosevelt, who dragged Insull into a biblical tale when he railed against "the Ishmael or Insull whose hand is against every man's."[2]

Insull traveled to Europe for a rest, and while in Paris he heard of federal and local indictments against him for mail fraud, violations of the bankruptcy laws, and embezzlement, so he fled to Greece to avoid extradition. The US Congress brought international pressure to bear on Insull, and he was brought back for trial in Chicago. He took the stand in his own defense and told the remarkable story of his life, including his early years with Edison, and his push to make his companies big enough to make electricity affordable to all. His story—the story retold below—won him acquittal on all counts.

SHOULDER TO SHOULDER WITH EDISON

In 1874, at age fourteen, Insull started his career as an office boy for an auctioneer in London. In preparing to give a talk, Insull read an article in *Scribner's Monthly* about a young inventor named Thomas Edison, and from that moment on Edison was Insull's idol. At nineteen, Insull was fired unfairly to make room for a client's son. He came across an ad in which an American banker was seeking a secretary. Insull won the position when he offered to do the job for the lowest salary. As it turned out, his new employer was Colonel George E. Gouraud, Edison's European representative.[3]

Not long after Insull started his new job, Edison's chief engineer, Edward Johnson, came to London. Insull impressed Johnson with his willingness to work hard and with his in-depth knowledge of Edison's European ventures. As a result, Johnson convinced Edison to hire Insull as his private secretary, and Insull began his new position in New York in 1881. Forrest McDonald, Insull's biographer, reports that at their first meeting, Edison and Insull were disappointed with each other. To Edison, Insull looked too young, and his accent made it difficult for Edison to understand him. Edison also looked askance at Insull's impeccable dress. To Insull, Edison was poorly shaven and shabbily dressed, and his midwestern accent was equally hard for Insull to understand. But regardless of his appearance, Edison had invented the stock ticker, the multiplex telegraph, the phonograph, and the telephone transmitter, all by the age of thirty-four, and Insull was impressed.[4]

Edison's dilemma was that he needed money to build out his electric system; his investors had put $500,000 into his effort, which he spent on the research and development phase, without offering further funds for manufacturing or equipment. To raise the cash, Edison decided to sell the most valuable of his European securities—those in the telephone business. Insull, having come from the London office, had all the necessary information in his head, and by 4 AM on his first day in New York, Insull had written up a detailed plan to sell these assets. With this inaugural effort, Insull became the Wizard's financial wizard.[5]

Another reason why Edison had to find his own cash for his work was that, while he firmly believed in larger, central station electric generation in which one large power plant would serve a neighborhood of customers, many banks thought it was a safer financial bet to build smaller, more isolated power plants for each customer; they were hesitant to fund Edison's larger central stations. Edison asked Insull to take control of this central station endeavor, and by 1886, there was an explosion of growth in central station service. Edison would need to manufacture a lot more equipment, and to that end Insull found an unused locomotive factory in Schenectady, New York. Edison moved much of his manufacturing there and sent Insull off to run the whole manufacturing effort with the admonition, "Do it big, Sammy. Make it either a big success or a big failure." Insull managed to make it a big success; he started with two hundred employees in 1886 and drove that to six thousand by 1892.[6]

INSULL'S INNOVATIVE PRICING

From the start, Insull made it clear that he wanted the lowest possible price so that electricity could become a product for all people. In 1892 Chicago had a population of one million, but only five thousand were electricity customers. Insull, more than anyone else at the time, understood the pricing problem faced by an electric utility that wanted to win over new customers. The issue stemmed from the physical fact that electricity could not be efficiently stored on a large scale, which made it different from other forms of energy commodities. For example, gasoline for a car can be produced from crude oil at an oil refinery and then stored to be used in the future, something that could not be readily done with electricity. With few exceptions, electricity must be used the moment it is produced, and it must be produced the moment it is needed. The consequence of the lack of effective storage is substantial. It means that a large number of power plants and extensive transmission line capability would have to be built to meet the maximum or *peak* demand for electricity, and much of that capability would sit idle at times other than at that peak moment.[7]

Still, Insull was determined to decrease the price of power. The best way to do this was to sell more electricity from the same power plants using the same transmission and distribution lines, so the fixed cost could be spread over more units of electricity and thus be cheaper for everyone. Selling more electricity *using the same equipment* was central to Insull's plan to decrease price, and having customers that needed electricity at different times of day or times of year was essential to using the same equipment. Electric-lighting customers needed electricity at night because of the dark. On the other hand, transit customers who operated the electric streetcars did so during the day, so they were perfect for diversification with residential customers. Large factories were the ideal customers when they used electricity to run motors around the clock. Insull pursued these customers with vigor to make his business model work. One challenge was that electric streetcar companies and large industrial customers often generated their own power; Insull worked hard to get these customers to buy electricity from the utility instead.[8]

Next, Insull had to price electricity in a way that encouraged customers to sign up with his utility and then to use more and more electricity over more and more hours of the day. On a trip to London in 1894 Insull learned of something called the two-part price from a man named Arthur Wright. One part of the price would recover the fixed costs of generating and delivering electricity, such as the monthly payments on loans that were used to build the power plants. The other part would recover the variable costs of electricity, like the cost of the coal burned to make the power plant run. Large industrial customers found the two-part pricing especially attractive. It allowed them to use as much electricity as they wanted to around the clock, and to spread the fixed cost over a substantial number of kilowatt-hours (kWh) of actual electricity use, thus making the average cost quite low. For this same reason, transit customers found two-part pricing attractive, too. Think of it in terms of an all-you-can-eat buffet. Insull was inviting customers to come on in and, for a fixed fee per month, eat as often and as much as they wanted.[9]

In assembling a diversified set of customers and by implementing an innovative, two-part pricing structure, Insull created two essential elements for his plan to make utility electric service available and affordable to all.

BIGGER AND BETTER EQUIPMENT

Insull then created the third essential element by buying and installing bigger and better equipment to generate and deliver the electricity he sold. One of Insull's first steps was to build an innovative power plant in Chicago, located where Harrison Street crossed the Chicago River. He increased the efficiency with which coal was turned into electricity, thereby cutting coal use in half.[10]

Next, Insull strived to change the fundamental technology of generation by switching from steam *engines* to steam *turbines*. Here, Insull truly pushed the technology envelope. General Electric agreed to build the 5,000 kW steam turbine Insull wanted, but if and only if Insull shared the risk of its development. Within ten years, he was pushing for even larger-scale generation with 35,000 kW steam turbines. By pushing larger-scale plants, Insull was achieving the economies of scale that would push electricity prices down.[11]

OLD POLITICS LEADS TO NEW REGULATION

Local politics in Chicago suffered from its share of corruption in Insull's day, and city council votes were routinely bought and sold. The city council was also known to extort funds from electric utilities by setting up fake, competing utilities that were secretly owned by the city council members. These fake companies were never operational, but they were granted long-term monopoly franchises, which meant the existing local utilities had to buy them out. The city council employed this tactic against Insull in 1897 when its members created a company called Commonwealth Electric and granted it a fifty-year franchise to serve the public. Had the extortion worked, Insull would have been compelled to buy Commonwealth Electric, but Insull had the upper hand. Unknown to the city council, Insull had won an exclusive right to buy equipment from every US manufacturer for the Chicago area. Since the city council's fake company was not worth much if it could not buy equipment, Insull bought out the city council members for a small sum and secured a fifty-year franchise.[12]

Insull then proposed that the prices charged by electric utilities like his be regulated by state governments. Yes, Insull actually asked for state government regulation and in exchange utilities would receive an exclusive "monopoly" franchise—that is, the exclusive right to serve a city or another local area. The state would set the prices or "rates" the utility could charge its customers based on costs actually incurred by that utility, plus a normal profit; that's why they are called *cost-plus rates*. Cost-plus regulation was the fourth essential element Insull set in place for the electricity business.[13]

Surprising as it may seem, Insull was not alone in his support for cost-plus regulation. The National Electric Light Association, the lobbying group for electric utilities, promoted regulation both as an antidote for political corruption and as an alternative to city takeovers of utilities. In 1907 the National Civic Federation, another high-profile group in which Insull was prominent, issued a three-volume study on the principles of regulation. The study became the general blueprint for state regulatory laws and led directly to the regulatory commission in Wisconsin, the Wisconsin Railroad Commission. John R. Commons, then a prominent economist from the University of Wisconsin and Insull's colleague at the National Civic Federation, wrote the law. After Wisconsin, Massachusetts and New York established laws for utility regulation in 1907; thirty more states followed within the next nine years.[14]

The creation of local monopolies governed by cost-plus regulation by the states was crucial to Insull's strategy. A monopoly allowed the utility to get big enough to achieve economies of scale. This, in turn, allowed costs to fall. Falling costs meant lower prices for customers, and allowed the monopoly to grow, which created even more opportunities to realize economies of scale and scope. It was a virtuous cycle of sorts. This was a beneficial strategy as long as economies of scale remained and the goal was higher and higher electricity sales.

Insull offered to lower electricity rates if Chicago would approve the merger of his two companies, Chicago Edison and Commonwealth Electric.[15] In 1907 Insull went forward with his merger to create the Commonwealth Edison Company—a company that today still serves 70 percent of Illinois's customers. (It is part of the Exelon family of companies.)

INSULL'S RISE

In 1907, just forty-eight years old, Insull had achieved a great deal. His own electric companies were a success, as was the electricity business across America. It was Insull's vision that served as the business model for this success. He pushed for monopoly regulation that allowed economies of scale to be realized, and in so doing he had started on the quest to make electricity available and affordable to all—"as natural and necessary as breathing."[16]

Insull pushed forward with his plan to expand service and cut prices. In 1910, through his Lake County experiments, he showed that he could cut the cost of power to even sparsely populated rural areas by using long-distance transmission lines to deliver power from central station generators. In 1911 Insull saw the potential for substantial cost savings for customers outside the Chicago area and he formed the Public Service Company of Northern Illinois to reach those households. Insull had calculated that, for Illinois outside Cook County with its diversified electric needs, a single company could serve the area with 38 percent less generating capacity, cutting costs in half.[17]

In 1912 Insull's companies had $90 million in assets centered on electric service in the Chicago area, but by 1917 his assets had more than quadrupled to $400 million and included electric, gas, and transit service in thirteen different states. In 1912 Insull formed his first holding company, the Middle West Utilities Company. In that same year, Middle West acquired Central Illinois Public Service Company, which operated a transit company and a small electric utility. Insull rapidly grew the company by providing service to communities that previously did not have any. Starting with about 15,000 customers who used 15 million kWh of electricity, the company expanded to 150,000 customers with 400 million kWh by the time Insull was finished. To be sure, holding companies brought with them complicated ownership and financial structures that could easily raise suspicions when scrutinized, and they would be put under a microscope during the Great Depression. Those suspicions included concerns that stock prices were inflated and fees charged by the holding company to its subsidiaries were too high. For Insull, however, in the early years the holding companies were simply a vehicle for growth that meant he could offer lower electricity prices.[18]

On top of his other successes, Insull was asked to serve as the chairman of the board of Peoples Gas Light & Coke Company to clean up the corruption within the company and restore its credibility.[19] Insull was also very active in community service during World War I and supported the war effort as chairman of the Illinois State Council of Defense, winning praise from the Illinois governor for making the organization innovative and effective.[20] Insull's actions during World War I helped to cement his reputation and shape his empire. Soon after the United States entered the war, the price of coal skyrocketed.[21] To his credit, Insull was insulated from this price spike and supply disruption because he had negotiated long-term contracts with Peabody Coal.[22] Again, Insull, a leading businessman, invited government regulation by encouraging the federal government to control coal prices, which it did by setting a price cap to cut the price of coal in half.[23]

In 1914 the Illinois Public Utility Commission was created as a key governing body in the state regulation that Insull had fought to put in place.[24] Almost immediately, however, Chicago and other municipalities started to lobby for a return to "home rule"—city regulation of utilities.[25] The new state commission soon was caught up in another sweeping change in politics once America went to war. Those leading the antiwar campaign turned it into a campaign against war profiteering, and utilities were at the top of the list of targeted companies.[26] Progressive leaders all over the Midwest—including well-known figures like Robert M. La Follette, a senator from and former governor of Wisconsin, and George W. Norris, a senator from Nebraska, joined in.[27]

Utilities of all sorts kept coming to the regulatory commission to win rate increases after the war; when denied, many went into bankruptcy. Calls to abolish the Illinois Commission grew, and in the face of home rule fervor in 1920 the Illinois Public Utility Commission was dismantled by the state legislature. Once the public denunciations ended, Insull worked quickly to reestablish state regulation, at first by simply renaming the commission; it would be called the Illinois Commerce Commission, a name that persists today.[28]

Insull's star rose in the postwar era. As McDonald puts it, "In the hero-worshipping postwar decade, Insull became the Babe Ruth, the Jack Dempsey, the Red Grange of the business world. The people—butchers, bakers, candlestick-makers who invested in his stocks—fairly idolized him, and even titans viewed him with awe. He measured up to America's image of itself: a rich, powerful, self-made giant, ruthless in smashing enemies, generous and softhearted in dealing with the weak."[29]

INSULL'S FINANCIAL FALL

The beginning of Insull's financial fall came when he responded to a threat from a corporate raider named Cyrus Eaton. Eaton started to buy Insull's operating companies' stock, and to defend against Eaton accumulating too many shares, Insull created a trust company in 1928 called Insull Utility Investments (IUI). Insull and his friends sold all their stock in the operating companies to IUI in exchange for its stock. They would then attempt to buy, through IUI, enough voting shares of the four big operating companies to achieve voting control and fend off Eaton. Insull set the initial value of the IUI stock at twelve dollars per share, and he then secured the option to buy a large block of additional shares at fifteen dollars per share. In the months that followed, the price of IUI stock skyrocketed to $150 per share. Stock prices for the operating companies soared, too. These high stock prices stifled Insull's plan to gain voting control; it became too expensive to buy enough stock. Insull set up a second trust in 1929, the Corporation Securities Company of Chicago, or simply Corp. IUI and Corp also owned each other's shares, and this became another level of protection against the corporate raider.[30]

But these measures reached their limits. Ultimately, Insull agreed to buy Eaton's shares and had to borrow money to do so. His bank loans were secured by IUI and Corp stock. Since the value of IUI and Corp stock was driven by the stock prices for the operating companies, the stock prices for IUI and Corp would fall accordingly if the operating company's stocks fell. This was exactly what happened as the Great Depression took hold, and since IUI and Corp stock were the collateral for the bank loans, as stock prices fell, more and more of the IUI and Corp shares became collateral. In this way, the banks slowly began to control more and more of Insull's companies. By mid-December 1931, all of IUI and Corp's stock had turned into collateral for Insull's bank loans.[31]

In April 1932, Insull went to New York to discuss refinancing $10 million of debt for Middle West.[32] According to McDonald, men from the House of Morgan entered the room and, after asking Insull to wait outside, declared that the banks would not refinance. For a paltry $10 million loan, Middle West, a $1.5 billion company, was forced into receivership, and IUI and Corp went under, too. By June 1932 Insull was forced to resign from his other operating companies. As Insull saw it, the banks "had ceased to find me of use to them, and had decided that the time had come to throw me overboard."[33]

INSULL'S POLITICAL FALL

After the banks were through with Insull, having taken his money and gained control of his companies, the politicians saw their chance to attack. Insull's political mishaps had originally begun much earlier, in the mid-1920s, when he gave a great deal of money to Frank Smith, head of the Illinois Commerce Commission, to finance Smith's campaign for the US Senate. Insull had given $125,000 directly and financed $33,735 in what today might be called "issue ads" related to the World Court. US Senator James Reed of Missouri subsequently opened an investigation into utility contributions to campaigns in Illinois and Pennsylvania. After joining forces with Senator La Follette, the Reed Commission held hearings in Chicago in 1926, and Insull was attacked almost daily in the press. Ultimately, the Senate declined to seat Smith because of the Insull contributions to his campaign.[34]

Worse yet for Insull was the fact that 1932 was an election year. John Swanson, Illinois' attorney for Cook County and a Republican, faced a tough reelection battle. Despite his apparent respect for Insull, Swanson announced an investigation into the collapse of Insull's business to help his reelection bid. In the wake of Swanson's investigations and the headlines it spawned, politicians nationwide jumped into the fray. The threat of federal indictments was added to Insull's burdens when US Attorney Dwight H. Green announced that the US Department of Justice had begun its own investigation.[35]

For his Cook County investigation, and to add a spark to his reelection campaign, Swanson released a secret list he had of who got to buy

IUI shares for the bargain price of twelve dollars per share. The list was pretty routine; mostly people known to be friends of Insull's, but it also included Anton Cermak, the Democratic mayor of Chicago. Others who appeared on the ever-growing list were the speaker of the state house of representatives, a former lieutenant governor of Illinois, and a number of state and federal judges. With the list and press reports as a dramatic backdrop, Swanson won grand jury indictments of Insull and his brother for "embezzlement, larceny, and larceny by bailee" in October 1932.[36]

Before the indictments, in June 1932, Insull had left quietly for Paris with the hope of finding some peace after his defeat by the banks. The financiers had cut a pension deal in which he would receive $50,000 a year for life. With news of possible indictments, Insull's son, Samuel Jr., pleaded with him to leave Paris, so he escaped to Athens because his lawyer told him that the United States did not have an extradition treaty with Greece. Samuel Jr., age thirty-two, did most of the fighting against both the Illinois and federal governments while Insull was abroad. His son bore the brunt of the anger and hatred from all sides, and later he, too, would be indicted for mail fraud and violation of bankruptcy laws.[37]

As Insull's lawyer had hoped, Greece refused to extradite Insull. To persuade the Greek government, however, the US government put pressure on the Greek American Merchants Association. These well-to-do merchants sent substantial funds home to Greece, so the federal government threatened to block the transfer of those funds. Faced with this ultimatum from the United States, Greece asked Insull to leave the country in March 1934. Both chambers of the US Congress passed a bill ordering Insull's arrest in any country with a treaty agreement that would allow them to do so. Insull chartered a ship and spent two weeks in the eastern Mediterranean, but when the ship docked in Turkey for provisions, the Turkish government arrested him and turned him over to the US embassy.[38]

Insull was jailed at the Cook County jail upon his return to America. His lawyer, former Illinois Supreme Court judge Floyd Thompson, had arranged bail at $100,000, but the amount was doubled when Insull actually got to the jail. In his memoirs, Insull noted that his bail was four times higher than that set for the infamous gangster Al Capone. Foreshadowing what would become the core of his defense at trial, in a brief statement to the press Insull said, "I have erred but my greatest error was in underestimating the effects of the financial panic on American securities and particularly on the companies I was trying to build . . . I worked with all of my energy to save those companies. I made mistakes—but they were honest mistakes. They were errors in judgment but not dishonest manipulations." From the start, Insull's defense was that he was being "persecuted for the sins of his generation," and that his trial would be "persecution" rather than "prosecution."[39]

THE TRIAL OF AN ERA

The federal indictment on mail fraud was the first to be brought to trial on October 2, 1934. The core of the fraud allegations was the claim that Insull and his codefendants induced buyers to purchase Corp stock at inflated prices. According to the federal government, this fraud was perpetrated primarily through a nationwide sales campaign in which the value of Corp holdings was deliberately and significantly overstated. Francis X. Busch, who chronicled the trial, reports that it was a complicated indictment that the jurors might have had trouble following in detail. However, it is certain that the jurors got the gist: the federal government was alleging that Insull had swindled the "little people."[40]

The allegations concerning Insull's Middle West Utilities are illustrative of the type of transaction causing scandal. Insull had fundamentally reorganized the financial structure of Middle West and two elements of that reorganization were relevant to the government's case. First, Middle West was aggressively selling stock, which was purchased by Insull, his agents, and companies he controlled, like Corp. Importantly, its value was then marked up in Corp's books to a value higher than the purchase price. For example, shares of Middle West Utilities were purchased for $12.7 million, but valued on Corp's books at a much higher price of $23 million, which was the value at the current stock market price. The government referred to the increase from $12.7 million to $23 million as a "write-up"; today, this would be called a mark-to-market valuation, which is widely used and still problematic at times (as will be seen in the story of Enron). Insull and his codefendants were always on both sides of the transaction, and the write-up looked suspicious, so it is understandable that people began to question him. The government argued that the efforts to sell Corp stock were fraudulent because the value of the assets in Corp were wildly overstated, and they pointed to the write-ups as evidence.[41]

The second element of the Middle West reorganization was that, going forward, Middle West stock would pay dividends in stock rather than in cash. The government charged that the way Corp treated the stock dividends from Middle West as current income was fraudulent because stock dividends could not be income until the stock was sold. If these stock dividends could not be counted as income, Corp was not only unprofitable, it suffered a loss as well. In 1930, for example, if stock dividends were erased from income and some losses on stock transactions were shown as an expense, the government alleged that Corp suffered a loss of $550,000 instead of the $6.8 million profit it showed on its books. This meant it was fraudulent to have claimed that Corp was a "sound" company when promoting stock sales, since technically it suffered a large loss.[42]

Once the government rested its case, Insull, at age seventy-five, had to testify to rebut the charges against him. The core of his testimony was simply the honest story of his life, and with that story he refuted every claim against him. Busch says that Insull captivated the courtroom with the narrative of his arrival in America and his role as Edison's most trusted adviser. He told of how he built the then-small Chicago Edison into a big company, and how he wanted to grow his companies in order to lower electricity rates for customers.

Insull spoke of how investment trusts like Cyrus Eaton's were a real threat to control of the operating companies he had built up to serve customers in Chicago and elsewhere in Illinois. Insull Utility Investment, he explained, was formed to combat the Eaton threat.

Insull went on to say that all the mechanics of the Middle West Utilities reorganization were sound and accepted accounting practices.[43] Defending himself more broadly, he conceded that stock prices of his companies had fallen with the Depression, but so had the prices of blue-chip stocks; the stocks for companies such as GE and GM lost 90 percent of their 1929 values. Insull had held on to his stock and lost everything as a result: his Lake County estate, his wife's dowry, even a life insurance policy.[44]

In his summation, Insull's lawyer stated there had been no fraud because all Insull's alleged fraudulent actions were shown clearly on the books of his companies. Surely, the lawyer argued, Insull's actions were motivated by attempts to save the operating companies from the Eaton menace and not to seek a profit. Insull and the others saw these as "good investments." Insull's lawyer argued that the federal government brought the case solely to lay blame for a "tragic era." In truth, said the defense attorney, it was only "Old Man Depression" who got all that money that was lost. Judge Wilkerson in his instructions to the jury said that the jurors must find intent to defraud, and that Insull could not be convicted for honest mistakes. In a bit over two hours the jury returned a verdict acquitting Insull and his codefendants of all charges.[45]

In the early years of the twenty-first century, Insull's trial found an echo—and a stark contrast—in the proceedings against Enron.

INSULL'S LEGACY

The editor of Insull's memoirs put it this way: "Single-handedly, Insull brought the industry from the status of an experiment that served the rich to a major industry and utility that served everyone." It is in this sense that Insull *democratized* electricity. He achieved this by relentlessly pursuing economies of scale and scope, which drove electricity prices down dramatically. And to create the incentives to pursue these economies, he invited cost-plus regulation of local monopolies by state governments. That regulation, with profits dictated by size, was a hugely effective incentive to get big.[46]

As long as the overarching goal was availability and affordability, and economies of scale were not exhausted, Insull's strategy was a remarkable success. The choice to invite state regulation was not ideological; it was a specific means to a specific end, and it worked. It is also worth noting that cost-plus regulation had a known downside: it diminished the incentive to take the risks necessary to make further advances on the science and technology that made the electricity business possible in the first place. In this sense, it killed the goose that had laid the golden egg.

Insull's career also set the pattern for the spectacular rise and igno-minious fall of a high-profile businessperson. With his rise from the small platform of Chicago Edison in 1892, Insull created one of the two larg-est conglomerates of electricity companies in America. In 1930, by one count, the Insull holding companies accounted for 12.3 percent of all elec-tric revenue in the United States. Insull's group then was equal in size to the conglomerate of companies controlled by Electric Bond & Share, General Electric's holding company. Adding in his gas utilities, at the time of his fall in 1932 the public held over $2.6 billion of Insull company stocks and bonds. His companies had more than 600,000 shareholders, 500,000 bondholders, and served 5,000 communities in thirty-two different states.[47]

With his fall, Insull himself lost everything, but how much was lost by the public who owned stocks and bonds in his companies? By one esti-mate, they lost far less than might be presumed based on the headlines of the day or the exaggerated references in the decades that followed. It is estimated that of the $2.6 billion in publicly held securities in 1932, investors lost $638 million or about 24 percent of the money they put into Insull's companies. These losses were not in the operating companies, the core of what Insull created; they were in the investment or holding com-panies, or in the transit and railway companies.[48]

Moreover, when the allegations of fraud and other wrongdoings were judged by a jury, Insull was acquitted in all three cases against him. The jurors' ultimate role was to weigh the good and bad of Samuel Insull. And there surely were both—the good being his democratization of electricity and the bad being his murky accounting and use of vast holding compa-nies. Also central to his acquittal was the fact that the jurors abided by the judge's instructions that, to find him guilty, there must be evidence of intent to defraud the public. No such intent could be proved.

In 1938, at the age of seventy-eight, about four years after his acquittal, Insull died of a heart attack in a Paris subway.

ELECTRICITY'S MARCH ACROSS AMERICA

Americans are greatly impressed by technology entrepreneurs and venture capitalists who invent, finance, and manufacture products that sell to huge numbers of people and businesses. They affirm the American Dream: that ingenuity and hard work, rather than inherited wealth and status, lead to worldly success. Electricity became one of those products. The electricity business used a brand-new technology to provide a brand-new product that sold spectacularly. In twenty years, from about 1902 to 1922, American electricity generation grew at a rate of over 12 percent per year. Another measure of success for this new product was market penetration in residential electricity use: only 8 percent of homes had electric service in 1907, but that grew to 35 percent in 1920, and then to 90 percent in 1948. Equally impressive was that each of these homes continually increased the amount of electricity it used. In 1912, the Census reported that, on average, a household used 264 kWh each year. By 1920 each household used 339 kWh per year on average, or about 28 percent more. By 1948 the average electricity use per household was 1,563 kWh, almost six times larger than use per residence in 1912. In sum, over the first half of the twentieth century, electricity became something that no home could do without, and each home found ways to use more and more of it.[49]

A dramatic decrease in the price of electricity to residences drove, or at least enabled, increased market penetration and use of electricity by each residence. In 1902 the average price of electricity to a residential customer was 16.2 cents per kWh. By 1920 the average price decreased by more than half to 7.45 cents per kWh. By 1948, the average price for residential customers was 3.01 cents per kWh, an 81 percent decrease from the price in 1902. Economies of scale and scope did, as Insull calculated, make electricity far more affordable.[50]

The new electricity business had a significant impact on the American economy because it drove what historians define as a second industrial revolution. The term is apt because the electric motor replaced the steam engine, which had powered the first Industrial Revolution. As was true with the steam engine, the payoff of the new technology of the electric motor was a surge in productivity throughout the manufacturing sector. The advent of electric motors allowed a dramatic redesign of the entire manufacturing process. With that redesign, the productivity of American workers and machines soared and helped make America a world power.

PART THREE

The Age of Big

Big Government, Big Power Plants

CHAPTER 8

FDR's New Deal (for Electricity)

I do not hold with those who advocate Government
ownership or Government operation of all utilities.
—Franklin D. Roosevelt (1932)[1]

In September 1932 the Democratic candidate for the presidency of the United
States, then New York governor Franklin Delano Roosevelt, traveled to
Portland, Oregon, to give a campaign speech entirely about electricity.
What was most surprising about Roosevelt's speech was his admission
that he did not believe the government should take over the electricity
business. In the midst of the Great Depression, which many viewed as a
great failure of capitalism, and after the financial collapse of giant utility
holding companies, the leap to government ownership was an obvious,
attractive option in the stressful context of the time. Instead, Roosevelt
went on to define and implement a three-prong New Deal policy for
electricity: "fund major federal [hydroelectric] power projects, bolster

the threat of municipal competition, and dismantle holding company systems."[2]

When he emphasized funding federal hydroelectric power projects, Roosevelt certainly was aware of his audience; he told the people of Portland that the "next great hydro-electric development to be undertaken by the federal government must be that on the Columbia River," in the Pacific Northwest. As will be discussed in subsequent chapters, two other federal hydroelectric projects with their roots in the pre-Depression 1920s came alive in the midst of the Great Depression: what Roosevelt called the Boulder Dam—now the Hoover Dam—and the Tennessee Valley Authority. Roosevelt would keep his promise for the Pacific Northwest with the construction of the Bonneville Dam in 1938 and the Grand Coulee in 1942.[3]

When Roosevelt addressed municipal development, the intent was to give the investor-owned utilities some potential competition. Roosevelt spoke of establishing a "yardstick" for the rates and other terms of service offered by private utilities through the development of federal hydroelectric projects. He also spoke of potential municipal utilities as a disciplinary force that could be brought out when a "mere scolding" was not enough to get private utilities to improve their service or rates.[4]

The dismantling of holding companies, meanwhile, began in dramatic fashion with the Public Utility Holding Company Act of 1935. One scholar concluded, "Except in wartime, the federal government never before assumed such total control over any industry." In 1935 a companion piece of legislation was passed by the US Congress, the Federal Power Act (FPA). At the time, it was likely considered the lesser of the two laws; however, it was put in place to address an additional problem created by the holding companies. The problem was that interstate holding companies were beyond the reach of state regulators. Under the FPA, the Federal Power Commission (FPC) would regulate interstate transmission and wholesale sales of electricity. Roosevelt understood this would trigger a political problem with states' rights.[5]

Today, the FPA is the primary source of the federal government's jurisdiction to regulate the electricity business. This federal regulation is a significant legacy of the Great Depression; it was born at a time in which most people wanted the federal government to intervene aggressively to stop the worst economic crisis in American history. National regulation born of crisis is not necessarily the regulation needed for the long term; it most likely isn't designed for when there is no crisis, or when there is a different crisis. And if national regulation is standardized, it runs the risk of undermining America's great strength: the diversity of ideas and opportunities across the states.

BACKDROP TO THE NEW DEAL

Two high-profile studies done in the 1920s provide an interesting backdrop to President Roosevelt's New Deal for electricity in the 1930s. The purported impetus for these studies was the near shortage of electricity during World War I and the concurrent coal price spike. Both studies also reflected Insull's call for economies of scale; bigger power plants would cut the cost of electricity (as well as coal use). And, to accommodate these bigger power plants, long-distance, high-voltage transmission lines would be built.

The two studies were the Super Power study and Giant Power study; their names alone reflect that they were dreamed up in the Age of Big. The two differed in their recommendations of who would control this network of big power plants and big transmission lines: the existing electric utilities (Super Power) or state government (Giant Power). That made a big difference in how the electricity business would operate. With utility control, the plan was that all the utilities would be interconnected, but the decision on whether to use that interconnection to buy power from neighboring utilities would be at the discretion of the utilities themselves. The alternative was to assure that interconnection meant integration—that is, all the power plants would be integrated and run as one system.

Super Power and Giant Power have historical importance for two other reasons. First, holding companies like Insull's Middle West Utilities were achieving lower prices through economies of scale and integrated operation in the 1920s and, despite this, concerns over large utility control of the business were already emerging. By 1935 the dismantling of those holding companies became the central goal of the federal government. Second, neither the Super Power nor Giant Power proposals made progress in either federal or state legislatures. However, together, they started a slow process toward the development of a template for how a competitive electricity business should work. Traces of Super Power and Giant Power proposals can be discerned in the gold standard for competitive markets today, the PJM Interconnection.

The acronym *PJM* is driven by the original geographic coverage, which included Pennsylvania, New Jersey, and later Maryland. Today the coverage extends west to include Chicago. PJM is one of the six large regional electric markets in the United States, called regional transmission organizations (RTOs) or independent system operators (ISOs). Within the boundaries of these markets, competition among electricity suppliers is promoted by assuring all suppliers have fair access to the transmission system and by operating transparent markets for electricity sales. While the promotion of competition is new, these RTOs and ISOs relate directly to Insull's vision in the sense that they also promote the capture of economies of scale by including widely dispersed customers and suppliers.

Super Power

The request for funding for the Super Power study covering power supply in the Northeast and mid-Atlantic first came in 1918. Congress ultimately provided $125,000 for the study and an advisory board was established, which included Herbert Hoover, then secretary of commerce.[6]

The rhetoric in the Super Power report was muted. Other than reference to extreme power shortages being narrowly avoided during World War I, there was little appeal to the reader's emotions to win support. The report made it clear that Super Power was not intended to compete with the existing electric utilities but was intended only to "coordinate and supplement these utilities."[7]

As is typical of proposals for power systems in any era, there is plenty of mind-numbing data describing the physical reality of the system as it was in 1919 and also as the proponents of Super Power hoped it would be by 1930. The Super Power "zone" covered an area from Maine to Maryland that included 25 percent of the American population. The report claimed that the effect of the Super Power plan would be to decrease the number and to increase the size of the power plants serving the zone. The report found that utilities had 558 power plants with a small average size of just 7.9 megawatts (MW). Regarding the future, with Super Power in place in 1930, the report claimed that a total of only 273 power plants would be needed, with 218 belonging to existing utilities. And the base load plants—those intended to run all the time—would be significantly bigger, ranging from 60 MW to 300 MW.[8]

For all the reasons Insull noted, the payoff of going big would be dramatic cost decreases. The Super Power study estimated that by 1930, there would be a 46 percent drop in cost, a savings of $278.5 million. With the importance of coal conservation in the wake of wartime coal shortages, Super Power promised a significant decrease in the amount of coal used to generate electricity. According to the proposal, new power plants were expected to cut coal use by 48 percent per kWh by 1930.[9]

Giant Power

In contrast to the muted rhetoric of the Super Power report, the tone of Giant Power was piercing, especially the sections by Gifford Pinchot, then governor of Pennsylvania and one of the proposal's originators. No less ardent in his support for the proposal was Morris Llewellyn Cooke, a prominent engineer. Drawing a distinction between his proposal and Super Power, Pinchot argued, "Giant Power and Super Power are as different as a tame elephant and a wild one . . . The place for the public is on the neck of the elephant, guiding its movements, not on the ground helpless under its knees." Carrying on with his case against Super Power, Pinchot wrote, "The main object of the superpower idea is greater profit to the companies. The main object of the Giant Power idea is greater advantage to the people."[10]

Pinchot went on to claim that the electricity business was a "natural monopoly"; that is, it could not be regulated by competition, and instead had to "be regulated by public control." And yet Pinchot was clearly not proposing public ownership when he wrote, "The Giant Power Plan takes no account of public ownership." However, Pinchot warned that public ownership would come if the utility companies "opposed and prevented reasonable and effective regulation by the states and by the nation."[11]

The substance of the Giant Power proposal began with Cooke's proposal to construct only very large power plants in an effort to conserve coal. To make his point that big power plants used less coal, he estimated that a 1,000-horsepower generator used 4 pounds of coal per horsepower, while a 50,000-horsepower generator used less than 1.5 pounds of coal per horsepower. Cooke called for power plants of no less than 650,000 horsepower or about 488,000 kW—a dramatic increase in size for the time. He also proposed that these power plants be built on top of the coal mines—called "mine-mouth" coal-fired generation—so that railroads would not be needed to deliver the coal; this would conserve coal, since the railroads would not need coal for fuel as they traveled to deliver the fuel itself. Cooke also argued that by electrifying all the rails in the United States, even more coal would be saved.[12]

In his proposed improvements to the electric transmission system, Cooke wanted to build a backbone, or "trunk line," electric transmission network. In this pattern, high-voltage lines would become a "common carrier"; that is, the lines would be open for use by any power plant, not just the power plants owned by the existing electric utilities who typically owned transmission lines. Giant Power, however, went beyond the interconnection. The proposal pushed forward to integration. The intent was to coordinate the operation of all power plants across the state to minimize the cost of power at each moment. This practice is referred to as "economic dispatch," in which, at each moment, as demand for electricity increases, the dispatcher surveys all the power plants that are not operating at full capacity to find the next lowest-cost kWh among them. By always choosing the next lowest-cost kWh, the dispatcher is guaranteed to minimize total fuel cost of electricity supply. As a payoff for all the important changes in his Giant Power proposal, Cooke promised "radical reductions in rates."[13]

Pinchot proposed a three-part regulatory jurisdiction. A Giant Power board would grant charters for the new large power plants envisioned, which would allow it to share regulatory authority with the state commission. This reflected Pinchot's and Cooke's views that the state commission was too close to the holding companies to be trusted with Giant Power's implementation. Also, the split of federal and state regulation would be clarified by allowing the federal government to intervene only when multiple states were implicated and those states could not agree on what to do.

Both Giant Power and Super Power were precursors to deregulation or, using today's apt terms, "competitive reform" or "restructuring" of the electricity business. Those reforms wouldn't begin to be seen until the 1990s, but, despite the distance in time, it is important to note that most of the foundational elements of competitive reform in the 1990s took root in the 1920s. Giant Power and Super Power's vision of regional interconnection came true. Giant Power predicted that there would be a single interconnected system with a single dispatcher for the area from "Chicago to the Gulf, and from the Atlantic Coast to the Great Plains." Today, the Eastern Interconnection is a single system serving much of the country east of the Mississippi River. Moreover, the roots of PJM Interconnection are in actions taken by utilities in the 1920s in response to the Super Power and Giant Power proposals; the first form of PJM Interconnection was created in 1927 as the Pennsylvania–New Jersey or PNJ Interconnection. Again the PJM Interconnection now sets the gold standard for designing competitive markets through integration and coordination.[14]

ROOSEVELT'S PLAN FOR POWER

Candidate Roosevelt in 1932

In his Portland speech Roosevelt made it clear that he saw power as a major issue. He began, "I have strengthened the belief that I have had for a long time and that I have constantly set forth in my speeches and papers in my work as Governor of the State of New York, that the question of power, of electrical development and distribution, is primarily a national problem." Taken alone, this opening statement might have caused some concern about a nationalization of the power business, but toward the end of his speech Roosevelt emphatically said, "I do not hold with those who advocate Government ownership or Government operation of all utilities. I state to you categorically that as a broad general rule the development of utilities should remain, with certain exceptions, a function for private initiative and private capital."[15]

Roosevelt simply did not trust Samuel Insull and the others who ran the large holding companies. His principle was that "when the interests of the many are concerned, the interests of the few must yield." It soon became obvious that "the few" he was talking about were those associated with the holding companies. He explicitly referenced a Federal Trade Commission (FTC) study, which he said revealed "a systematic, subtle, deliberate and unprincipled campaign of misinformation, of propaganda, and, if I may use the words, of lies and falsehoods." Roosevelt used the FTC study as evidence of an increasing concentration of control of power plants and power sales by a few big companies, and in the end that motivated legislation to dismantle the holding companies.[16]

Roosevelt opined that while regulation had been imposed by state public utility commissions, too many state commissions had failed. He felt that state commissions saw themselves as an "arbitrator" or "umpire" when they should be an "agent of the public" that proactively and aggressively ensured "adequate service" and "reasonable rates."[17]

Continuing with his speech, Roosevelt made mention of Senator George Norris, who had used the FTC study to show how the holding companies overstated their investment in order to charge excessive rates. Taking a shot at Roosevelt's opponent in the upcoming presidential election, Norris said that President Hoover wanted to shut down the FTC study. Roosevelt also poked at his opponent for saying, back in 1925 when he was secretary of commerce, that "nothing could be [a] more hideous extension of centralization in Federal Government than to undermine State utility commissions and state responsibility." President Hoover believed that there had been "outrageous exaggeration of the probable extent of interstate power." Again, in sharp contrast, Roosevelt believed state regulation had failed to control the holding companies effectively.[18]

Roosevelt's stance made it clear that "electricity is no longer a luxury. It is a definite necessity." Noting that Canadian homes used twice as much electricity as American homes, he concluded that American power was "certainly backward." Roosevelt believed that "selfish interests" in the power business had "not been sufficiently far-sighted to establish rates low enough to encourage widespread public use." He targeted the "Insull monstrosity" and went on to say that the "Insull failure has done more to open the eyes of the American public to the truth than anything that has happened." Ironically, as we saw in the previous chapters, Insull was indeed establishing "rates low enough to encourage widespread public use."[19]

With respect to his reference to Canada, that nation had created large public utilities early on and per capita electricity use was higher. However, the difference could have causes other than backwardness. For example, by 1920, Canada produced 97 percent of its power from a single, often less expensive source—hydroelectric power. The same opportunity did not exist in the United States. Differences in weather and industrial use could explain higher use, too. Note that Canada's higher per capita use persists today.

Roosevelt called for a "new deal" for the "relationship between the electric utilities on the one side, and the consumer and the investor on the other," and then listed what he saw as the necessary reforms. Business must become more transparent and "turn on the light" on all financial information from public utilities, on who owned the stock in these companies, and on all intercompany contracts. He called for regulation of holding companies by the FPC, and for cooperation between the FPC and state commissions. Regulation was also needed for all issuance of stocks and bonds by the holding companies, and consumer-paid rates should be based on the actual dollar value of the investment made—the so-called "prudent investment" principle. And he wanted to ensure that it was made "a crime to publish or circulate false or deceptive matter relating to public utilities."[20]

Roosevelt said he was not an advocate of government ownership but that there were exceptions. If a locality felt it was not getting adequate service or fair rates, that municipal government should be free to hold a referendum asking for a municipal utility. He argued that this right would allow the locals to create a yardstick by which other utilities' service and rates could be judged. He also called the right "a 'birch rod' in the cupboard to be taken out and used only when the 'child' gets beyond the point where a mere scolding does no good."[21] Thus, the "birch rod" was a tool of last resort for municipal governments when the "child"—the private utility company—was not acting fairly.[22]

Large hydroelectric power projects were another exception to his hesitance on government ownership. He pointed to the work already under way on what he called the Boulder Dam on the Colorado River. In this context, he foresaw four yardsticks to "prevent extortion against the public": "the St. Lawrence River in the Northeast, Muscle Shoals in the Southeast, the Boulder Dam project in the Southwest, and . . . the Columbia River in the Northwest."[23]

In closing, Roosevelt told the crowd, "Judge me by the enemies I have made," referring to those in the investor-owned utility business, like Insull, who many believed were corrupt and greedy.[24]

A contemporary account reported that Governor Roosevelt was cheered by the Portland crowd both before and after his speech on utilities. Big crowds and loud cheers, however, were the norm for Roosevelt no matter what topic he chose. He drew thirty thousand at Columbus,

Ohio, for a speech that included proposals for more federal control over stock exchanges and banks. He drew one hundred thousand at a rally in Sea Girt, New Jersey, where he spoke about ending Prohibition. Cheering crowds greeted him in Topeka, Kansas, where he spoke of reforming agricultural policy and tariffs. In the wake of economic crisis, Americans were ready for change and ready for the federal government to take control to end that crisis.[25]

President Roosevelt's Report in 1935

On January 4, 1935, Roosevelt—now President Roosevelt—delivered his State of the Union address to Congress. He said, "We have undertaken a new order of things"—although he clarified that this new order was pursued "under the framework and in the spirit and intent of the American Constitution." To justify his proposals for change and renewal, he said that "masters of old practice and theory were unprepared" for the economic problems confronted around the world and that "a distinction between recovery and reform" should not be made. Much as one would treat a sick man, there was the need to seek not only a "cure" for the symptoms but "also removal of their cause." Among the major programs, Roosevelt listed "the restoration of sound conditions in the public utilities field through abolition of the evil features of holding companies."[26]

Was the president right? The United States was in an unprecedented economic crisis and aggressive action by a federal government could readily be justified. What might be questioned is the president's apparent presumption that holding companies caused the Great Depression. It is probably closer to the truth that holding company practices magnified the harm of the Depression, and that high levels of debt and other tactics made these companies especially vulnerable to failure in a financial downturn.[27]

In March of the same year, Roosevelt transmitted to Congress "A Report of the National Power Policy Committee with Respect to the Treatment of Holding Companies." In his transmittal letter, he noted that many of the recommendations in the committee's report had already been incorporated in the House and Senate versions of the Public Utility Holding Company Act. Not mincing words, the president pushed back against "the propaganda" with the intent to "make the investor believe that the efforts of Government to protect him are designed to defraud him." The holding company law would "not destroy a penny of actual value of those operating properties which holding companies now control." If holding companies could not "prove to the Securities and Exchange Commission that their existence is necessary for the achievement of the public ends which private utility companies are supposed to serve," then their "disappearance at the end of 5 years" was the best end result. Regulation, he said, had "small chance of ultimate success against the kind of concentrated wealth and economic power" that the holding companies had created. Stating that he was against socialism of any kind, he concluded that the "destruction of private socialism [of the holding companies] is utterly essential to avoid governmental socialism."[28]

To show that there was a precedent for the federal government intervening in other businesses, the committee report drew parallels between the holding companies in the power business and those addressed in the railroad industry by the Interstate Commerce Act, and in the banking industry by the Banking Act, both passed in 1933. The committee's report went on to list the holding companies' wrongdoings, as well as the actions that had to be taken to rein them in. In so doing, it was laying out the content for legislation Roosevelt wanted in order to dismantle the holding companies.[29]

THE BIRTH OF BIG GOVERNMENT

The FTC's Report on Holding Companies

Edwin L. Davis, then chairman of the FTC, submitted the summary report on holding companies to the US Senate in June 1935. Davis reminded the Senate that the report was prepared pursuant to a Senate request that had been approved in February 1928, more than seven years earlier. The early pages of the report documented the rapid growth and increasing importance of electric (and gas) utilities. This was a golden age for the electricity business, with sales increasing rapidly while cost and, consequently, prices fell substantially.[30]

It was the concentration of control by a few holding companies, however, that became the headline of the FTC report, and that was clearly the major problem the FTC believed had to be addressed. By 1932 just three corporate groups controlled 44.5 percent of all electricity generation in America. Largest among the top three was the United Corporation Group, created by Morgan interests; United controlled 19.8 percent of all generation in 1932, through five affiliated holding companies. In second place was Electric Bond and Share, created by General Electric; it controlled 13.6 percent of all electricity generation in that year, through four affiliated holding companies. In third place were the Insull interests, which controlled 11.1 percent of nationwide generation through seven holding companies, including Commonwealth Edison and Middle West Utilities. Another thirteen holding companies controlled another 32 percent of generation, bringing the total control for the top sixteen holding companies to over 76 percent.[31]

The FTC then turned to the question of how much of the holding companies' electricity sales were interstate; this evidence on interstate sales would lead to the significant increase in federal regulation of both electricity and gas over the long term. The FTC found an "almost unbroken chain" of interconnected electricity companies stretching across ten states, from Boston, Massachusetts, to Tampa, Florida; thirteen companies controlled this chain along the Eastern Seaboard. Another chain went from Chicago to Mississippi.[32]

The report then went into more detail on holding company control of interstate sales. Notably, the FTC reported that 98.5 percent of all interstate transfers were controlled by twenty holding companies. Electric Bond and Share alone accounted for 21.9 percent and the Insull interests accounted for another 19.7 percent. The United companies controlled 28.3 percent—bringing the top three to a total of almost 70 percent.

The FTC also reported the number of states in which holding companies distributed electricity in the year 1930, and the geographic breadth is compelling. General Electric's Electric Bond and Share distributed electricity in thirty-one states, while Insull's Middle West Utilities did so in twenty-nine.[33]

While the finding on control of interstate sales of electricity by holding companies was persuasive on its own, the share of all electricity sales traced to interstate transactions was not. The FTC reported that interstate trade of electricity accounted for just 14.4 percent of all public and private electric generation in 1930. In other words, holding companies handled the majority of interstate sales, but those sales were not a particularly large part of the market. This is telling because interstate sales (of both electricity and stock) were the principal justification for the federal government to expand its regulatory role substantially and thereby diminish the role of state regulation.[34]

The Public Utility Act of 1935

To address concerns about the concentration of ownership and the control of interstate sales by the holding companies, Congress passed the Public Utility Act of 1935. As often happens, the Senate passed one version and the House passed another, so the final conference report reflects how the two chambers of Congress came together. Title I was the Public Utility Holding Company Act (PUHCA). Title II was the Federal Power Act. This is the moment when aggressive federal regulation began. The consequence was a proliferation of laws, regulations, reports, and acronyms . . . lots of acronyms. Congress gave the legal basis for federal government intervention by stating that holding companies were "affected with a national public interest" for several reasons. Primary among these was the fact that their securities were sold through the mail—the same fact that sparked Insull's indictment for mail fraud—and through the "instrumentalities of interstate commerce." Moreover, the conference report found that holding company "activities extending over many States are not susceptible to effective control by any State and make difficult, if not impossible, effective State regulation of public-utility companies."[35]

The FTC report to Congress had listed both pros and cons for holding companies and thus gave a sense of what Congress had to address. Holding companies did help with beneficial integration, said the FTC; once formed, they extended service to share cheap electricity. Because of their large size, holding companies could give small operating companies access to lower-cost financing, but that also meant that a holding company could force a deal on smaller utilities. So, too, operating companies got better price and contract conditions for new equipment because holding companies bought on a larger scale and were experienced buyers. Holding companies gave smaller operating companies the constant attention of in-house experts, rather than the intermittent attention of outside consultants; in this way, operating companies got better access to technical and managerial expertise. However, the fees for these services were too high: the FTC found that fees charged by the biggest holding company, Electric Bond and Share, constituted up to 30 percent of its gross income.[36]

The FTC pointed to additional cons, such as writing up asset values to be greater than actual costs; in a sample audit, the FTC found a 22 percent write-up on average. These write-ups increased rates to consumers. In addition, holding companies understated depreciation, thus understating expenses and overstating profits. This detailed evidence from the FTC was used both to prove the allegations of wrongdoing by the holding companies and to dictate what practices would be prohibited by the law itself.[37]

Dismantling the Holding Companies

With the case made and the law passed, implementation began. It was expansive and successful—so successful that anyone who came to the electricity business in the 1970s or later wouldn't know of any great controversy surrounding PUHCA. Most would not have been aware of just how aggressively the government intervened through PUHCA into one of America's most important industries. That is because by the 1970s, holding companies were either completely gone or completely regulated. In this context, the most important action was the implementation of section 11 of PUHCA, which required the "physical interconnection" of holding companies and the "simplification" of "corporate structure." Three moments in history illustrate the implementation of PUHCA over the 1935 to 1959 period.[38]

On June 5, 1941, as World War II raged in Europe, Securities and Exchange Commission (SEC) chairman Edward C. Eicher spoke at the annual convention of the Edison Electric Institute, a trade association for electric utilities. He intended to push back on what he called "loose and dangerous talk in holding company circles" about the commission's implementation of section 11 of PUHCA. That talk had been "calculated to generate the fear that the enforcement of Section 11 would result in the distress sale of assets." Calling all that propaganda, he declared that he had come to tell the truth.[39]

The first part of the truth was that the SEC intended to enforce section 11, about which Eicher said, "let there be no question about that." He quickly added, however, that there was nothing in the law that required a sale of holding company assets at "unfair or inequitable prices." The third part—one that contributed to the second—was that the SEC would be flexible on the timing of the sales. The fourth, said Eicher, was that the sale of assets was not "a losing proposition" for the holding companies because, in the eyes of the financial community, the "break up" value of many holding companies is substantially greater than the "'present going' value." In short, Eicher said that the holding companies would not be "smashed by sudden explosions of dynamite."[40]

There were protests that section 11 enforcement should be slowed due to the war effort, but Eicher made it clear that that would not be the case. He spoke of the need to clear the "debris" of holding companies. Operating companies must be freed of holding companies, he said, so "that they can go full speed ahead in serving the nation's needs." What was needed was "all-out production" by operating companies to meet the "all-out-war-making of the totalitarian states."[41] He added that it "might be fatal to our national existence if we should experience the power short-ages" of World War I.[42]

In March 1950, almost a decade later and with the war over, SEC commissioner Donald Cook spoke to the New York Society of Security Analysts to express the view that the culprit in the bankruptcies of hold-ing companies in the 1930s was "frenzied finance," not the fundamental reality of the electric utility business. Cook believed that the PUHCA was "one of the great and enduring New Deal reforms" and that it was opposed by those in favor of "radical finance." He cited several metrics of success for PUHCA implementation. For example, fixed payment coverage—the amount of operating income available to pay fixed-loan payments and preferred dividends—went from 1.9 to 2.7 times the necessary amount. The higher this ratio, the more open banks would be about loaning money to a utility. And, most important, Cook reported that, in the 1937 to 1948 time frame, generating capacity increased by 42 percent and kWh gener-ation was up 107 percent—all during the time of PUHCA enforcement.[43]

Because the implementation of PUHCA made utility investments far more stable and secure, new sources of investment came to the electricity business. Specifically, Cook reported that as a result, "capital from savings banks, life insurance companies, and personal trusts is becoming available for investment in this industry in ever-increasing amounts." All this flow of capital was needed to finance the growth industry the electric utility business had become.[44]

The scale and aggressiveness of the federal government's effort to dis-mantle holding companies under section 11 was documented in a report by the SEC staff in 1959. The staff found that 2,387 companies had been "subject to the act as registered holding companies or subsidiaries thereof during the period from June 15, 1938, to June 30, 1959." Over that twenty-year period, 2,064, or 86 percent, had been "released" from regu-lations under the act or ceased to exist. Of these, 924 were divested. Of the rest, 777 were released from regulation because of "dissolutions, mergers and consolidations," and 363 were granted exemptions from PUHCA. Of the 323 companies not released, 176 were part of the 18 active holding companies at the time and the remainder was made up of parts of three small holding companies. In sum, PUHCA restructured the electricity business, company by company.[45]

PUHCA marked a significant change in government involvement in the electricity business. The government's role became politically charged, with a focus on allegations of bad behavior that were blamed as a root cause of the Great Depression. In the eyes of big business, meanwhile, government began to be seen as an existential threat—it could determine whether a corporation existed or not. And it was the federal government, not the states, driving all this intervention.

THE FEDERAL POWER ACT: THE LESSER PART

The FPA can be judged to be the lesser of the two major titles in the Public Utility Act of 1935. While its importance faded from memory in the 1970s, and it was actually repealed in 2005, PUHCA was the more important provision at the time, since dismantling the holding companies was Congress' primary goal. The FPA addressed the related problem that multistate holding companies were beyond the reach of state regulation; in general, the reason the courts gave was that state regulation was judged to impede interstate commerce, which is not allowed under the Commerce Clause of the US Constitution. For that reason, the law gave the FPC regulatory authority over wholesale power sales in interstate commerce.

Federal Power Commission Responsibilities and States' Rights

In its annual report to Congress in 1935, the FPC explained that it was first created under the Federal Water Power Act of 1920 and given authority over hydroelectric power. The Federal Water Power Act was renamed the Federal Power Act as part of the Public Utility Act of 1935, and the FPC noted that this law "greatly enlarged [the commission's] functions and duties." The FPC described its new roles as follows: "The Congress has established regulatory control over the interstate activities of electric utilities which were beyond the constitutional or effective administrative control of the States, and has set up machinery for Federal assistance to State regulatory bodies in their efforts to accomplish effective public regulation."[46]

The commission explained further that it would have "the authority to fix rates over that part of interstate energy which is sold wholesale for resale." In this way, the 1935 FPA would, in effect, "fill in the gap" the Supreme Court found in its 1927 decision on *Public Utilities Commission v. Attleboro Company*, the gap, of course, being that interstate sales were beyond the reach of state regulators. It is interesting to consider that Westinghouse and Tesla provided the technological underpinnings of federal regulations; only with alternating-current systems would the geographic reach of a utility be likely to cross state lines—to be an "interstate sale."[47]

Adding to the established theme that the FPC would only supplement, not replace, state regulators, the FPC emphasized that its intent was to cooperate with the states. The FPC quoted a commissioner, saying, "This legislation sets up a new milestone on that highway which leads to complete cooperation between State and Nation in the solution of one of the most serious problems that has thus far presented itself to our dual system of government." Congress, in the FPA itself, went out of its way to say it would not have the federal government intrude on state regulations. The FPA stated, "The Commission shall have jurisdiction over all facilities for such transmission or sale of electric energy, but shall not have jurisdiction . . . over facilities used for the generation of electric energy or over facilities used in local distribution or only for the transmission of electric energy in intrastate commerce, or over facilities for the transmission of electric energy consumed wholly by the transmitter." Despite this plainly stated deference to the states, the battle between state and federal jurisdiction rages on even today.[48]

The FPA gave the commission other specific regulatory powers over electric utilities. The FPC had to approve the sale, lease, or other disposition of facilities under its jurisdiction; however, the FPC could not duplicate the actions of the SEC in oversight of these transactions. Other sections established crucial policy standards for rates. All rates, for example, had to be "just and reasonable"; this is a standard that still motivates most federal regulatory actions today. Further, the FPC could not give any "undue preference or advantage to any person." This later was phrased as a prohibition on "undue discrimination," again, a term used to motivate major federal action in the electricity business today.[49]

Implementation of the FPA: In Depression and War

Five years after the 1935 Act was passed by Congress, the Federal Power Commission's annual report to Congress added more context and perspective on the emerging role of the federal government in regulating America's electricity business. The FPC started with data showing the huge growth in electricity sales over the twenty years from the FPC's creation in 1920 to the time of the report's publication in 1940. It then explained how, in the five years from 1935 to 1940, the commission played its substantially expanded role under the FPA.[50]

The commission also noted its greatly expanded role in the natural gas business, instituted by Congress' Natural Gas Act of 1938. The FPC stated, "The Commission has authority to regulate rates charged for the transportation and wholesaling of natural gas in interstate commerce, to grant certificates of convenience for proposed new pipe lines extending into the market area of existing companies, to direct extensions of existing pipe lines where existing markets will not be adversely affected, and to control the flow of natural gas over the nation's boundaries." This expansion to natural gas is not an unimportant tangent. To this day, the fortunes of the electricity and natural gas businesses are joined at the hip. This is because natural gas is an important fuel for electricity generation, and because regulatory practices in one industry set a standard for the other.[51]

In another major theme that has grown even more prominent today, the FPC argued that regulation was a substitute for competition.

Specifically, it stated, "The regulation of privately owned public utilities came into being in large measure as a substitute for competition in a field which had come to be considered as adapted to 'natural monopoly.'" A natural monopoly occurs when "a single firm is able to serve the market at less cost than two or more firms." Persistent economies of scale are presumed to be the reason average cost continues to fall until only a single utility remains. Decades later, deregulation or "competitive reform" would be proposed by the FPC's successor precisely because such economies of scale were no longer thought to prevail in the generation segment of the electricity business. [52]

Reflecting the demands of war mobilization, a letter from President Roosevelt was appended to the FPC's 1940 report. It asked that the FPC, working with the National Power Policy Committee and the Advisory Commission to the Council of National Defense, undertake efforts to be up to date on the full power needs of the military, the supply of power available to meet those needs, and the equipment needed to build new power stations. Six years later, the FPC documented that a boom in both electricity and natural gas production was necessary to fit the needs of a wartime economy. Thus, on the heels of an expansion of the government's role necessitated by the Great Depression came a new motivation for expansion: World War II. [53]

THE NEW DEAL'S LEGACY
(FOR ELECTRICITY)

Federal regulation of the electricity business was born in the midst of the collapse of capitalism during the Great Depression. If ever there was a moment to consider government ownership of power plants and transmission lines, this was it. And, yet, public ownership was not proposed by President Roosevelt or by anyone with the political sway to make it happen. From this perspective, one legacy of the New Deal for the electricity business was an explicit choice of regulation of private business over public ownership. This reveals a certain restraint.

There was nothing restrained, however, about the big-government regulation that came out of the Great Depression. The dismantling of the holding companies under the Public Utility Holding Company Act of 1935 must be one of the largest interventions in any American business. Big government set its sights on the big holding companies—they were public enemy number one—and ripped them apart. In addition, these new regulations plugged a hole in state regulation—the federal government would regulate the 14 percent of electricity sales that crossed state lines. This was the start of ongoing federal regulation of interstate sales and, though written with great deference to the states at the time, the new policy triggered a fight over states' rights that continues undiminished today.

From the start, big-government regulation had an awkward, love-hate relationship with big business. That awkwardness might best be reflected in what happened to Samuel Insull. Insull's big business was admired because it dramatically reduced electricity prices through economies of scale. Insull himself, however, was blamed for the financial and accounting scandals that contributed to the collapse of the stock market. This love-hate attitude has grown more moderate because circumstances are less dire, but it persists today. For example, mergers that create bigger companies and promise big savings are rarely rejected, but allegations of market power abuse by big companies are still common.

FDR's presidency caused a significant change in the government's role in the electricity business and added a new twist to the narrative. FDR created far-reaching federal regulation, which focused on the fundamental structure of the business. His actions were politically charged, driven by allegations of bad behavior leading to the financial crisis—not allegations of inadequate electricity supply or cost overruns. FDR's regulation was born out of crisis—the Great Depression, but also, later, World War II. Since these were national and international crises, actions by the national government were warranted at the time. Aggressive intervention is to be expected in times of crisis or war, but action inspired by these exceptional and urgent situations is not necessarily the best policy for the long term.

Some in the industry would argue that the real danger of national regulation—federalization—is that it leads to standardization. America is a big, diverse nation, and diversity is a strength. The future is uncertain and different states have different views on how to mitigate the risk. Much like a diversified portfolio of stocks, a diverse set of state policies assure America as a whole that someone will try all the options and that, out of this variety and innovation, the best options will emerge. Diversity also allows actions to be tailored to local needs and opportunities. There are concerns that federalization may undermine the strengths of diversity.

However, others will say even the two most visible symbols of FDR's Depression-era actions—Hoover Dam and the Tennessee Valley Authority—reflect the value of diversity regardless of their federal origins. As we will discover in the next chapter, Hoover Dam was an investment that took advantage of a unique regional opportunity, which itself grew out of a bottom-up compact among the western states. The Tennessee Valley Authority, meanwhile, was a first-of-a-kind experiment with state capitalism, but, it, too, was tailored to unique regional circumstances.

CHAPTER 9

Herbert Hoover's Dam

This morning I came, I saw and I was conquered, as everyone
would be who sees for the first time this great feat of mankind.
—President Franklin D. Roosevelt (1935)[1]

Hoover Dam has been called "the great pyramid of the American West." Like the
pyramids, Hoover Dam rose out of the bleak desert, and it is a monument
to the ingenuity and skill of the Americans who built it. As Hoover Dam
historian Joseph Stevens writes, it was built at "a time when the engineer
and the builder were romantic figures, when taming a wild river was a
heroic endeavor, when holding a job and feeding one's family was a feat
of honor."[2]

Hoover Dam is the perfect physical symbol of the Age of Big. Located
southeast of Las Vegas on the Arizona-Nevada state line, Hoover Dam
was a big project in every sense of the word. When it was completed in
1935, it was the tallest dam in the world, at a height of 726 feet, more than
twice as tall as the next tallest dam, the Arrowrock Dam in Idaho. The
scale of electricity generation was big, too; from 1938 to 1948, Hoover
Dam was the biggest producer of hydroelectric power in the world. Most

150

important, success at Hoover Dam proved to many that big government was the only entity that could successfully complete such a big project and produce big benefits like flood control, water supply, and electric power production on such a scale. This point was buttressed by the fact that the private sector had tried over many years to achieve at least the first two of these benefits, and not only did it fail but it caused one of the largest environmental disasters in history.

The timing of this big project by big government could not have been better. The political compromise and engineering design needed to make it happen had been achieved over the preceding decades, so Hoover Dam was ready to start construction in 1931, just as the Great Depression deepened in severity. Three Republican presidents—Teddy Roosevelt, Calvin Coolidge, and Herbert Hoover—contributed significantly toward making the dam a reality, but it was President Franklin D. Roosevelt who claimed it as a symbol of his New Deal. On September 30, 1935, with his next presidential election looming, FDR traveled to the newly built Hoover Dam for its dedication.[3] Historian Michael Hiltzik writes, "He aimed to appropriate as a tangible symbol of the New Deal, his Democratic administration's economic recovery program, a great dam on the Colorado River conceived and launched by Republicans."[4]

The story of why and how Hoover Dam was built is fascinating in and of itself, but its telling in detail here is justified for a deeper reason: Hoover Dam raised and resolved many of the issues that confront the choice of electricity technology even today. It shows how the culture of the day drove technology choice, just as it does now. It also resolved the ideological debate on federal versus state government control by involving both in the ways that best ensured the project's success—an ability to coexist and cooperate that is sometimes forgotten by the federal courts today. Hoover Dam was built by the federal government, but it was tailored to regional opportunities for power production, water supply, and flood control. It resolved the debate of government versus private enterprise by assigning responsibility to each in a way that best served the project. It also used straightforward tools such as pay-for-performance contracts to give the right incentives to private enterprise—something that had to be relearned four decades later. And, perhaps because it got the incentives right, Hoover Dam was a first-of-a-kind project that came on line on time, and on budget—in sharp contrast to first-of-a-kind projects in more recent times.

Hoover Dam also fit well into Franklin Roosevelt's three-prong plan for electricity; big hydro was the third prong. And it helped fulfill the manifest destiny of electricity: to cover all of America by pushing into the West.

EARLY EFFORTS TO TAME THE COLORADO RIVER

Success at Hoover Dam came via many failures. Geologist William P. Blake wrote of the potential to use Colorado River water for irrigation in 1853, when he was a member of a survey team for the Pierce administration that was sent to find a southern route for an intercontinental railroad. His most notable discovery, however, was the then-dry bed of a large ancient sea in Southern California. He said the soil could grow anything if a supply of water was found, and that supply could come from the Colorado River, which was over one hundred miles away. The next man to pursue this dream was Oliver M. Wozencraft, a physician from Ohio. In 1859 Wozencraft sought and won a grant from the California legislature for ten million acres of land in the area, but he also had to win the approval of Congress. Indeed, Wozencraft made his last journey to Congress in 1887 only to see the bill he was counting on lose another vote in the House. After that vote, Wozencraft returned to his rooming house and died a few days later, having lost both his wealth and health to his dream for the Colorado River.[5]

Charles Robinson Rockwood was next. He told his story of great hope and disappointment in his 1909 book *Born of the Desert*. Like the others, Rockwood wanted to make the California "desert bloom" with irrigation water from the Colorado. Soon after he attracted an investor, as Stevens put it, "[The] sinister names Colorado Desert and Salton Sink were replaced with the grandiose title Imperial Valley." Rockwood's idea was that settlers would buy land at a very low price per acre but would also be required to buy the irrigation water needed to make that land worth anything. In 1901 the first settlers came to the Imperial Valley, and the first water flowed to them from the Colorado River through the Mexican Canal. Despite all that was working against it, by about 1904 the Imperial Valley's population numbered about seven thousand and by 1905 the land produced dairy products and crops, such as barley, alfalfa, and cattle feed, valued at over $2 million in total.[6]

Unfortunately, Rockwood's firm, the California Development Company, was built on a rickety financial structure. It did not directly own the land the settlers lived on because that was sold to them by the federal government, so the company didn't make money on land. What the company did own was the canal and the water it carried. Unfortunately, the Colorado River carried so much silt that locals joked it was "too thick to drink and too thin to plow," and it clogged the canal that carried water to the Imperial Valley. The best way to take care of this problem would have been through constant dredging of the canal, but Rockwood's company did not have the money for that. Instead, their remedy was to divert the river farther south from the original diversion in Mexico at Pilot Knob. With no cash, Rockwood cut the new diversion site at what would become known as Disaster Island.[7]

In 1905 a series of heavy rains and floods on the Colorado River changed everything. Rockwood could not control the diversions or control the river water, which flooded into the Salton Sink, the ancient basin that Blake had discovered—a basin that had been dry for four hundred years. Rockwood inadvertently transformed the dry Salton Sink into the 60-foot deep, 150-square-mile Salton Sea. He realized he had created an environmental disaster and was overwhelmed, so he turned to Edward H. Harriman, the powerful owner of the Southern Pacific Railroad, which was profiting from shipments out of the Imperial Valley. Thinking only of the theoretical value of irrigated land, and not aware of the environmental crisis, Harriman demanded a controlling 51 percent share of the company in return for a $200,000 loan. With this uninformed negotiation completed, Harriman unwittingly took on the full liability of the environmental disaster and eventually the full brunt of President Teddy Roosevelt's wrath. Only when Epes Randolph, one of Harriman's chief engineers, finally got to the site did he see the scope of the disaster. It would take the Southern Pacific more than eighteen months and more than $3 million to attempt to fix it.[8]

Rockwood's mistakes had created the largest lake in California, and they also set in motion all that was needed to motivate the Hoover Dam. The dream to make the desert bloom was the same, whether a private enterprise or the government was in control. Private enterprise had been given its chance, and it failed. President Theodore Roosevelt would keep the dream alive but made it clear that a comprehensive plan to tame the Colorado River was now a matter for the federal government. Electricity would be an important part of that plan. The science Benjamin Franklin had discovered through study of one untamed natural force—lightning—would be used to tame another: the raging Colorado River.

THE FIRST STEPS TOWARD HOOVER DAM

In 1920 Philip D. Swing, a Republican, was elected as a congressman from California. He sought a $30 million loan guarantee for the Imperial Irrigation District to cover the cost of building an All-American Canal. Around the same time, Arthur Powell Davis, head of the Bureau of Reclamation, sought congressional approval for a study of a comprehensive flood control program for the Colorado River. Davis had his geologists search for sites suitable for a high dam—the term *high* indicated it would be suitable for electricity generation. Three sites at Boulder Canyon were proposed, as well as two at Black Canyon. All five were on the Colorado River, at the border between Nevada and Arizona.[9]

In 1922 Davis presented his report to Congress, where it became known as the Fall-Davis report. The report proposed the Boulder Canyon Project, which would provide flood control and water storage, and concluded that it should be financed with electric power sales; an all-American canal would also be built. Oddly, the Bureau of Reclamation's survey of the sites ultimately concluded that the Black Canyon sites were superior to the Boulder Canyon ones, and Hoover Dam was actually built at Black Canyon, despite still being called the Boulder Canyon Project.

The essential political foundation for Hoover Dam would be agreement on the allocation of the water diverted from the Colorado River. President Warren G. Harding appointed his secretary of commerce, Herbert Hoover, as chairman of the Colorado River Commission. This was the first step in the journey to naming the Hoover Dam. Political heavyweights from seven states were sent to the commission; the four "upper basin" states were Colorado, Utah, Wyoming, and New Mexico, and the three "lower basin" states were Nevada, Arizona, and California. The commission's objective was to establish an interstate compact that would distribute Colorado River water equitably to all seven. The US Constitution had always allowed interstate compacts subject to congressional approval, but such compacts had never involved more than two states at a time.[10]

At first, a wide range of allocation methods to determine how to distribute this common resource made sense to the different states. Pressure mounted for a compromise when a Supreme Court ruling seemed to favor the lower basin states. In addition, if federal funds were won, the federal government might impose an allocation. The crucial first compromise came with the proposal to have the upper and lower basins split water rights fifty-fifty. The allocation among the states in each basin would come later. Finally, in 1922, representatives from the seven states signed the Colorado River Compact. After fraught negotiations, this agreement was still only a first step: the seven state legislatures as well as Congress would have to approve it, too.[11]

THE STARS ALIGN FOR LEGISLATION

As the agreement sought approval by the state legislatures, five of the states were relatively easy; the real battles were in California and Arizona. Hoover lobbied hard and finally won approval from the California

Legislature. In Arizona the opposition was led by a party motivated by a competing project.

In Congress, Swing joined forces with the powerful Senator Hiram Warren Johnson, also from California, to introduce legislation: The Swing-Johnson bill would fund the construction of a high dam in the vicinity of Boulder Canyon as well as the All-American Canal. Historian Michael Hiltzik points to two factors that pushed the Swing-Johnson legislation toward approval. The first factor, sadly, was the Midwest flood of 1927, which killed 246 people along the Mississippi River. As Congress got ready to vote on the Flood Control Act of 1928, Swing made it clear that the same kind of risk loomed for his constituents from the untamed Colorado River. The second factor was a report showing the full extent of the campaign against municipal power by private electric utilities. Hiltzik writes, "In all, the utilities were revealed to have spent $1 million a year to turn public opinion against public power, and another $400,000 specifically to torpedo the Boulder Canyon bill."[12]

Pushing against approval was another tragic event: in 1928, at a critical moment for Swing-Johnson, the Saint Francis Dam burst, killing over four hundred people in Southern California. William Mulholland, head of the Los Angeles Department of Water, took the blame. By many accounts, he was indeed largely responsible, but no charges were made. The House of Representatives had trusted his engineering endorsement of the site and the design of the Boulder Canyon Project. When Mulholland's reputation was undermined by the Saint Francis collapse, the judgment of his political supporters was called into question, too. To mitigate this and avert future disasters, the Swing-Johnson bill added a requirement for a "blue-ribbon" panel of experts to review engineering designs. Swing-Johnson also added provisions to win support in California on the water allocation issue: California would get 58 percent of the allocation of the three lower basin states and would also be allowed to use Arizona's and Nevada's allocations until their water needs grew. The House passed the Swing-Johnson bill in May 1928. The Senate then passed it by a vote of sixty-four to eleven and, on December 21, 1928, President Coolidge signed it into law.[13]

To determine the allocation of the electricity to be generated at Hoover Dam, Secretary of the Interior Ray Lyman Wilbur conducted a competitive solicitation: in other words, he requested bids. But the bids to buy electricity from Hoover Dam far exceeded the amount available. Indeed, two entities—Southern California Edison, a private utility, and the city of Los Angeles—each offered to buy 100 percent of what was available. The project had an ambitious cost-recovery plan, so maximizing sales was absolutely key. The revenue from power sales had to be sufficient to repay the federal investments as well as to pay all expenses plus interest over fifty years.[14]

"SIX COMPANIES"

President Herbert Hoover saw an urgent need for public works spending, and so he pushed for the start of Hoover Dam. By 1930 the effects of the Great Depression were catastrophic and Hoover was looking for what today would be called "shovel ready" public works projects; Hoover Dam was one of the few megascale projects that was shovel ready. Its total estimated cost of $165 million was a big investment, given the fact that the *total* budget for federal construction was typically $150 million a year. The scope of the federal government was much smaller than it is today, with its spending equaling 3.4 percent of gross domestic product in 1930, compared to 20.7 percent in 2015.[15]

The Bureau of Reclamation next invited bids from private companies to build Hoover Dam, surely the largest such effort in American history. The winning bidder would have to submit a $5 million performance bond—a substantial sum to put at risk at any time, made even bigger by the context of the Great Depression. This approach assumed that "the government would provide all the materials entering into the completed work, such as cement and steel, but it was up to the contractor to furnish all the machinery, tools, vehicles, and supplies needed to carry out construction." Bidders also took on the risk of pay-for-performance penalties for missing deadlines. For example, a $3,000-a-day penalty would be imposed if the four diversion tunnels were not completed by October 1933, and a further penalty would be imposed if the dam's final height for power generation was not achieved by August 1936.[16]

There were only three compliant bids. In the end, the winner was a consortium of bidders calling themselves Six Companies. Incredibly, Six Companies would complete the Hoover Dam two years early. Every leader in the Six Companies team brought something important to the effort; most had built their companies from the ground up by successfully completing major projects in the West. Utah Construction, headquartered in Ogden, Utah, was created by the Wattis brothers. As its business grew, it formed a partnership with Harry Morrison of Morrison-Knudsen. All would agree that, for the Hoover Dam bid, Morrison-Knudsen's most desirable feature was that its team included America's foremost dam builder, engineer Frank T. Crowe.[17]

Morrison approached the J. F. Shea Company in Portland, Oregon, the biggest tunnel and sewer builder on the West Coast. Shea suggested Portland's Pacific Bridge Company as another member of the consortium. Morrison then went to San Francisco and won over McDonald & Kahn, another successful construction firm. Next was Henry Kaiser and his firm Kaiser Paving in Oakland, California. Kaiser's big break came when Warren Bechtel, an influential San Francisco contractor, became Kaiser's mentor and invited him into joint ventures. Kaiser and Bechtel were in, and they brought along the only eastern firm in Six Companies, Warren Brothers.[18]

When it came time to bid, the consortium took Crowe's cost estimates, added a profit of about 25 percent, and bid $48.9 million. Six Companies won handily. Stevens, the Hoover Dam historian, paints the scene with the press crowded into William Wattis' room at the St. Francis Hospital, where he was being treated for cancer. His wife made him presentable, sitting in his robe puffing a cigar. The victory had put him in a positive frame of mind. Wattis, one of the brothers behind Utah Construction, quipped, "Now this dam is just a dam but it's a damn big dam."[19]

CHALLENGES AND INNOVATIONS

The start of the project brought a new set of challenges. It was difficult to get workers to the construction site, as well as to get materials and electricity into and out of the site. To meet the challenge, three major parts of the infrastructure were under construction by the time Crowe got there: the Union Pacific railroad spur; Southern California Edison's 200-mile, 90 kV power line from Victorville, California, to Boulder City; and the road and rail lines from Boulder City to the work site.[20]

After the Saint Francis Dam catastrophe, a blue-ribbon panel had recommended tougher standards. One of these required that the maximum stress allowed on the dam be cut. In the end, the bureau sidestepped this standard. On many other technical design decisions, however, the bureau was methodical and precise. Expert science and engineering lay behind many of the decisions made for the Hoover Dam. The bureau conducted substantial practical research on several elements, most notably on the concrete used and the shape of the dam. Between 1931 and 1933, the bureau experimented with as many as fifteen thousand concrete samples in ninety-four different formulations. When a major design issue was discovered, the bureau pursued a solution with real intellectual vigor. A Swedish engineer argued that an "arched dam" would be more stable and also would be much thinner; it would be more cost effective because less concrete would be needed. The bureau looked seriously at both the stability issue and the promise of cost reductions claimed for the arched dam. In partnership with Southern California Edison, it built the Stevenson Creek test dam on the San Joaquin River in California. The test dam showed that huge savings in concrete could be achieved safely with the arched dam, but still the bureau did not go that way in the end. One major reason was that the Saint Francis Dam had been an arched dam, and memory of that dam's failure was still fresh.[21]

Crowe was under pressure to get the four diversion tunnels for the dam completed by October 1933. Practical innovations let him meet that tight deadline. The best single innovation was the "William Jumbo." This was scaffolding mounted on a truck along with drilling equipment, an efficient way to make scaffolding mobile. It could quickly be pulled away when dynamite was to be detonated, and just as quickly put back into the tunnel to continue drilling.[22]

Hoover Dam reflects America's scientific and engineering might. Surely the successful design and construction of this dam is among the great engineering accomplishments, in a time rife with great engineering feats. This America might surface again and perhaps it already has in the shale gas and oil revolution at the turn of the twenty-first century.[23]

HARD WORK AND HARDSHIP

The Great Depression created a nation of desperate people seeking work. Before Hoover Dam even began construction, hundreds of workers and their families moved to the Nevada desert and set up camp to be first in line for jobs. They formed what was known as "Ragtown," a small, improvised settlement on the scorched landscape. Once construction began, Boulder City housed the workers, but it was only a marginal improvement from Ragtown. Housing was often unfinished, lacking plumbing and, ironically, electricity; rules were strictly enforced, including prohibitions on gambling, liquor, and dance halls within city boundaries. Of course, all these vices were readily available outside Boulder City, especially in Las Vegas.[24]

Amenities were scarce. Residents pushed the federal government and Six Companies to build a school, and Boulder City School was completed in 1932. It was given some guidance by the Las Vegas School Board, although because it was on federal land it could not be funded by the state. A hospital was set up a year after construction began, but only dam workers could get treatment—their families had to "fend for themselves." Still, workers clamored for a job on Hoover Dam because the average annual salary was $1,825. According to the Bureau of

Reclamation, that meant dam builders were paid better on average than steel workers ($423 a year), coal miners ($723 a year), and civil service employees ($1,284).[25]

Although the pay was good for the times, workers were pushed to the limit while building Hoover Dam. Most worked eight-hour shifts seven days a week, and enjoyed only two optional, unpaid vacation days a year: the Fourth of July and Christmas Day. But while living conditions were difficult, the weather and working conditions were extreme; Hoover Dam workers were true unsung heroes. Men generally worked from either 4 AM to noon or from 4 PM to midnight, since by noon it was so hot that they would pass out from the heat. "It was so hot we used to soak sheets and crawl inside them to try and get a little sleep," one man reported. Night workers who slept during the scorching-hot day had to endure suffocating heat in their dormitories. The desert heat was a leading cause of death at Hoover Dam. In 1931 alone, sixteen died of the heat as temperatures rose to 120 degrees. One worker recalled that in one month temperatures never went below 100 degrees at night, and at times topped out at 138 degrees. In the summer months, it was normal to send ten to twelve men to the hospital each day, "unconscious and incontinent," their body temperatures measuring up to 112 degrees. Sadly, these men had about a 50 percent chance of survival until doctors adapted and began to immediately submerge the overheated men in ice baths and then give them intravenous fluids.[26]

The heat was rivaled by the dangerous work conditions. Industrial accidents, rock slides, drownings, blasting accidents, truck accidents, canyon walls collapsing, electrocutions, and accidental falls were just some of the causes of death recorded at the time. Oddly, pneumonia shows up as one of the leading causes of death among workers at Hoover Dam. It is likely that the hospital recorded pneumonia as the cause of death when the real culprit was carbon monoxide poisoning, which resulted from machinery being operated in improperly ventilated tunnels and other tight spaces. Perhaps the most dangerous of all jobs was that of the "high scalers." These courageous men, often Native Americans, would descend canyon walls on ropes and use forty-four-pound jackhammers and dyna-mite to strip away loose rocks. This job was so dangerous that a special hat called a hard-boiled hat was created for them by coating their cloth hats with tar. This crude hard hat proved highly effective; men struck by fall-ing objects did not suffer skull fractures or death quite as frequently—just broken jaws.[27]

The hard work and hardship of these workers is an important part of the story of Hoover Dam. The workers were physically responsible for the successful construction, and it is they and their families who are at the heart of this accomplishment.

PROTECTING PROGRESS ON THE DAM

Crowe's speed earned him the nickname "Hurry-Up Crowe," but it also created another kind of challenge. The workforce was much larger than the bureau had expected, and it quickly ran through its budget alloca-tion for the year. It would be a great challenge to get more money from Congress in the middle of the Great Depression. Indeed, the federal gov-ernment was determined to balance its overall budget, and it had already planned to cut funds in 1932. Crowe pushed the speed because he knew he would be at the mercy of the spring floods, and if he did not complete the work before the floods there was the danger that the partially completed work would be swept away.[28]

Elwood Mead, head of the Bureau of Reclamation, testified to Congress that he needed $7 million immediately or the work would be shut down, but unfortunately Mead was an unconvincing witness and may have done more harm than good. It was Henry J. Kaiser from Six Companies who saved the day with an argument focused on job losses. He immediately pointed to the damage that would result from a failure to provide the $7 million: "It will kill the possibility of the Government [con-tinuing] to employ the 3,000 men who are now employed . . . It will kill the possibility of the Government to profit from the extent of six or more millions of dollars through a 1-year earlier completion . . . It will kill the opportunity of the Government to remove the hazard to life and property for one whole year of the thousands and thousands of people living in the Imperial Valley." If Congress were to shut down Hoover Dam, Kaiser went on to say, "You will force about 7,000 citizens of the United States out on that desert."[29] Soon after Kaiser made his case, Congress restored all the money it had planned to cut from the Hoover Dam budget and added more.[30]

Despite the jobs Hoover had created, during the election of 1932, Franklin D. Roosevelt criticized him for "wasteful spending on public works." Roosevelt soundly beat Hoover in the contest, winning forty-two states to Hoover's six. Even in Boulder City, Roosevelt easily beat Hoover by 1,620 to 454 votes. After Roosevelt's election there would be no more coddling of Six Companies. Harold Ickes took over at the Department of the Interior, and one of his first moves was to remove any possibility that the dam would be credited to Hoover; he ordered that it be named Boulder Dam. He also took aim at the cement cartel. Ten suppliers had colluded and come in with exactly the same price, which was 20 percent higher than it had been a few months earlier. Ickes negotiated a winning price at a 10 percent discount. Next, he ended the Six Companies' practice of paying the workers in company scrip, which forced the workers to buy necessities at the company store and denied independent stores a fair chance to compete. Finally, Ickes promoted hiring black workers to combat racial discrimination.[31]

Still, Six Companies continued to play hardball with the workers. A battle over the diagnosis of carbon monoxide poisoning, for example, continued into 1935, the year in which President Roosevelt would give his dedication speech. To rebut claims of disability, Six Companies' experts testified that "victims of carbon monoxide poisoning 'either die or get well.'" Given the "die or get well" standard, the very presence of the complainant in court would rebut their claims of harm. The human costs mounted as workers hammered the dam into existence.[32]

PRESIDENT ROOSEVELT'S DEDICATION SPEECH

On September 30, 1935, President Roosevelt gave the dedication day speech at Hoover Dam. He was eager to claim the dam as a success for his New Deal, despite the fact that three other presidents over thirty years had helped the project reach important milestones. He wanted to show that progress was being made in economic recovery through big public works projects like the dam, and he reflected the attitude of his time toward the physical environment, which was that nature exists for mankind to make it productive. Electricity was central to both showing progress and making nature productive. Sales of electricity would pay for the dam and powerhouse, and electricity would pump the water to make the California desert bloom. Electricity was both tangibly and symbolically the source of the progress called for by the times.

Roosevelt began his speech with a twist, more clever than apt, on Julius Caesar's *veni, vidi, vici*: "This morning I came, I saw and I was conquered, as everyone would be who sees for the first time this great feat of mankind." He noted how barren the site had been: "Ten years ago the place where we are gathered was an unpeopled, forbidding desert." He called what had been done at Boulder City "a twentieth-century marvel." Appropriately for the Age of Big, he celebrated Hoover Dam's size. Most directly relevant to the history of electricity, he told his audience that the Hoover Dam power system, yet to be completed at that point, "will contain the largest generators and turbines yet installed in this country," capable of generating "nearly two million horsepower of electric energy"—that is, about 1,500 MW of electricity.[33]

Emphasizing the efficiency and hard work that went into the project, he noted that the work had been completed two years ahead of the contract deadline and expressed the nation's "gratitude to the thousands of workers who gave brain and brawn to this great work of construction." Going beyond the impressive feat of construction, Roosevelt celebrated the "agricultural and industrial development" that the dam would bring and the contribution to "health and comfort" to citizens in the Southwest. He noted the federal and state cooperation that had settled the issue of water allocation, and he gave credit to Congressman Phil Swing and Senator Hiram Johnson, but he made no mention of Herbert Hoover, who had negotiated the all-important deal on water allocation. Again, reflecting the cultural attitude toward the physical environment, he said that the Colorado River had been "an unregulated river" that added little value to the region. Indeed, said the president, the "people of Imperial Valley had lived in the shadow of disaster." Roosevelt credited the dam with stopping a flood in June 1935 because the water was "safely held behind the Boulder Dam" and believed that if the dam had been completed one year earlier, it could have prevented the drought suffered then. Rightfully, jobs were center stage and he celebrated that in the midst of the Great Depression, the project hired "4,000 men, most of them heads of families," plus "many thousands more were enabled to earn a livelihood through manufacture of materials and machinery."[34]

Absent from Roosevelt's speech, however, was a big-picture view of what Hoover Dam meant to America's future. Back on October 30, 1932, President Hoover had pushed back on his opponent, Roosevelt, with such a view. President Hoover said, "This campaign is more than a contest between two men . . . It is a contest between two philosophies of government." Roosevelt and his big-government philosophy won in the 1932 election in the depths of the Great Depression. To the man on the street, there could be no better evidence that Roosevelt's philosophy worked than the jobs, the water, and the electricity brought by the newly completed Hoover Dam.[35]

After the speech, construction on Hoover Dam moved quickly to cross the finish line. In December 1935, a 1,180-foot roadway built over the dam was first opened to traffic. In March 1936, the federal government took possession of the dam from Six Companies. Change Orders added about $5.8 million to what Six Companies was paid—a relatively small increase of about 12 percent above the bid. Six Companies earned a profit of $10.4 million after taxes. The first electric power was delivered to California in October 1936.[36]

THE LEGACY OF HOOVER DAM

Hoover Dam was a big project by any measure: it was the tallest dam in the world when it was completed in 1935, and it was the largest hydroelectric power generator in the world for many years thereafter. Because its large scale made Hoover Dam a first-of-a-kind technology, its success was never guaranteed. The fact that it began generating power on time and on budget, and performed as promised, makes it a significant success in the history of the electricity business. First-of-a-kind nuclear plants did not have the same success in the 1980s, nor did today's clean coal projects.

Hoover Dam's construction nicely balanced the competing ideologies of big government and big business. The dam could have been all about big government with big business left out of the equation. After all, the dam was built after private business failed to provide much-needed water supply and flood control. Indeed, private investors had caused a massive environmental crisis—the inadvertent creation of the Salton Sea. Big business could have been pushed aside by the broad brush of concern over the collapse of capitalism in the Great Depression. It was not. A partnership between government and business was sought from the start.

Too often, business and government are portrayed as black-and-white alternatives. Hoover Dam was built under what today would be called a public-private partnership. As it should, the government tackled the thorniest political issue, water supply allocation, through an interstate compact initiated by all seven affected states; there was no pitched battle over states' rights. Roosevelt rushed to the dedication in 1935 to claim Hoover Dam as the symbol of the success of his New Deal, but as we have seen, business played a major role. More broadly and more important to history, having a precedent for successful public-private partnerships must have given Roosevelt confidence when he turned to big business to produce the armaments that would be used by American soldiers in World War II. These experiences would persuade Roosevelt to shift to the right politically by teaming with big business—a shift with significant consequences.

Another part of the legacy of Hoover Dam is the guidance it gives to those looking to understand the past or predict the future of any power-generation technology. Hoover Dam was driven by overarching goals well beyond the existing demand for electricity. The purpose of Hoover Dam was to provide water supply and flood control for the western states—power generation was part of the equation mainly to pay the bills. America must look for this overarching goal in every technology going forward, from nuclear power to wind power to solar power and beyond. The example of Hoover Dam also inspires us to look for the cultural beliefs that support or undermine new technologies and projects. Hoover Dam would not have been built except by a culture that celebrated the use of science and technology to serve human needs and tame the environment.

The culture of the time was deeply affected by the need for job creation. The harsh treatment of workers on Hoover Dam reflected how desperate many workers were. Notably, the harshness was not mitigated by government's involvement. What historian Michael Hiltzik calls "mordant book ends" illustrate the sacrifices of the workers. The first bookend was the first person to die for the project; J. Gregory Tierney, part of a survey crew, drowned in the Colorado River on December 20, 1922. The second bookend was that his son, Patrick, was the last to die, thirteen years later on exactly the same day. Physical risk is part of progress, but the consequent human losses should never be forgotten.[37]

Hoover Dam might best be seen as the third of a four-stage manifest destiny for electricity, each with its own impetus and imprint. In the East, Franklin, Edison, and Westinghouse brought both the concept for and commercialization of electricity technology. In the Midwest, Insull paved the way for electricity to be available and affordable to all. In the West, Hoover Dam showed how electricity could tame nature and make it productive in the eyes of that era; the giant dams in the Northwest followed. The fourth and final stage would take place in the South, with Roosevelt's Tennessee Valley Authority.

CHAPTER 10

David Lilienthal's Tennessee Valley Authority

> A Promised Land, bathed in golden sunlight, is rising out
> of the grey shadows of want and squalor and wretchedness
> down here in the Tennessee Valley these days.
> —Lorena Hickok (1934)[1]

As he had done with Hoover Dam, President Franklin Roosevelt furthered the manifest destiny of electric service by creating the Tennessee Valley Authority (TVA) early in his presidency. Where Hoover Dam had pushed electricity service to the West, TVA pushed service into the South. The two furthered the same national goal, yet they were distinctly different in their methods. The details of that difference are essential to the history of electricity. They show that, while federal action was central to both Hoover Dam and TVA, federalization *did not* mean a one-size-fits-all template was imposed, nor did it mean local or regional voices were silenced.

Roosevelt was just thirty-seven days into his first term when he sent Congress his proposal to create TVA. Roosevelt signed the TVA Act soon thereafter, on May 18, 1933. The president's New Deal led to a torrent of new legislation, which created new government agencies that aggressively intervened in the American economy and American lives. No legislation, however, better reflected the president's faith in the good that might be done by the right kind of government agency with the right kind of planning. Roosevelt claimed TVA would be "a corporation clothed with the power of government but possessed of the flexibility and initiative of a private enterprise." The TVA Act reflected faith in the good electricity could do, too; electricity would make life easier and attract new business with new jobs.[2]

Later, TVA's electricity was crucial to America's effort to win World War II. Well established by 1940, TVA was in an ideal position to expand its generation to provide electricity at Alcoa, Tennessee, for the aluminum that went into aircraft, and at Muscle Shoals, Alabama, for the nitrates that went into munitions and fertilizer. TVA also powered the Manhattan Project, the research and development project that created the first nuclear weapons, by supplying electricity to the lab in Oak Ridge, Tennessee. With this surge in electricity production during the war, TVA transformed itself from a regional agency focused on waterway navigation and agricultural development—with electricity supply as a third priority—to a very large electric power monopoly. Its interests went well beyond the hydroelectric power that was envisioned at the time of its creation. TVA aggressively moved to produce electricity from coal, and it was among the first to build nuclear power plants. In time, TVA grew to be the largest public power supplier in the United States.[3]

While it took Roosevelt's political mandate and skill to launch the idea of TVA, it was David Lilienthal who made the idea a reality. Even today, TVA itself refers to Lilienthal as the "Father of TVA." Author Steven Neuse, meanwhile, offers another apt characterization in the subtitle for his biography of Lilienthal: *The Journey of an American Liberal* (1997). Neuse found that, at TVA, Lilienthal was transformed from a typical progressive calling for regulatory reform for the electricity business into a state capitalist calling for government ownership and operation. In this sense, Lilienthal's TVA was a unique, bold experiment on the governance of an industrial democracy. That is, the TVA experiment addressed this question head on: on the spectrum from complete private enterprise to complete government direction, how should the American economy, in general, and the electric power business, in particular, be driven? Only with the right person—Lilienthal—at the right moments—the Great Depression and World War II—could America have conducted such an experiment.[4]

A SEED PLANTED BEFORE THE NEW DEAL

TVA's origins lay in the turmoil of World War I. In 1916 the National Defense Act authorized the production of nitrates for munitions development at Muscle Shoals, Alabama. By the end of the war, the federal government had spent $100 million on a yet-to-be-finished facility with two nitrate plants, for use in munitions or fertilizer production, and a hydroelectric dam. Roscoe Martin, a professor of political science at Syracuse University, writes that after 1918 there ensued a fifteen-year battle over the disposition of the government's holdings at Muscle Shoals; after the war, President Harding wanted to privatize federal projects and the Muscle Shoals facility was put out for bid. None other than Henry Ford offered $5 million for Muscle Shoals and went to visit the site with his equally famous friend, Thomas Edison. A not-so-famous Senator George William Norris, a Republican from Nebraska, put a stop to Ford's bid in 1924.[5]

Norris' interest in the Tennessee Valley was piqued, and in 1926 he introduced a bill to have the federal government expand Muscle Shoals and to build other dams along the Tennessee River. The bill made no progress under President Coolidge and a later bill was vetoed by President Hoover. In 1929 Theodore Roosevelt, then the governor of New York, proposed the construction of publicly owned dams and power plants along the St. Lawrence waterway in New York; momentum for Norris' idea was building. In January 1933, President-Elect Roosevelt and Norris went together to Muscle Shoals. The president-elect spoke of Muscle Shoals being part of a greater effort for the Tennessee River that would benefit "generations to come" and "millions yet unborn." When the senator asked Roosevelt how he would describe the nature of this federal corporation he intended to create, Roosevelt said, "I'll tell them it's neither fish nor fowl, but whatever it is, it will taste awfully good to the people of the Tennessee Valley."[6]

THE TVA ACT OF 1933

Roosevelt's famously productive Hundred Days started on March 5, 1933, the day after his inauguration. He signed the TVA Act on May 18, 1933. The act set both a direction for and the boundaries of TVA's actions. The long-term goals for TVA in the "Tennessee drainage basin and adjoining territory" were listed, in order, as maximum flood control, maximum development for navigation, and "maximum electric power generation consistent with flood control and navigation."[7]

Crucially, TVA would be run by an independent board of directors consisting of three members, all appointed by the president with the advice and consent of the Senate. To eliminate any conflicts of interest, board members were required to have no financial interests in public utilities and other related companies. TVA was given the authority to construct dams and reservoirs in addition to "power houses, power structures, transmission lines, navigation projects, and incidental works in the Tennessee River and its tributaries, and to unite the various power installations into one or more systems by transmission lines." TVA also was given the authority "to produce, distribute, and sell electric power." It had the right to sell "surplus power . . . to States, counties, municipalities, corporations, partnerships, or individuals," though a preference was to be given to non-profits. The right to raise funds with power sales or the sale of bonds also was included. The act made it clear (at least implicitly) that TVA did not have to limit itself to hydroelectric power; it spoke of raising funds for the "construction of any future dam, steam plant, or other facility."[8]

LILIENTHAL'S PATH TO TVA

Early Years

David Lilienthal's *New York Times* obituary summed him up this way: "Keen but relaxed, Mr. Lilienthal was an athlete, intellectual and executive. A professional boxer trained him when he was in high school. He left college with a Phi Beta Kappa key and a reputation as a light-heavyweight boxer." He continued to fight throughout his high-profile career because he was pressured by opponents at every stage. He did not fight, of course, with his fists; instead he deployed the spoken and written word. Lilienthal's lifelong passion for public life began in high school in Michigan City, Indiana, where he gave a speech titled "The Predatory Rich." Later, Lilienthal attended DePauw College in Indiana, where his academic record was satisfactory but not exceptional. However, he kept sharpening his speaking skills by entering oratory competitions and seeking out opportunities to speak.[9]

Lilienthal became impressed with the role lawyers played in progressive politics and decided he wanted to join that profession. He went on to Harvard Law School, where he did well in Felix Frankfurter's public utility law course. Frankfurter, a Roosevelt adviser and eventual Supreme Court justice, became one of Lilienthal's key mentors. Louis D. Brandeis was also among Lilienthal's heroes. In 1923 Lilienthal wrote his first article in the *New Republic*, titled "Labor and the Courts," in which he picked up Brandeis' call for "socially-minded, creative judge[s]." Frank Walsh, another Harvard law professor, encouraged Lilienthal to go to Chicago after graduation and work with Donald R. Richberg. Richberg was a labor lawyer who worked on behalf of the railroad unions and helped write the Railway Labor Act in 1926. He later became a prominent adviser to President Roosevelt and helped write the National Industrial Recovery

Act. Lilienthal's practice of cultivating and working with prominent progressives soon put him on a path to meeting Roosevelt.[10]

At Work

Lilienthal worked for Richberg for three years and contributed to two labor cases that Richberg won at the US Supreme Court. He also got exposure to important utility issues, such as how to value assets when determining rates, including the "prudent investment" versus replacement cost debate. If the original cost of an asset—say a small power generator—was $100,000, the prudent investment method would base the rate customers pay on this original cost. In contrast, if the cost to replace this small generator rose over time to $110,000, the replacement cost method would base rates on the $110,000.[11]

With the courts taking aim at labor, and his interest in utility issues whetted, Lilienthal decided to drop labor law and move more fully into public utility law. Around the time he made the transition, in 1929 and 1931, he wrote two articles on state regulation of public utility holding companies, both published in the *Columbia Law Review*. Reflecting the concerns of the time, he called for closely monitoring contracts between the holding company and the utilities it owned as well as closely scrutinizing the securities issued by them. The Progressives in Wisconsin then provided him with a high-profile opening to public utility issues: in 1931, Governor Philip La Follette asked Lilienthal to join the Wisconsin Railroad Commission.[12]

Although Wisconsin was among the first states to establish utility regulatory commissions in 1907, Neuse, Lilienthal's biographer, concludes that La Follette saw big opportunities for reform. In the midst of the Depression, prices and profits were falling everywhere, but utility profits remained high. One of the reasons was that replacement cost had been widely used as the basis for setting rates and profits; that would be challenged now. Reformers were elected in many states across the country, and Lilienthal became the de facto chairman of the Wisconsin Commission.

Under Lilienthal, the commission would no longer wait for a complaint to be filed but rather would investigate any utility's rates at its own initiative, and the utility would pay for the investigation. The commission got a new name as well: the Wisconsin Public Service Commission.[13]

Lilienthal promoted new ways to set the price, or rates, for products from utilities. In a telephone utility case, Lilienthal championed a rationale for cutting rates that pragmatically reflected the reality of the Great Depression. His core argument was that because prices were falling economy-wide, holding electricity rates steady meant that those rates were increasing in real terms—that is, they were higher after adjusting for the deflation. Prominent economists, like James Bonbright from the Columbia University School of Business, one of the foremost experts on utilities, supported Lilienthal's arguments.[14]

Lilienthal became nationally known when he served on the Committee on Public Utilities in 1931, which was initiated by, among others, Senator Robert La Follette Jr. and Senator George Norris. By 1932, however, Lilienthal was in political trouble because of his aggressive style and the declining popularity of Governor La Follette, who had appointed him. In stepped Justice Brandeis, who recommended Lilienthal to President Roosevelt. FDR nominated him for the TVA board in June 1933.[15]

BREAKING THE STALEMATE

As Lilienthal took up his job at TVA, he noted that "in many states instead of the regulators regulating the utilities, the utilities are regulating the regulators." TVA would regulate public utilities in a fundamentally different way. Lilienthal believed that TVA "represents an attempt to regulate public utilities not by quasi-judicial commissions, but by competition. The act definitely puts the Federal Government into the business of rendering electric service." The results would be "intended to serve as a 'yardstick' by which to measure the fairness of rates of private utilities, and to prevent destructive financial practices." With TVA in operation, Lilienthal was saying that government would not restrict itself to courtroom debates about what honest electricity rates should be, but would define honest electricity rates in the real world by producing electricity itself.[16]

Lilienthal saw a major obstacle holding back progress on lowering electricity rates. He wrote, "Here we have a picture of a complete business stalemate. The electric utilities' position is that until the use of electricity is greatly increased, the rates cannot be drastically decreased. Nor will the use of electricity be greatly increased, they urge, unless electric-using appliances come into general use." To combat this catch-22, the Electric Home and Farm Authority, a subsidiary of TVA, would get appliance manufacturers to produce lower-cost appliances and the new Authority would subsidize the purchase of these appliances by electricity customers.[17]

AT TVA

Lilienthal sat alongside the two other TVA board members, who, coincidentally, shared the same last name. A. E. Morgan was the appointed chairman. He had been the president of Antioch College in Ohio and was a follower of both the utopian thinker Edward Bellamy and the management guru Frederick Taylor. In 1888 Bellamy wrote a novel that promoted his utopian ideals titled *Looking Backward, 2000–1887*. It was a direct, strong indictment of capitalism and strongly favored government management of all aspects of life. A. E. Morgan's interest in Bellamy was strong enough that he wrote a biography, but in practice he seemed to take Bellamy's utopian ideals with a grain of salt. Neuse characterizes Morgan as "a master engineer, dam builder, and social visionary" who had little interest in bureaucratic order and political maneuverings.[18]

A. E. Morgan and Lilienthal were both outsiders to Tennessee, but H. A. Morgan, the third of the three board members, had been a local for fifty years. He had been the dean of Tennessee's School of Agriculture, and his broad vision centered on the interconnectedness of the "soil, air, water, and all of life." Lilienthal joked that he was the only board member that was not named Morgan, did not hold a doctorate, and had not been a college president.[19]

The three board members allocated responsibilities: A. E. Morgan would be responsible for dams, navigation, "social experiments," and overall "integration" of the disparate goals of the TVA. H. A. Morgan would be responsible for agricultural programs, while Lilienthal would be responsible for electric power and legal issues. From the start, there was tension among the board members, with many meetings enlivened by "shouting matches." To Lilienthal, TVA was all about electric power, and his goal was to make power cheap and abundant. He saw the opportunity to make life easier for people who couldn't afford much; for example, he said the introduction of public power would mean a 60 percent rate cut in Tupelo, Mississippi. Interestingly, this motive was much like Insull's, in the sense that he saw low prices leading to higher use, which in turn led to even lower prices as the fixed costs were spread over a larger number of customers. With this theme, Lilienthal was heading toward a confrontation with A. E. Morgan, who envisioned a lesser role for power and wanted to limit the sale of TVA power to private electric utilities rather than compete with those utilities for customers.[20]

Lilienthal clashed with A. E. Morgan over power sales, but he also had to battle Wendell Willkie, the president of the Commonwealth and Southern Holding Company, the private electric utility that served the Tennessee Valley. Willkie was in a tough spot. If he sold local electric transmission and distribution facilities to TVA, as Lilienthal proposed, then TVA would get bigger and be an even more formidable competitor. If Willkie did not sell, however, TVA would use federal money to build competing transmission and distribution facilities anyway. Moreover, while Lilienthal's (and Roosevelt's) notion of having a public yardstick for comparison to the prices of private utilities sounded right, there were several reasons it was not a fair comparison. Willkie could not compete on price because TVA enjoyed subsidies from the federal government that lowered its costs. In addition, TVA used mostly low-cost hydroelectric power, and TVA's price did not include distribution costs, as Willkie's did.[21]

TVA AND THE COURTS

Lilienthal had to fight the private electric utility companies in court. The courts were seen as a place to fight by those who opposed Roosevelt's New Deal programs in general and that was true with TVA, too. The constitutionality of TVA power sales and the use of federal government money to help communities build transmission and distribution facilities were challenged in the case *Ashwander v. Tennessee Valley Authority*. TVA lost the challenge in Alabama courts, but the Alabama ruling was overturned by the federal court. In 1935 the US Supreme Court upheld the federal court's decision. That decision was narrowly focused on allowing power sales from just the Wilson Dam at Muscle Shoals.[22]

The constitutionality of TVA power sales was challenged again in a case involving the Tennessee Electric Power Company. The US Supreme Court eventually ruled that the utilities had no standing to sue because they had not been guaranteed a monopoly. While the court did not rule on the constitutionality of TVA, per se, the cumulative impact of the losses in court meant that Willkie finally realized that TVA was "there to stay." He agreed to sell to TVA substantially all of Commonwealth and Southern's facilities in Tennessee as well as some facilities in Alabama and Mississippi. Congress appropriated funds for the purchase and, significantly, did not put any territorial limit on TVA.[23] Cases like this are important because the Supreme Court allowed state capitalism to displace regulated private capitalism.[24]

The very public battle between Lilienthal and A. E. Morgan continued, with dueling articles in the *Atlantic Monthly* and the *New York Times*. In his *Atlantic Monthly* article A. E. Morgan defended public power, but said that some—namely Lilienthal—actively hated private utilities and wanted nothing short of an all-out war. An article by Wendell Willkie appeared in the same publication before A. E. Morgan's, and Lilienthal used this fact, along with a detailed comparison of the two articles, to imply that A. E. was somehow controlled by private utilities.[25]

The battle came to a head after A. E. Morgan accused Lilienthal of improprieties, such as alleged bias against private utilities regardless of the circumstance, and called for a congressional investigation. Roosevelt grew tired of A. E. Morgan's allegations and demanded that he substantiate any charges. When he could not, or would not, Roosevelt fired him. Still, the congressional investigation went on with seventy days of hearings in the period from May to December 1938, but in the end six of ten people on the panel exonerated TVA and Lilienthal of any wrongdoing.[26]

TVA GOES BIG AND GOES TO WAR

TVA was big from the start simply because of its geographic footprint. It served the Tennessee Valley that drains the Tennessee River and its tributaries, covering parts of seven states: Tennessee, Kentucky, Virginia, North Carolina, Georgia, Alabama, and Mississippi. In his landmark book *TVA: Democracy on the March* (1944), Lilienthal reported that TVA then controlled twenty-one dams, sixteen of them newly built, and that $700 million had been invested in the TVA system. Of that total, he attributed 65 percent to electric power generation, 20 percent to flood control, and the remaining 15 percent to navigation. Lilienthal then ticked off what he saw as the important impacts the TVA investment had had on the Tennessee Valley over the ten years from 1933 to 1943. Given all this investment, how did TVA now compare to the rest of the country? Lilienthal proudly concluded that TVA was the United States' second largest electric generator, providing 1 billion kWh per month in 1944. Ten years earlier, per capita electricity production in the TVA footprint had fallen well short of the US average. After a decade of investment, that per capita electricity production was about 50 percent higher than the national average. Lilienthal had made electricity the priority of TVA, and these results were ample evidence of his success.[27]

Looking at all that had been accomplished by TVA, Lilienthal could not help but compare his dams to the great dams of the West: Hoover and Grand Coulee. In terms of the concrete used alone, he found that

TVA, in building all its new dams, used four times as much concrete as Hoover Dam had. Moreover, he recalled that it took six major private firms to build Hoover Dam and ten such firms to build Grand Coulee, but it was just a single agency, TVA, that built all the dams in the Tennessee Valley system.[28]

Even more noteworthy was TVA's importance to America's victory in World War II. TVA's key role in the war developed in four ways, according to Lilienthal. First, TVA supplied plentiful electricity to the Alcoa aluminum plant south of Knoxville; the Alcoa plant was the largest aluminum plant in the world, and it was central to achieving the fifty-thousand-plane air force President Roosevelt had called for. Second, it was the assured supply of electricity, as well as its ample supply of water for cooling, that enticed the federal government to secretly build an entire city called Oak Ridge in Tennessee. Electricity supplied to Oak Ridge was used to produce nuclear material to make the atomic bombs that would ultimately be dropped on Japan. Specifically, the electricity was used to run the "centrifuge and the electromagnetic and gaseous-diffusion processes"; such equipment is central to the development of nuclear weapons, as evidenced by the fact that it is part of the debate on the Iranian nuclear deal today. Third, the nitrate output at Muscle Shoals was used both as material for making munitions and as material for making fertilizer, much of it sent to America's allies in Europe. Fourth, TVA had advanced aerial reconnaissance and other mapping techniques in the Tennessee Valley, and these techniques were implemented in Europe to create maps for Allied aviators.[29]

While TVA was invaluable to the war effort, Erwin C. Hargrove, professor emeritus at Vanderbilt, shows that the war was equally important to TVA. It was the war that allowed TVA to go far beyond the navigation and flood-control mandates in the original TVA Act. TVA became the largest integrated electricity system in the United States, and at one point 75 percent of its power went toward the war effort. Employment there grew significantly, from fourteen thousand in 1940 to forty-two thousand in 1942. The war gave justification to TVA building its first nonhydro unit, a steam unit, at Watts Bar in Tennessee. With this steam plant, TVA was no longer solely focused on the Tennessee River and hydroelectric power; it was transformed from a regional river basin manager to a major power monopoly. It used the power needs at the federal facilities in Oak Ridge, Tennessee, and Paducah, Kentucky, devoted to atomic development to justify building many more steam generation facilities. Most important, it was TVA's contribution during the war that gave it the reputation as a government agency that worked.[30]

TVA'S IDEOLOGICAL FOOTPRINT

For many people, TVA was always about much more than concrete, water, and electricity. The project had a spectacular impact on the well-being of people in the Tennessee Valley. TVA was a symbol of American progress and expertise in the field of electricity and could serve as a model for economic and political development around the globe. Publicity about TVA's spectacular impact started not more than a year after TVA began. In the depths of the Great Depression, Harry Hopkins, Roosevelt's trusted adviser, sent Lorena Hickok, a pathbreaking journalist and close friend to Eleanor Roosevelt, to report on the effects of the New Deal.

In 1934 Hickok wrote to Hopkins to tell him what she had found out about TVA. She wrote, "A Promised Land, bathed in golden sunlight, is rising out of the grey shadows of want and squalor and wretchedness down here in the Tennessee Valley these days." She continued, "Ten thousand men are at work, building with timber and steel and concrete the New Deal's most magnificent project, creating an empire with potentialities so tremendous and so dazzling that they make one gasp." She visited the sites of the Norris and Wheeler Dams as well as the Wilson Dam. She marveled that the men working on the Norris and Wheeler Dams were "working five and a half hours a day, five days a week, for a really LIVING wage. Houses are going up for them to live in—better houses than they have ever had in their lives before. And in their leisure time they are studying—farming, trades, the art of living, preparing themselves for the fuller lives they are to lead in that Promised Land." Contrast this with the hardship at Hoover Dam. Hopkins deployed correspondents like Hickok to keep him informed about the needs and mood of Americans. Although they had an audience of one, Hickok and the others gave eyewitness accounts that are a valuable part of the historical record.[31]

Lilienthal also wrote passionately about TVA's impact. He modestly described his book as merely a report to the stockholders, but he went on to claim that all Americans owned TVA. He said, "This is a book about tomorrow." In assessing TVA's impact, Lilienthal said that "men and science and organizing skills applied to the resources of waters, land, forests, and minerals have yielded great benefits for the people," and chief among those benefits was a change of attitude for the future: "No longer do men look upon poverty as inevitable, nor think that drudgery, disease, filth, famine, floods, and physical exhaustion are visitations of the devil or punishment by a deity." Lilienthal wrote that, after ten years, the Tennessee Valley and surrounding areas had become "an entirely different region."[32]

These powerful pronouncements by Hickok and Lilienthal go far beyond political spin. They reflect a fundamental cultural view: electricity was a good thing because it modernized and civilized the Tennessee Valley. In the end, success or failure can be judged only in the context of a nation's stated goals. With a goal of making electricity available and affordable to all, including rural areas, TVA was an important success. When the goal changes, however, so can the judgment on success or failure.

Lilienthal also freely referred to TVA in terms of a moral or spiritual purpose. He wrote, "The physical achievements that science and technology now make possible *may bring no benefits* . . . unless they have a moral purpose." And that moral purpose must be reflected in unified resource development "governed by the unity of nature herself." Plus, he put forth that resource development must ensure grassroots participation by all the people affected. Speaking of meetings held by electric cooperatives for the people who were both consumers and owners, he claimed that they often had an "emotional overtone" and "spiritual meaning." Here we might compare Lilienthal's secular faith to the spiritual faith of Faraday, both centered on a unity found in nature.[33]

TVA is a government-owned entity; it is not a private utility regulated by government. Its unique status would become ever more pronounced during the Cold War, in which free market capitalism was at war with the centrally planned, state-run industries under communism. As if anticipating this, Lilienthal said he feared overcentralization in both government and business, and rather than aligning with big business or big government, he instead offered the TVA model, which he believed balanced the two by establishing centralized electric power production with decentralized delivery through the local distributors TVA served. He argued that control from distant Washington, DC, by the federal government would have been no better than control from New York City by electric holding companies. And anticipating curiosity about his politics, Lilienthal said plainly that politics had to be kept out of TVA; as he put it, "A river has no politics."[34]

Lilienthal's *TVA: Democracy on the March* got plenty of good reviews, but there were also skeptics. One of the old guard of the New Deal, Rexford Guy Tugwell, a famous member of Roosevelt's brain trust, referred to the project as "democracy in retreat." Donald Davidson, poet and founding member of the Southern Agrarians, pointed to the seventy thousand people displaced by TVA's dam building, and said TVA was more "paternalistic" than "participatory." William Chandler, author of *The Myth of TVA* (1984), believes that, because Lilienthal's book was so widely used as *the* source of information on TVA, his book kept honest critiques of TVA at bay until the 1970s. Only then did problems with TVA's nuclear energy program cause scholars to look more critically at Lilienthal and his model of government ownership.[35]

ASSESSING TVA AFTER TWENTY YEARS

In the early 1950s, two respected academics looked back at the original goals for TVA compared to its actual achievements. Each came up with somewhat different assessments of the path TVA had taken, but Norman Wengert of the City College of New York and the aforementioned Roscoe

Martin of the University of Syracuse came to similar conclusions. In short, TVA had accomplished a lot, but by the end of World War II, it had evolved into simply another large electric utility.

The title of Wengert's article hints at his conclusion: "TVA—Symbol and Reality." Wengert concludes that TVA's "symbolic significance may be more lasting and more important than its actual accomplishments in the development of the Tennessee Valley, however impressive the latter might be." To document the high hopes for TVA, he quotes the noted historian Henry Steele Commager, who wrote, "Here [at TVA] were tested the broad construction of the Constitution, large-scale planning, the recasting of federalism along regional lines, new techniques of administration, and new standards of civil service, the alliance of science and politics, and the revitalization of democracy through a calculated program of economic and social reconstruction."[36]

After the "collapse of the 'free enterprise system,'" the hope for broad-based government planning was understandable—though Wengert concludes TVA did not fulfill that hope. Wengert offers many reasons for the narrow planning actually done at TVA. The first is that the early battles of the board of directors hurt such efforts. The congressional investigation triggered by A. E. Morgan's allegations hurt, too, as did the hard-won litigation victories. Another factor was that electric power outsized the other TVA functions. Moreover, according to Wengert there was no political support for broad, region-wide economic planning by TVA because such planning was essentially a political goal-setting process and there was no regional political body to handle such goal setting; TVA could not fill this political role because its board and its managers were not elected. Thus, TVA settled into a narrow "preoccupation with power and chemical development."[37]

Martin comes to essentially the same conclusion: TVA achieved a lot in flood control and navigation early on, but looking back from the 1950s, it had become just another large electric power utility. He sees TVA's roots in the fifteen-year battle over what to do with the Muscle Shoals munitions plants and power facilities—essentially an ideological battle between public and private development. Norris tried and failed seven times with legislation to initiate TVA, but, finally, the Great Depression and Roosevelt's election created the context in which Norris' dream could come true in the form of the TVA Act. Still, the act was blunt in its narrow focus on flood control, navigation, and power production; Supreme Court rulings furthered this narrow focus by establishing TVA's right to sell power.[38]

LILIENTHAL'S (AND TVA'S) LEGACY

Like Hoover Dam, TVA furthered electricity's march across America. It did so, however, in a way that reflected local or regional aspirations and opportunities. At the start, TVA was meant to be all about an ambitious effort to jump-start economic development in the impoverished Tennessee River Basin—with electricity playing an important but not dominant role. Such an effort would have reflected the desires of Roosevelt and Norris as well as those of the two Morgans who served on the original TVA board of directors. Quickly, though, factors conspired to turn the project into an ambitious effort to make TVA-generated electricity available and affordable to all the businesses and homes in the Basin, and this created far-reaching consequences for economic development. The most influential "factor" driving TVA in that direction was Lilienthal himself. Under his direction, TVA became a rare and large-scale experiment in state capitalism in America.

The TVA experiment could have significantly changed the path forward for the American electricity business. What emerged was a head-to-head battle between Lilienthal's state capitalism and regulated, private capitalism, as embodied in Wendell Willkie's Commonwealth and Southern Holding Company. Private capitalism, even of the regulated variety originated by Insull, did not really have a chance to win in the Basin. It could not duplicate the advantages bequeathed by the federal government and, without federal financing, private capitalism surely would not have moved as fast given the risks. And if it did not move fast, it would not have achieved the economic impact that Lorena Hickok wrote about and thousands more experienced.

Notably, the goals and tactics of Lilienthal's state capitalism and Insull's private capitalism were the same: increase the scale and scope of the electricity system to capture economies of scale and lower the price of electricity. That lower price would increase demand, which would set in motion another round of cost and price declines. Attracting new customers was central to the effort, and TVA was in a perfect position to do so during World War II.

It is surprising that Roosevelt let the experiment go forward, given his words and actions on other occasions. In his Portland speech, he said electricity should be mostly a matter for private business. And with Hoover Dam, there was no broad effort at state capitalism: the power was sold to other private and public utilities. The TVA experiment might indicate that Roosevelt was no ideologue but rather a pragmatist looking for what worked.

Had TVA shown itself to be a reliable, beneficial alternative to private capitalism, the American electricity business would look quite different today. In the end, however, TVA was (and is) a large public utility with all the attendant opportunities to do good or bad, just like a private utility. Like many private utilities, TVA's later failure with nuclear power would undermine its political support because it lost touch with its pursuit of abundant and affordable power. Also like any private utility, TVA would face charges that it caused environmental harm, ranging from threatening endangered species (such as the snail darter, a fish discovered during TVA's construction of the Tennessee Dam) to causing global climate change with carbon dioxide emissions from its large coal-fired power plants. By the time this ultimate equivalence of TVA to a large private utility became clear, however, TVA was simply too big to go away. Despite its emphatic rhetoric, and the substantial belief engendered by that rhetoric, TVA did not prove to be either transformative or replicable as a new form of governance for an industrial democracy in general, or for the electricity business in particular.

CHAPTER 11

King Coal's Reign

> For within coal is man's industrial holy
> trinity of light, heat, and power.
> —Saul Alinsky (1949)[1]

The term King Coal *seems to stick no matter the actual conditions of the coal* industry. In 1917 Upton Sinclair wrote a novel titled *King Coal* that details the harsh conditions coal miners faced; his sequel *The Coal War* was published posthumously in 1976. Even today, with vastly different conditions in the coal business, King Coal is a frequent and prominent characterization in the newspaper headlines from the *New York Times*, the *Economist*, and many others.[2]

Coal was not always king. By 1850 coal production and delivery had become a fully commercialized business in America, but coal still faced substantial competition from the primary fuel of the day, wood. Coal's share of all energy use in 1850 was just under 10 percent, with wood accounting for much of the rest. Coal eventually pushed wood aside so that, by 1920, coal accounted for three-quarters of all energy use. That rise to dominance is what earned it the title of King Coal. Coal's rise was driven by falling coal prices and the fact coal proved to be more effective in the industrial processes of the day.[3]

Writing in 1949, community organizer and writer Saul Alinsky eloquently described the breadth of coal's reign: "Coal is the prime mover of our life. In these black chunks of the earth's history is the energy that pours power into our gigantic industrial empire . . . This Gargantuan industrial scene reveals interlaced speeding railroads, giant whirring dynamos lighting up the nation, and overwhelming surges of power spun from steam . . . For within coal is man's industrial holy trinity of light, heat, and power." Alinsky's powerful writing reveals how coal was used to power the mighty steam engines that, in turn, powered many of the mightiest industries. Coal was essential to iron and steel making as both a fuel and a catalyst, and was also used to power America's railroads and to heat homes.[4]

And, of course, coal was used to generate electricity. However, the crowning of King Coal cannot be attributed to the use of coal in the electricity business alone. In 1920, with coal's market share at its peak, only 7.3 percent of all coal went to electric utilities. As the coal business lost its broad customer base, however, it grew more and more dependent on the electricity market. By 1955 the share of coal sold for electricity generation had increased to 32 percent. By 2015 the electricity business had become coal's primary customer: 93 percent of the coal consumed in America was used to generate electricity. While the electricity business is certainly important to the coal business, there is another side to the coin, too: coal has been and, for the moment, still is quite important to the electricity business. By 1950 coal was used to generate 46 percent of all electricity in America. Even as recently as 2015, coal was used for 33 percent of all electric generation.[5]

The history of electricity, then, is intertwined with that of coal, and the story of one cannot be told without the story of the other. Coal *directly* fueled the industrial and subsequent military rise of America. Then, although it lost much of its direct sales, coal continued to fuel America's rise *indirectly* by emerging as a preferred fuel for electric power generation. It achieved that preference by virtue of the cost-plus regulatory scheme chosen for electricity.

So, too, the intertwined history of electricity and coal cannot be told without some perspective on the headline-grabbing drama of violence and labor unrest in the coal business. Central to that drama, in turn, is the larger-than-life character of John L. Lewis, the famous (or infamous) labor leader.

THE CROWNING OF KING COAL

Coal Caused a Revolution

Business historian Alfred Chandler posed an interesting question that illuminates how coal transformed American life. He begins with a premise that most historians agree on: factories, and the division of labor they employed, were essential for early industrialization. Further, there were many factories in England by the late 1700s. Why then, he asked, was there a delay until the 1840s in the widespread development of factories in America? A clue to the answer, Chandler said, could be found in Pittsburgh, Pennsylvania. Pittsburgh was the exception, and the reason was that it had access to sufficient reserves of coal to meet both home and business needs. That, in turn, gave Pittsburgh manufacturers an inexpensive source of steam power and an inexpensive way to manufacture iron, which were necessary for the prolific rise of factories.[6]

In time, coal suppliers finally delivered those same two resources—inexpensive steam and inexpensive iron—to the rest of America's big cities in the Northeast. Coal created America's early Industrial Revolution, after which factories quickly proliferated. Metal works began to produce "shovels, saws, scythes, stoves, pots, pans, wire, plows," and many other products that were necessary for homes, farms, and other businesses. The revolution then extended to other industries, from glass and paper to wood and leather making. And the revolution was so effective that eventually the British came to America to study American manufacturing techniques, not the other way around.[7]

King Coal by the Numbers

By 1850 coal was a fully commercialized business, and two surges in coal production growth ensued. The period from 1850 to 1885 was the first period of rapid growth for coal production. From just 8.4 million tons in 1850, coal production grew in America to 110 million tons by 1885—a thirteenfold increase over thirty-five years. In the second period, from 1885 through World War I to 1920, growth spiked again, with coal production increasing sixfold to 658 million tons.[8]

The reason for this rapid growth might aptly be termed the *iron-rail-steam interdependence* on the basis of research by Sam Schurr, the former director of the Energy and Mineral Resources program at Resources for the Future, and his colleagues. Coke, which is almost entirely carbon, is made by heating coal to high temperatures to drive out impurities. Coke and iron ore are mixed in a blast furnace to create molten or pig iron. That pig iron is the primary material from which a range of steel products are made. Early on, the railroads needed the rails that were made from pig iron. Then, as the Bessemer process for making low-cost steel from iron came into wider use, it provided another market for the coke made from bituminous coal. Railroads, however, were even more important to coal's rise, because coal was used to make the steam that powered locomotives.[9]

In 1850 coal's market share was just about 9 percent of all energy, and wood was still the dominant fuel with a 91 percent market share. By 1885 coal's share had increased to 50 percent and wood's share had decreased to 47 percent; the small, 3–4 percent remaining share was served by oil and natural gas. By 1920 King Coal was dominant with a 75 percent share of all energy use; wood's share had fallen to just 7 percent, while oil, natural gas, and hydro took about 18 percent. King Coal was crowned.[10]

STAGNATION

The crowning of King Coal, surprisingly, also marked the beginning of a decline for coal in the face of competition. American coal production had reached its peak in 1918 during World War I, when annual production reached 678 million tons. With the onset of the Great Depression, coal production fell. Then, during World War II, coal production resumed its rise and reached a slightly higher peak of 683 million tons in 1944. In the post–World War II period, coal production kept level at first, in part because of coal exports to a devastated Europe. By the mid-1950s, however, it had declined again.[11]

Crucial competitive changes largely explain this decline. Railroad demand for coal dropped precipitously from 135 million tons in 1920 to just 15 million tons in 1955. Schurr concludes that coal's decline was the "direct result of oil's invasion of the railroad fuel market." Oil's rise and coal's fall can be seen more broadly, too. Indeed, the fortunes of oil and coal rose and fell in inverse proportion to each other. In 1920 coal's share of all energy use stood at 75 percent and the oil and natural gas share was just 15 percent, but by 1955 coal's share fell to 32 percent, while oil and natural gas rose to 62 percent.[12]

Another important competitor of sorts was the efficiency of coal use. As time and technology progressed, America got more and more useful work out of each ton of coal. Schurr points out, "Thirty years earlier at least five pounds of coal had to be burned in a steam boiler to produce one horsepower hour of mechanical work, whereas the most modern internal combustion engines run with coal gas could produce one horsepower hour with as little as one pound of coal." Even as electric utilities became coal's most important customer, they did so despite steady, substantial improvements in efficiency. Schurr reports that 7 pounds of coal were needed to produce 1 kWh of electricity in 1900, but by 1950, that had been cut to just 1.19 pounds. Still, rapid growth in coal use by electric utilities offset to some extent the loss of important customers like railroads. To wit, coal use by electric utilities rose from 37 million tons in 1920 to 141 million tons in 1955.[13]

With the two peak-coal years (1918 and 1947) looking like equal-sized bookends, it could be said that coal was left behind by a growing America. In that same time frame, America's population increased by 31 percent and its gross national product increased by 130 percent, while manufacturing

activity had increased by 225 percent. Despite economic growth, coal production, even looking at just the peaks, had stayed flat.[14]

COAL'S DRAMA: CULTURE AND POLITICS

While the numbers are essential to the story, they cannot capture all
its dimensions. In particular, the numbers do not describe the culture
and politics that caused much of the drama and garnered most of the
headlines about coal in the first half of the twentieth century. This side
of the story is best discussed in the context of the life of John L. Lewis,
the leader of the United Mine Workers, because he acted out the drama
on a prominent national stage. Saul Alinsky published an unauthorized
biography of Lewis in 1949, and it is a uniquely valuable source about
Lewis' life and times—which were so well publicized then, and are
little-known today.[15]

"Where Death Never Takes a Holiday"

On no topic is Alinsky more blunt than on the culture of the coal miners.
He writes that the miners are a "mystery to the overwhelming majority
of our nation." The miner has a "bitterness bordering on hatred for the
outside world" and has the "indifference of men who work down below
where death never takes a holiday." This is why it is said that Lewis and
his miners cared little about what anyone said to them or about them.
Alinsky drives home his point about miner culture by vividly evoking their
daily life. He describes the breakfast a miner eats compared to what a city
worker might eat. Most of us, he writes, eat the "morning ceremonial of
toast and coffee." In sharp contrast, "the miner's breakfast may be fried
chicken or ham with potatoes, hot cakes, biscuits, and endless steaming
coffee, or as they say, 'the kind of eats that sticks to a man's ribs.'" The
miners need this hefty meal to beat the physical and mental exhaustion
their works brings.[16]

Alinsky describes the difficulty and danger of a miner's day in detail. A miner might have to travel miles underground to get to the "face" of the coal, where the coal seam is sandwiched in between layers of rock above and below. A coal-cutting machine could rip the coal from the seam. At the time, explosives were used to break the coal up into pieces. Those pieces were loaded—by hand or by machine—onto a railcar to be carried to the surface. Alinsky describes the mining process as "the ripping of the undercutting machine, the whir of drills, the blast of explosives, the crash of the coal face, and the clatter of loading." Collapsed roofs and errant explosives were frequent killers. The Mine Safety and Health Administration reports that 68,835 miners were killed and over two million injured in the thirty-five years from 1910 to 1945.[17]

Those who use the electricity made with coal, then and now, should never forget the human cost. Later, this same point would apply for the environmental impact of coal-fired electric generation. Those who use electricity should be aware of the consequences along the full path of production.

"One of the Bleakest and Blackest Episodes of American Labor History"

Sadly, the history of coal contains frequent reference to massacres. A prime example of this was the infamous Ludlow, Colorado, massacre in 1914, which novelist and historian Wallace Stegner called "one of the bleakest and blackest episodes of American labor history." In September 1913 miners working at the Rockefeller-owned Colorado Fuel and Iron Company went on strike to protest low pay and deplorable working conditions. Striking miners were evicted from their company-controlled homes, so they built tent encampments as temporary housing. Everything was calm until the National Guard was brought in to break up the strike once and for all. In spring 1914 a war broke out between the strikers and the National Guard.[18]

That April, the National Guard enticed labor leader Louis Tikas to come out of his tent to discuss a truce. An eyewitness recounted seeing

Tikas walk out of his tent when "suddenly an officer raised his rifle, grip-ping the barrel, and felled Tikas with the butt. Tikas fell face downward. As he lay there we saw the militiamen fall back. Then they aimed their rifles and deliberately fired them into the unconscious man's body. It was the first murder I had ever seen, for it was a murder and nothing less." The National Guard did not stop there; witnesses said that the "militia dragged up their machine guns and poured a murderous fire" into the camps and then set fire to the tents. In total, sixty-six fell as a result of the massacre and ensuing riots. The stage had been set for a bloody twentieth-century power struggle between mine owners and striking workers. Eventually this conflict would reach the highest office in the land, as John L. Lewis and Franklin D. Roosevelt squared off over strikes that could have under-mined America's efforts in World War II.[19]

Lewis' United Mine Workers

With the constant tension between miners and mine operators, a strong union would seem to be inevitable. Unions and union leaders, however, did not have an easy rise to power. In 1890 the United Mine Workers of America (UMW) was created in Columbus, Ohio, through the merger of two other labor organizations. The UMW then affiliated itself with the nationwide craft unions under the American Federation of Labor (AFL). UMW members were not easy to lead, or to please. Even John Mitchell, a founding member of the UMW and its president, considered a "patron saint" of organized labor, was pushed out by the rank and file in 1908. The UMW needed a strong leader and they got one in John L. Lewis, who took the reins in 1920 and kept them firmly in his hands for the next forty years.[20]

Lewis had solid roots in mining. His father, Thomas, was among the Welsh miners who immigrated to America in the 1870s. Thomas Lewis settled in Iowa, then a hotbed of populism, and joined the Knights of Labor. The family tale is that much later Thomas led and won his first strike against the coal companies. The price he paid was being blacklisted by those companies and not being able to find work—a difficult situation

with, by then, eight children. John L. Lewis had no chance for even a
high school education. He began working in the mines at age fifteen and
inherited his father's hatred for mine owners.[21]

Chronic Turmoil and Disorder

As Lewis began his career in the 1920s and early 1930s, the coal business was in a period of upheaval and anxiety. Writing in 1935, the *Yale
Law Journal* spoke of the "chronic state of turmoil and disorder" in the
American coal industry. This was the case despite the hopeful claim
by the journal that America had huge advantages with coal. In 1929
the United States accounted for 40 percent of the world's coal production, and it was believed that America held within its borders half of
all the coal that remained to be mined in the world. The core problem,
wrote the journal, was excess mine capacity; the capacity to produce
coal persistently exceeded the demand for coal. The journal listed the
varied causes of this excess, including, among others, seasonal demand,
railroad subsidies for long-distance transport of coal, and the increasing
productivity of mine workers.[22]

The journal reported that the excess capacity was consistently in the
40–50 percent range in the years that followed. Equally important was
the fact that, to compensate for the decline in coal prices due to the excess
capacity, coal mine operators chose cuts in wages as the primary business tactic. That, consequently, led to strikes in both union and nonunion
mines. In 1919 and 1922 the strikes caused coal shortages, and several
unions lost their gamble. By 1930, 80 percent of coal mines were nonunion, up fourfold.[23]

The Great Depression: Great for the UMW

The Great Depression completely reversed the prospects of the UMW. A
major reason for that was the passage of pro-labor legislation such as the

Norris-LaGuardia Act, which dramatically curtailed the use of court injunctions to combat strikes and said that companies could not ask their employees to sign a contract that prohibited union membership. The short-lived National Recovery Act also gave labor a shot in the arm and helped bolster unions before being found unconstitutional in 1935. Section 7 of the National Recovery Act said, "Employees shall have the right to organize and bargain collectively through representatives of their own choosing . . . employers shall comply with the maximum hours of labor, minimum rates of pay, and other conditions of employment." After section 7 became law in 1933, UMW membership exploded to over five hundred thousand in just two years from a low of about seventy-five thousand.[24]

Although the UMW would be his top priority throughout his life, Lewis had big plans for unions nationwide. In 1921 Lewis challenged Samuel Gompers for the presidency of the AFL. Gompers was the first and, ultimately, the longest-serving president of the AFL. With his leadership the group grew to be the biggest labor organization in the world, with three million members in 1924.

Lewis lost the election but garnered one-third of the votes. He kept up his criticism of the AFL into the 1930s. The UMW soon was out of the AFL, and Lewis began his effort to create a rival nationwide union organization named the Congress of Industrial Organizations (CIO). Through the CIO, Lewis took the dramatic step of taking on General Motors—a part of Detroit's auto industry long considered the "greatest genii of industrial power known to mankind." The General Motors strike showed just how influential Lewis had become on a national stage. It also showed that, beyond coal, the war between capital and labor had many fronts, and that Lewis' powerful tactics and personal charisma significantly affected every one of them.[25]

The result of the General Motors battle in particular depended on whether Michigan governor Frank Murphy would call the National Guard to evict the sit-down strikers from General Motors factories. Governor Murphy told Lewis that, indeed, he would call the National Guard because he had a duty to uphold the law. Lewis fired back that if Murphy used the National Guard to evict the strikers he would be "giving complete victory to General Motors" and would betray the workers. He reminded Murphy of his own family's plight in Ireland. Lewis asked Murphy if he would choose to "uphold the law" when his father was imprisoned by the British, or when Murphy's Irish grandfather was hung. The inspired arguments worked. After that conversation, the governor did not order the troops to evict the strikers. GM signed a contract with the United Automobile Workers to end the sit-down strike, which had lasted over forty days.[26]

Lewis' Break with FDR (and America)

Lewis had doubled down on his bet on Franklin Roosevelt in the 1936 presidential election; UMW contributed more than $500,000 to the Roosevelt campaign. Trouble began to brew early in Roosevelt's presidency, however; in Lewis' mind, labor had not yet gotten any quid pro quo for their money or their votes. The Lewis-Roosevelt rift began when, in the midst of the General Motors strike, FDR told the Michigan governor to "disregard whatever Mr. Lewis tells you." Lewis said he would support the Republican presidential nominee, Wendell Willkie, in the 1940 election—the same Wendell Willkie who had battled David Lilienthal at TVA less than ten years earlier. Lewis pulled no punches. In a major radio address, he said FDR "means War! War! War!" and was trying to create a "political dictatorship." Finally, Lewis threatened to quit the CIO if the members did not turn away from Roosevelt. He said, "Sustain me now, or repudiate me." Roosevelt did win, of course, and Lewis did stay true to his word—he resigned the presidency of the CIO.[27]

After resigning, Lewis returned to the UMW, saying, "Coal miners of America have the first call and the first right on whatever John L. Lewis has to contribute." As his first big issue there, Lewis chose captive mines; these are the mines run by the big consumers of coal such as steel mills. Captive coal mines employed fifty-three thousand mine workers and 95 percent of them already were unionized. Still, Lewis wanted to raise that proportion to 100 percent. In 1941, to that end, Lewis ordered a strike. FDR came out against the strike, claiming he was saying no to a "dangerous minority of industrial magnates" and to "a small but dangerous minority of labor leaders." On December 7, 1941, Lewis won his campaign for the captive coal mines. And on that day of infamy, the Japanese bombed Pearl Harbor.[28]

War economics created the backdrop for Lewis' next strike. His miners faced flat wages, and yet the prices they paid at the stores were rising because of wartime shortages. Lewis argued that mine workers were struggling to pay food prices that had increased 125 percent since 1939. In a full-throttle attack, he called the current pay level "substandard starvation wages," and he claimed the miners' children "cry for bread." Tension mounted again between Lewis and President Roosevelt. In April 1943 Lewis threatened that his 450,000 miners would walk out if a wage increase was not achieved by May 1. Shockingly, on May 1, 1943, in the middle of the war, 450,000 coal miners went on strike in defiance of the president. They achieved little popular support; on the whole, the American public took offense at Lewis' strike. In response to theater newsreels about the strike, audiences would hiss and boo. High school students picketed Lewis' home in Alexandria, Virginia, with posters calling him "Hitler's helper." One political cartoon showed Lewis "stabbing the American soldier in the back." The military newspaper *Stars and Stripes* said, "Speaking for the American soldier, John L. Lewis, damn your coal black soul!"[29]

On May 2, President Roosevelt delivered his fireside chat "On the Coal Crisis" to plead for the miners to return to work for the sake of America's war effort. Addressing the striking miners directly, he said, "No matter how sincere his motives, no matter how legitimate he may believe his grievances to be—every idle miner directly and individually is obstructing our war effort." More bluntly, the president said, "A stopping of the coal supply, even for a short time, would involve a gamble with the lives of American soldiers and sailors and the future security of our whole people." He then spoke to the miners about their own sons. He cited stories he had heard of miners' sons from Pennsylvania, Kentucky, and Illinois who were serving in the military. He said he was appealing to the "essential patriotism of the miners" and made clear that the "three great labor organizations"—the AFL, CIO, and the Railroad Brotherhoods—had a no-strike pledge and that the UMW "was a party to that assurance." On June 23, 1943, government control of the mines ended as the coal strikes that had started on May 1, 1943, were tentatively resolved.[30]

But the wave of strikes was not over. On April 12, 1945, President Roosevelt died. Nineteen days after the president's death, on May 1, 1945, Lewis called another strike. In March 1946, Lewis again called a strike and then, because of the shortage of coal, railroads laid off 51,000 workers and Ford, with 110,000 employees, began to shut down. Lewis had not been tamed by powerful politicians. He would be tamed somewhat by the technological progress and geographic shift in the aftermath of these strikes.[31]

THE AFTERMATH BY NUMBERS

Major shifts in the coal business diminished Lewis and the UMW, and these reveal themselves in industry statistics from the 1950s into the 2000s. One shift shows the decline in the percentage of all coal mined underground. In 1950, 75 percent of American coal was mined underground; by 2010 that was cut to 31 percent. The alternative to underground mining is surface mining, which requires fewer miners. If only 31 percent of total coal production is mined with underground technology, then by definition the rest of the coal—69 percent—was mined with surface mining technologies in 2010. The shift to surface mining also meant a change in location to the West. In 1950 mines east of the Mississippi accounted for 94 percent of all coal production. By 2010 eastern coal accounted for just 41 percent. Western coal then accounted for 59 percent of all coal in 2010. Most of the western coal mining is in Wyoming, which alone accounts for 40 percent of total US production.[32]

Labor productivity also rose steadily with the move to surface mining out in the West. Even in 1950, before the shift, surface mining was almost three times more productive. By 2010 a miner in an underground mine produced 2.89 tons of coal an hour, on average, as compared to 9.46 tons an hour in a surface mine. In surface mines in the western United

States alone, a miner could produce over twenty tons per hour. All of this reduced the need for mine workers. In 1920 employment of coal miners was at 784,621, but by 1950, that number had dropped to 483,239, and by 2010 it had plummeted again to just 135,500.[33]

This drop in employment occurred despite the fact that there was substantial growth in coal use and in coal prices. In the 1950 to 2010 time frame, American coal production almost doubled from 560 million tons to 1.08 billion tons. Average prices in nominal terms increased at a pace of about 3.26 percent a year. At the same time, the electricity business became coal's primary customer. In 1950 electric power accounted for 19 percent of all coal consumption, while in 2010, it accounted for 93 percent.[34]

Looking at this plentiful data with a wide-angle lens reveals all that has changed as a consequence. The state-by-state political center of gravity shifted to the West. Wyoming, as noted, accounts for 40 percent of total US coal production while West Virginia accounts for just 11 percent. The shift to the West meant long-distance rail transport was required; rail transport prices now drive coal prices and, thereby, electricity prices, as much as coal mining costs do. Western coal has a significantly lower sulfur content, so sulfur dioxide emissions are lower when that coal is burned, garnering it an environmental preference. Most notably, the drop in total employment has diminished the political power of the coal business and its unions. The significance of this last point has become clear more recently, as concerns about global climate change have led environmentalists to push for the shutdown of all coal-fueled power plants.[35]

Sadly, one circumstance has not changed. Coal mining, despite all its progress as an industry, is still dangerous. On April 5, 2010, at Massey Energy Company's Upper Big Branch Mine in West Virginia, an explosion killed twenty-nine men. In 2016 Massey Energy's chief executive, Don Blankenship, was convicted of conspiracy to violate federal mine safety laws, but was acquitted of more serious charges. He received a sentence of one year in jail, the maximum under the law.[36]

As Mr. Blankenship left the court, a family member of one of the victims said to him, "You don't miss your kids like we miss ours."[37]

COAL'S LEGACY

Clearly, the history of electricity is inextricably intertwined with that of coal. It is important, however, to get the story straight in terms of when and how the electricity business and the coal industry started to matter to each other. This dynamic was hardly expected in the early days as coal was being crowned king. Coal's rise to provide 75 percent of all energy in 1920 was driven by a broad base of customers: Schurr's iron-rail-steam interdependence involving the iron and steel industries, railroads, and the many other industries using coal-fired steam engines. The electricity business played only a small role in coal's initial rise because it accounted for only 7.3 percent of use.[38]

The symbiotic relationship between the coal and electricity businesses began in a time of relative decline for coal, in terms of its economy-wide market share. The loss of railroads to oil and the competitive inroads made by both oil and natural gas elsewhere explain this decline. But how did it happen that, in 2015, 93 percent of all coal was used to generate electricity and from the 1950s on, coal often accounted for more than 40 percent of all fuels used to generate electricity? Why did coal rise in the electricity market while falling elsewhere? Why did the necessarily staid and stable electricity business latch on to a fuel source with a history of dramatic instability?[39]

The answer is that each got what they wanted most from the other. Because of the incentives of cost-plus pricing, electric utilities wanted to invest in big, capital-intensive generation technologies: coal-fired power plants represent exactly that. The arithmetic of this incentive to build big coal power plants is clear. A great deal more equipment is needed to burn coal to produce electricity than, say, for natural gas. If it takes $3,000 of equipment to turn a unit of coal into electricity, and the allowed return on investment is 10 percent, Insull's cost-plus pricing gives the utility $300 of profit—10 percent of $3,000. If, instead, it takes only $1,000 of equipment to turn natural gas into electricity, then the profit is just $100. From a profit perspective, it follows that a utility would prefer building coal-fired power plants.

The utility practice of economic dispatch gives further incentive to build coal power plants. Once a power plant is built, its equipment cost is put out of consideration; these costs are said to be "sunk costs." Now only the fuel cost is considered when determining how often to run a power plant. Since the cost of the coal itself historically was lower than other fuels, coal plants would be run 24-7 if possible. That meant the equipment costs could be spread across many more kWh of electricity, thereby driving down the per unit cost of coal power. Cost-plus pricing and economic dispatch worked together to favor coal.

Lewis' coal miners, however, had to provide stable prices for coal. The most important, long-term impact that Lewis and his union had on the price of coal was by virtue of *something they did not do*: they did not block the use of new technology or the "mechanization" of coal mines. Mechanization meant there were fewer and fewer miners to pay at each mine, which, in turn, helped mine owners keep coal price increases to a minimum. At the same time, mechanization meant each of the miners who remained produced more and more coal. So, even with stable coal prices, there was more money to pay each worker higher wages and benefits.

It might be expected that the controversial history of the coal business—the horrific violence and strikes, even in a time of war—would have a detrimental effect on the industry, but in the end, it did not. Instead, the drama hastened the development of new mining techniques (surface mines) and the move from the eastern to the western coal fields. It must also have given some pause to electric utilities and their government regulators. However, stable coal prices—and the ability to store great piles of coal in case of a strike—perhaps made them willing to tolerate the threat of further strikes. Another reassuring factor was the potential for a competitive fuel. If another energy source could bring both low fuel cost and low capital cost, then coal was vulnerable to being displaced. As will be seen later in the book, that is exactly what natural gas did in the throes of the shale gas revolution.

As for Lewis himself, his commitment to wartime strikes makes for a mixed legacy. In 1960 Lewis retired as president of the UMW after forty years at that post. In 1964 President Johnson awarded Lewis the Presidential Medal of Freedom, the highest civilian honor in the United States. Writing at Lewis' death in 1969, Saul Alinsky felt compelled to try to rationalize the 1943 strike in opposition to the president in the middle of World War II. Alinsky wrote, "His defiance of every power was a note of reassurance for the security of the democratic idea, that his dissonance was part of our national music."[40] The American warriors who fought in World War II, and their families, could be forgiven for not liking that particular music.

As a fitting tribute, upon Lewis' death, UMW members walked off their jobs for four days.[41]

CHAPTER 12

Albert Einstein's Equation $(E = mc^2)$

> According to the theory of relativity, there is no essential
> distinction between mass and energy. Energy has mass
> and mass represents energy. Instead of two conservation
> laws, we have only one, that of mass-energy.
> —Albert Einstein (1938)[1]

Albert Einstein, like Thomas Edison, is a name that just about everyone recog-
nizes. While *Edison* has become a synonym for inventiveness, *Einstein* has
become a synonym for genius. The four portraits Einstein hung on the wall
of his home office in Princeton provide a clue to the influences on his genius.
They featured Isaac Newton, Michael Faraday, James Clerk Maxwell, and
Mahatma Gandhi. Einstein's scientific breakthroughs were the culmination
of Newton's work on gravity and Faraday's and Maxwell's work on electricity and magnetism. And, like Gandhi, Einstein was a pacifist—despite the
enduring connection between his science and nuclear weapons.[2]

In 1905, what scientists collectively termed his "miracle year," Einstein revolutionized the science of physics with five technical papers. At just twenty-six, he wrote these revolutionary papers not as a professor at one of Europe's prestigious universities but rather while working as a patent officer in Bern, Switzerland. Through his papers he demonstrated that Newtonian physics, the theory accepted worldwide for more than two centuries, was a special case within the broader framework built on the great work of Faraday and Maxwell.[3]

In his fifth paper Einstein laid out the science for his famous equation: $E = mc^2$. The significance of this equation is that a small amount of mass—say, a handful of uranium fuel—could be converted into a massive of amount of energy because, in the course of the conversion, the mass would be magnified by a factor equal to the speed of light squared. It is this equation that led European scientists to conclude that the energy released by a nuclear reaction could be used for the purposes of war.[4]

Einstein's famous equation also links him to the modern electricity business. If a small amount of matter can be converted into a massive amount of energy with a nuclear reaction, that energy could also be used for peaceful purposes, such as the generation of electricity. Einstein is relevant to the story told here because he shows once again that the technology used to generate electricity—in this case, a nuclear power plant—is inextricably linked to the most advanced science of the age.

The link between electricity and advanced science, however, goes far beyond nuclear power technology alone. The study of the science of electricity played a leading and enduring role in the great scientific advances of the nineteenth and twentieth centuries: from Newton's mechanics, to Faraday's electric and magnetic fields, to Maxwell's electromagnetic waves, to Einstein's theory of relativity to quantum theory. As Einstein himself explained, the study of the science of electricity was not just an event on the path to scientific breakthroughs, but was itself that path. Much of this chapter is dedicated to describing the journey Einstein and these other great scientists made along this path.

THE EARLY YEARS

Albert Einstein was born on March 14, 1879, in the city of Ulm in Germany; the town motto, presciently, was "the people of Ulm are mathematicians." As explained in Walter Isaacson's 2007 biography, early encouragement in science came from a young medical student, Max Talmud, who dined with the Einstein family once a week. Talmud introduced Einstein to a series of science books written by Aaron Bernstein, a popular science writer of the day, which seemed to have made a lasting impression. One influential Bernstein story involved the reader imagining he was on a speeding train and a bullet was shot through the window. Einstein later used thought experiments similar to Bernstein's railcar.[5]

Once Einstein outran Talmud's knowledge of science, Talmud switched to philosophy, giving Einstein, at age thirteen, his first taste of Immanuel Kant through the *Critique of Pure Reason*. This early exposure to Kant may help in understanding the philosophical bent of Einstein's later writing for the public. As was true with Faraday, there are differences of opinion on whether Einstein's revolutionary science was influenced by his reading of Kant. Did Kant's view of unity in nature drive Einstein's search for unity of forces and of the sciences? Or was it driven more by Maxwell's work unifying electricity, magnetism, and light, which in turn was inspired by Faraday? The most intriguing speculation is that Kant's secular philosophy of science led Einstein to the same view—a unity of forces—that religious faith had defined for Faraday.[6]

Einstein tried for around two years to gain admission to Zurich Polytechnic. While he easily passed the parts of the entrance exam on math and science, he failed other subjects such as literature and French. In 1896 Einstein finally succeeded and enrolled there, though he did not immediately impress anyone; one professor referred to him as a "lazy dog" for his lack of hard work at mathematics. Einstein wanted to learn more about James Clerk Maxwell, but such material was not covered in any course of study there. Instead, Einstein read the latest in physics with Marcel Grossmann, a classmate who would become a lifelong friend, and his other friends while often skipping class. In 1900 Einstein graduated fourth out of five graduates.[7]

At twenty-one, Einstein had achieved nothing extraordinary and given no indication of genius. However, he did pursue science and began to publish papers. His first published paper concerned the interaction of atoms and molecules. He addressed the issue of what force explained the attraction of molecules in a liquid and wondered whether the attraction among molecules might be analogous to the attraction of planets in the universe—that is, directly related to mass and inversely related to distance, as Newtonian physics held. In 1901 Einstein submitted a proposal for his dissertation to the University of Zurich; it was rejected. All in all, Einstein's early career hardly presaged his later fame.[8]

THE SWISS PATENT OFFICE

Marcel Grossmann helped Einstein get a job at the Swiss patent office in Bern in 1902. Einstein was at his desk for eight hours a day but accomplished his work so quickly that he could spend over half of his time working on science. He tutored for another hour each day and did additional scientific research at night. Although this schedule would seem to be a disadvantage for one trying to make scientific breakthroughs, Isaacson writes that it had one big advantage: it made it easier for Einstein to be unconventional. Had Einstein been in an academic setting, his published work would have been vetted—and perhaps changed or discouraged—by those who established the conventional wisdom of the time.[9]

Einstein and his friends read widely and thought deeply about the philosophy of science. Together they read important works such as David Hume's *A Treatise of Human Nature* and Baruch Spinoza's *Ethics*.

THE MIRACLE YEAR (1905)

In his article "Five Papers That Shook the World," published in *Physics World* in 2005, Matthew Chalmers reported the awe-inspiring fact that a century earlier, in 1905, given "little more than eight months" and in his "spare time," Einstein "completed five papers that would change the world forever." That is why 1905 is considered Einstein's "miracle year." His work covered a remarkable breadth of material—his papers spanned "three quite distinct topics—relativity, the photoelectric effect and Brownian motion." Chalmers adds, "Perhaps even more remarkably, Einstein's 1905 papers were based neither on hard experimental evidence nor sophisticated mathematics. Instead, he presented elegant arguments and conclusions based on physical intuition." Chalmers quoted Gerard 't Hooft, winner of the 1999 Nobel Prize in Physics, saying, "Einstein made the world realize, for the first time, that pure thought can change our understanding of nature."[10]

The first of the five papers, "On a Heuristic Point of View Concerning the Production and Transformation of Light," was submitted to the leading German journal *Annalen der Physik* on March 17, 1905. This paper contained Einstein's contribution on the photoelectric effect, and it is cited as the basis for his 1921 Nobel Prize. A photoelectric effect occurs when light hits one side of, say, a thin metal sheet, and electrons are ejected out the other side. Thus, the photoelectric effect converts light into electricity—a flow of electrons. The effect is used in a range of technologies including light detectors for alarms, in which a steady light shines on a surface and produces electricity until an intruder blocks the light and the lessening of the electric current triggers the alarm. More consequential for the electricity business is the use of photovoltaic or solar cells to convert light to electricity.[11]

His second paper was based on his successful dissertation at the University of Zurich, which was accepted at last on April 30, 1905. It addressed topics including how to calculate "the size of molecules by studying their motion in a solution." Next, Einstein submitted a third paper on Brownian motion, which refers to the constant movement of particles suspended in a liquid. Einstein concluded that the movement resulted from the random collisions of atoms; the broad importance of Einstein's work is that, once confirmed by experiment, it convinced scientists that atoms were real.[12]

Einstein's fourth paper was received by the journal on June 30. "On the Electrodynamics of Moving Bodies" detailed what would become known as his special theory of relativity.[13] The word *special* was used because his theory applies "only to non-accelerating frames" of reference. Among the implications of the theory were that independent time and distance did not exist. As Chalmers puts it, "The length of an object becomes shorter when it travels at a constant velocity, and a moving clock runs slower than a stationary clock."[14]

The fifth paper was a three-page "afterthought" drawing out another implication of the special theory of relativity. The article "Does the Inertia of a Body Depend upon its Energy Content?" is the source of the most famous equation in all science: $E = mc^2$.[15]

How did the world react to the results of Einstein's miracle year? Isaacson reports that the five papers now seen as miraculous and revolutionary were initially met with "icy silence." The German scientist Max Planck supplied one exception, as he actually began to build on Einstein's special theory. Remarkably, after all that he had accomplished in 1905, Einstein continued working in the Bern patent office. Even after 1905, Einstein could not win a bottom-rung job at the University of Bern. In 1908 he was even rejected for a position as a high school teacher. Einstein's experience shows the path to scientific breakthrough can be long and difficult; it takes time to overturn entrenched paradigms. Patience and a commitment to the scientific process are essential.[16]

Then, in 1909, four years after his revolutionary papers, Einstein won a post as a junior professor at the University of Zurich. In that same year, he gained recognition with a speech at a Salzburg conference in which he fused the competing theories of light as particles and light as waves.

Einstein's reputation grew at an accelerating pace thereafter, as reflected in his many professional moves. In 1911 Einstein accepted an appointment as a professor in Prague. He reached out to meet with scientists of the caliber of Ernst Mach and Hendrik Lorentz. Mach was an Austrian physicist who studied supersonic motion; today, his name is the metric for the speed of sound. According to Einstein, Mach helped inspire the theory of relativity. Lorentz was a Dutch physicist who shared the Nobel Prize in 1902 and focused on extending Maxwell's theories.

In 1915 Einstein generalized the special theory of relativity he had developed in 1905; this important advance took a decade of effort. In 1919, with the war over, there was an opportunity to test Einstein's relativity theory's prediction that gravity bent light. Arthur Eddington, the director of the Cambridge Observatory, took up the cause, focusing on the next solar eclipse, which would be on May 29, 1919. The results showed Einstein was right, and Britain's Astronomer Royal, Sir Frank Dyson, then confirmed them. There were more than six hundred books and articles written on the theory of relativity in the six years following that first proof. At age forty, Einstein became an international celebrity.[17]

In 1921 Einstein traveled to the United States for the first time. On that trip, the press began the habit of asking interviewees if they understood the special theory of relativity. Einstein had traveled by ocean liner with Chaim Weizmann, then president of the World Zionist Organization, and Weizmann was asked whether he understood the theory. Weizmann joked in response, "During the crossing, Einstein explained his theory to me every day, and by the time we arrived I was fully convinced that he really understands it."[18]

Einstein traveled to America again in 1930 and 1931. In 1933 Hitler became chancellor of Germany. The Nazis raided Einstein's Berlin apartment and his second home at Caputh in Germany. Fortunately, his papers had already been secretly taken to the French embassy. That same year, Einstein traveled to America once again, accepting a position at Princeton's Institute for Advanced Study; he would never return to Europe.

ELECTRICITY AS THE PATH TO DISCOVERY

In 1938 the world received a unique opportunity to read the story of the exciting breakthroughs in physics by one of the field's most important revolutionaries—Einstein himself. *The Evolution of Physics: From Early Concepts to Relativity and Quanta* was coauthored with Leopold Infeld. It is unique in its accessibility to the layperson. Indeed, Einstein and Infeld described it as a "simple chat" between the reader and themselves. While the authors did not shy away from the more challenging parts of the story, they narrated them via thought experiments or analogies rather than with mathematical equations. The book was unique, too, because Einstein and Infeld were explaining not only what was happening but also why. That is, the book was about how the new theories evolved from problems or shortcomings in the previous theories as they were used to explain reality. In taking this step, Einstein and Infeld revealed a great deal about the intellectual process of scientific advance.[19]

The story has four parts. It begins with the origins of modern science founded in Newton's mechanical theory, about which Einstein wrote, "The reality of our outer world consisted of particles with simple forces acting between them and depending only on the distance." The next part came with Faraday and Maxwell's electromagnetic field theory. Einstein summed up field theory by stating a "courageous scientific imagination was needed to realize fully that not the behavior of bodies, but the behavior of something between them, that is, the field, may be essential for ordering and understanding events." Then came Einstein's own special relativity theory; it destroyed old concepts such as "absolute time" and created new concepts, including the view that the observer's frame of reference mattered in the description and explanation of events. Quantum theory was the fourth part; "instead of laws governing individuals," Einstein wrote, it brought "probability laws."[20]

Newton's Mechanics (and His Shadow)

Newton's quest was to explain motion; it was a more complex task than he perhaps anticipated. Einstein noted, "A stone thrown into the air, a ship sailing the sea, a cart pushed along the street, are in reality very intricate." For a body at rest, say a cart at a standstill, it takes some action or force to move it; that is, someone must "push it or lift it, or let other bodies . . . act upon it." Reflecting this need for force, Newton's law of inertia asserts, "Every body perseveres in its state of rest, or of uniform motion in a right line, unless it is compelled to change that state by forces impressed thereon."[21]

One of the early and most important intricacies Einstein attributed to Newton is this: a force is reflected not in the speed (or velocity) of the body observed but rather in the change in that velocity. Put another way, observing the velocity of the cart does not show that a force is acting on the cart. Rather it is an observation of a change in velocity that shows the presence of that force. If a stone is dropped from a tower, the velocity increases as it falls toward the ground. If this stone is thrown into the air, its velocity decreases—the moment it leaves the hand the upward force is very strong, but as the stone proceeds upward that velocity slows as gravity pushes the stone back toward the ground. From those changes in velocity one can conclude that a force—gravity—is acting on the stone. Einstein said the intricate or subtle point that force is reflected in a change in velocity, and not in the velocity itself, is the "basis of classical mechanics as formulated by Newton."[22]

Newton's laws are meant to apply to all phenomena, but gravity is the most prevalent force in daily life. Newton's law of universal gravitation states that between any two objects there is an attractive force called gravity. This force is proportional to the masses of the objects and inversely proportional to the square distance between the two objects. The bigger the object, the stronger the force will be, and the more distant the object, the weaker the force will be. This is the law that casts Newton's shadow. That is, it would incline all scientists after him to think of attractive force depending on the object's heft (mass) and the distance between the objects. What is so inspiring about Faraday and Maxwell is that they questioned that inclination and went a different way when explaining the attractive forces of electricity and magnetism. Einstein followed their example.

Faraday's and Maxwell's Waves

The hardest test for classical mechanics was to explain "all phenomena," not just gravity. When Newton's mechanical view failed to explain electricity, magnetism, and light, new science was needed. Einstein recalled Coulomb's attempt to make Newton's mechanics work for electricity by concluding that electrical force depends on distance and mass. Electrical force between two objects is stronger if the distance between them is shorter and the charges are larger. However, Einstein found that Coulomb's effort did not fully explain electrical phenomena because of the apparent differences in electrical force and gravitational force. Gravitational force already exists, but electrical force must be created. Further, gravitational force manifests as attraction, but electrical force manifests as both attraction and repulsion. Einstein showed that the link discovered between electricity and magnetism ultimately undermined the Newtonian mechanical explanation of electricity. For example, the magnetic force created by the electrical charge is perpendicular to the charge—the force does not act along a straight line as in mechanical theory. Further, the magnetic force created increases with the velocity of the electrical charge, so the force is more than a function of distance and mass (or charge) as Newton and Coulomb suggested.[23]

Einstein highlighted that "revolutionary ideas were introduced into physics" by Faraday, Maxwell, and Hertz. To illustrate the essential focal point for these ideas, Einstein concluded, "The new concepts originated in connection with the phenomena of electricity." Not only did electricity revolutionize how modern society worked, it also revolutionized man's basic understanding of how nature works.[24]

Next, Einstein explained why field theory was superior to the mechanical view when explaining electricity and magnetism. He recalled the effect of the electric current on a magnet and asked, how does this current exert its force on the magnet? In answering this question, he suggested, "The field proved a very helpful concept. It began as something placed between the source and the magnetic needle in order to describe the acting force." And if the empty space between objects was not empty but rather packed with lines of force of electric and magnetic fields, how could Newton's mechanics be right? How could the attraction depend only on the mass of objects and the distance between them?[25]

The ultimate payoff to field theory comes through its link to wave theory. Specifically, Einstein reports that the "outcome" of "mathematical deduction" from Maxwell's equations "is the *electromagnetic wave* . . . Every change of an electric field produces a magnetic field; every change of this magnetic field produces an electric field . . . and so on." Finally, Einstein asks another question: "With what speed does the electromagnetic wave spread in empty space?" The answer is: "*The velocity of an electromagnetic wave is equal to the velocity of light.*" Einstein states, "The theoretical discovery of an electromagnetic wave spreading with the speed of light is one of the greatest achievements in the history of science." Electricity is unique in its ability to serve the practical day-to-day needs of man while at the same time serving as the focal point for revolutionary scientific advances.[26]

Einstein's Relativity

Einstein's theories of relativity were, of course, landmark scientific breakthroughs. The crucial link to the study of electricity is that Faraday's vision and Maxwell's equations led Einstein to consider the implications of a constant speed of light. The special theory of relativity was one result of that consideration.

Einstein illustrated his science with a number of simple cases that each show how an observation can vary by what Einstein called the coordinate system, or, more simply, the frame of reference of the observer. In one illustration, Einstein puts this idea in terms of a person walking on a moving ship at a pace of three miles per hour. If asked, an observer *on the ship* would confirm that the man's speed is three miles per hour. What about an observer *on shore* rather than on the ship? If the ship itself were moving at thirty miles per hour, then the on-shore observer would say the man walking on the ship is moving at a pace of thirty-three miles per hour—the sum of the speed of the walking man and that of the ship itself. In Newton's classical mechanics, the difference between the two observations can be fully understood. Indeed, the observation from one observer can be "transformed" into the other observation as has already been done here—just add the two speeds.[27]

But problems arose when scientists turned to measuring the speed of light. Did the speed of light vary with the frame of reference? If the man on the ship were walking at the speed of light, would his speed be the speed of light plus the thirty miles per hour speed of the ship from the frame of reference of a man onshore? If Maxwell's equations are to apply, the answer must be no. As Einstein put it, "*The velocity of light is always the same in all CS* [coordinate systems or frames of reference] *independent of whether or not the emitting source moves, or how it moves.*" If the speed of light must be the same in all frames of reference, Einstein concluded, "[Newton's] transformation law is in contradiction."[28]

The implications of this are startling because if the speed of light is constant across frames of reference, then the measurement of time and distance must vary by frame of reference. Einstein wrote, "If the velocity of light is the same in all CS, then moving rods must change their length, moving clocks must change their rhythm." This is the famous outcome of relativity: the measurement of time and distance varies with the motion related to the observer. The consequent bedrock of Einstein's view, then, is that the speed of light is a constant that does not change.[29]

Next, Einstein compared the concept of force under Newtonian laws to that under his relativity theory and thus moved quickly toward his famous equation. With Newton, force was indicated by a change in velocity and that change in velocity would be the same whatever the starting point—that is, whatever the original speed. With relativity, that is no longer true. The change in velocity for a given force varies depending on the original speed: it takes more force to change the velocity from 100 miles per hour to 101 miles per hour than from 10 miles per hour to 11 miles per hour. That is, it takes more force to increase speed when starting from a higher speed. This allows for the rule that the closer a speed is to the speed of light, the harder it is to increase its velocity, and if the speed of light is achieved, the object cannot go faster.[30]

Einstein then turned to mass and considered the effect of motion and frame of reference. The Newtonian view was that more mass meant more resistance to change in motion: put simply, it takes more force to push and move a cart that is fully laden than it does to move an empty cart. Relativity contributes the concept that resistance to a change in motion also increases with speed. This allowed Einstein to link energy to mass. He wrote, "A body at rest has mass but no kinetic energy, that is, energy of motion. A moving body has both mass and kinetic energy. It resists change of velocity more strongly than the resting body . . . If two bodies have the same rest mass, the one with the greater kinetic energy resists the action of an external force more strongly." He pushed forward to a dramatic conclusion by stating, "All energy resists change of motion; *all energy behaves like matter.*"[31]

Finally, he contrasted the fundamental view of Newton's mechanics with his own relativity when he wrote, "Classical physics introduced two substances: matter and energy . . . In classical physics, we had two conservation laws: one for matter, the other for energy . . . According to the theory of relativity, there is no essential distinction between mass and energy. Energy has mass and mass has energy. Instead of two conservation laws, we have only one, that of mass-energy."[32]

For the science of electricity, the direct implications of this mass-energy equivalence were big. Out of the special theory of relativity came the famous equation that enabled nuclear power technology, $E=mc^2$. Einstein's path to the scientific breakthrough was equally important. It was founded on Faraday's and Maxwell's studies of electricity and magnetism, which led to the discovery of electromagnetic waves that required a constant speed of light. The constant speed of light, in turn, is the basis for finding the equivalence of energy and mass that is embedded in $E=mc^2$. This equation had the potential to change the world, and the electricity business along with it.

A Quantum Leap

At the end of his section on field theory and relativity, Einstein suggested the beginning of the fourth part of his story of the revolution of physics. He wrote, "Fundamental problems are still before us. We know that all matter is constructed from a few kinds of particles only. How are the various forms of matter built from these elementary particles? How do these elementary particles interact with the field? By the search for an answer to these questions new ideas have been introduced into physics, the ideas of *quantum theory*." Quantum mechanics is the study of the motion of objects that come in tiny bundles called quanta.[33]

What Einstein wrote about quantum mechanics in 1938 further supports the view that electricity played a central role in this major advance in physics. For Einstein, the fundamental links between electricity and quantum mechanics came from the "realization that the electron, the elementary quantum of the negative electric fluid, is also one of the components of the atom, one of the elementary bricks from which all matter is built." He noted that the renowned scientist J. J. Thomson proved that "the negative electric fluid is constructed of grains, just as the beach is composed of grains of sand, or a house built of bricks." For electricity, the grains of sand, or, more precisely, the elementary quanta, are called electrons. He emphasized the importance of electricity once again by writing, "This result closely connecting the problem of the structure of matter with that of electricity follows, beyond any doubt, from very many independent experimental facts."[34]

THE PATH TO NUCLEAR WEAPONS

Einstein's equation has a biography of its own; one told well by science writer David Bodanis in *E=mc²: A Biography of the World's Most Famous Equation* (2000). In setting the course for that equation's influence, Einstein made one very important cameo appearance in 1939, but other scientists forged the path from Einstein's great insight in the fifth of his papers in 1905 to the development and use of nuclear weapons in the 1940s.

The journey to nuclear weapons began, at least in part, with research on atoms in the early twentieth century. Around 1910 Ernest Rutherford's research found that atoms were mostly empty, with a nucleus at the core and negatively charged electrons orbiting at the outer edges. The nucleus had positively charged protons and, as later discovered in 1932 by James Chadwick, it also contained neutrons with a neutral charge. Chadwick bombarded the nucleus with neutrons in an attempt to see what happened when more neutrons were added, but bombardment at high speeds did not work. In 1934 Enrico Fermi found that slower-moving neutrons could be pushed into the nucleus.[35]

The next major discovery in this line of study was made by Lise Meitner in 1938. Meitner had been one of the leading scientists in Germany, where she had formed a productive scientific partnership with Otto Hahn at the Kaiser Wilhelm Institute. When Hitler rose to power, Meitner was first pushed out of the University of Berlin in 1933 because she was Jewish, and later pushed out of the Kaiser Wilhelm Institute, sadly with the help of Hahn. Still, Meitner continued collaborating with Hahn from her exile in Stockholm.[36]

Meitner focused on the heaviest natural element, uranium, which has a nucleus already brimming with neutrons. She wanted to know what happened when the uranium nucleus was bombarded with neutrons. What was created? Would it be something much heavier? On a Christmas holiday in Sweden, Meitner received a letter from Hahn reporting some unusual results. Meitner's nephew, Robert Frisch, a physicist working with Niels Bohr in Denmark, was with her for the holiday. Hahn reported that the bombardment of the uranium nucleus had resulted in radioactivity—a "spraying out of energy streams." It was not entirely a surprise result. Bohr's view of the atom was that the nucleus was always ready to burst, since the positively charged protons would repel each other—like repels like. The question was whether the addition of neutrons could tip the balance in favor of the nucleus splitting apart.[37]

What Meitner and Frisch discovered was that the bombardment of neutrons actually split the nucleus. Adding neutrons did not make the nucleus bigger; it broke the nucleus apart. Borrowing a word from biology, Meitner and Frisch termed this process *fission*. The link to Einstein's famous equation was crucial. In the creation of two smaller nuclei, the mass that was lost was converted into energy as Einstein's equation predicted. Bodanis explains how, at another time, this finding might have taken decades to work its way through the scientific community, "but in 1939, the world had just begun its largest war ever. The race was on to see in which country the equation's power would emerge first." The race was on to build the first nuclear weapon.[38]

In 1939 Einstein was approached by an old friend, Leó Szilárd, a Hungarian physicist then at Columbia University. Szilárd readily convinced Einstein of the dangerous potential of the discovery of fission using uranium. With some constructive guidance from Alexander Sachs, an economist at Lehman Brothers, Einstein wrote a letter to President Roosevelt. Einstein's letter stated in part: "This new phenomenon would lead also to the construction of bombs, and it is conceivable—though much less certain—that extremely powerful bombs of a new type may thus be constructed. A single bomb of this type, carried by boat and exploded in a port, might very well destroy the whole port together with some of the surrounding territory." Sachs was unable to deliver the letter for almost two months, and when he did, President Roosevelt did not make it a top priority, despite the fact that Einstein was the most famous scientist in the world. After reading the letter, the president set up an ad hoc committee headed by Dr. Lyman Briggs, director of the Bureau of Standards, the country's physics laboratory.[39]

By 1940, however, Meitner's nephew, Frisch, had convinced the British of the potential for nuclear weapons. The Germans believed it, too. They picked Werner Heisenberg, the second-most-famous scientist in the world, to lead the effort. In 1940 Heisenberg conducted his first tests using uranium from the Czech mines Hitler had captured. Fermi had already found that the trick was to slow down the neutrons, if the goal was to split the nucleus to convert mass into massive amounts of energy. But what would slow down the neutrons? Water would help, but heavy water—water with more deuterium—would be best. Deuterium is a variant, or isotope, of hydrogen that has twice the mass of ordinary hydrogen. The supply of heavy water was a weak spot for the Germans. They relied on a facility in Norway that became the target for Allied sabotage attempts; some were successful in slowing the supply of heavy water to German scientists.[40]

Heisenberg's first test failed, but by 1942 he had succeeded in finding a way to release the energy Einstein claimed was trapped in mass. Meanwhile, after the Pearl Harbor bombing on December 7, 1941, the United States pushed forward with what became known as the Manhattan Project. In August 1945 nuclear bombs were loaded on a plane destined for the skies over Japan. It was there that Einstein's famous equation had its first impact on man, with devastating consequences.

Einstein was not informed of nor invited to participate in the Manhattan Project because the FBI considered him to be a security risk. J. Edgar Hoover, the FBI director, wrote that Einstein posed a potential liability because of his actions supporting pacifism. Still, *Time* magazine later ran an issue with Einstein, his famous equation, and atomic bombs on its cover. *Time* acknowledged that Einstein did not work directly on the atomic bomb, but argued that the project had started upon his initiative, and it was based on his equation. Einstein said that, had he known the Germans would not succeed with nuclear weapons, he would not have helped. He made a plea for a world government that would have a "monopoly on military power." Einstein noted that Alfred Nobel "atoned" for his discovery of dynamite with the Nobel Prize, and Einstein felt that he and the other contributors must atone for making atomic weapons. Einstein said, "I do not know how the Third World War will be fought . . . but I can tell you what they will use in the Fourth—rocks."[41]

EINSTEIN'S LEGACY

Einstein's most direct legacy for the history of electricity began in the last of his five papers from 1905, the miracle year. That fifth paper gave the world $E = mc^2$, which was the scientific underpinning of the technology for both nuclear weapons and nuclear power. Einstein does not deserve full blame for the bad or full credit for the good done by these technologies, but he certainly enabled their emergence.

Even more profound for the history of electricity is that the great scientific advances in physics in the nineteenth and twentieth centuries, in substantial part, came out of the study of the nature of electricity and its intimate links to magnetism and light. Newton's shadow inclined scientists to use his pathbreaking understanding of gravitational force to explain the other major forces of nature. It was the failure of Newtonian mechanics to explain electricity, magnetism, and light that opened the door to Faraday's and Maxwell's major advances with field theories. Not only did Maxwell's work improve on the past but by showing that the speed of light must be constant, he set the stage for the next great advance—Einstein's theory of relativity. Remarkably, electricity continued its central role in scientific advances by virtue of the central role of the electron in quantum mechanics.

Sadly, efforts to further explore the scientific path of electricity have given way to efforts focused on other key drivers in the electricity business, including law, regulation, and, most notably, politics. Listen in on conversations about electricity today and few references will be heard to the great scientific revolutions born of the study of electricity, much less any mention that those revolutions, in turn, have advanced the technology for the production and use of electricity. Surely, the potential for science to lead to new revolutions has not been exhausted. It is essential for mankind to get back on this path of discovery.

Einstein's legacy also encompasses the standard he set for how a professional scientist should act. He was respectful, accessible, and humble. He appreciated the scientists who came before him and those who were his contemporaries. And he showed a deep regard for the scientific process, saying, "To raise new questions, new possibilities, to regard old problems from a new angle, requires creative imagination and marks real advance in science."[42]

He was down to earth, as evidenced by the book he coauthored with Infeld. In 1916 he wrote another layperson-friendly book titled *Relativity:*

The Special and General Theory, and he was also the author of a collection of writings titled *Ideas and Opinions.* Einstein wrote widely, not only on science but also on politics, government, pacifism, education, friends, and religion. He was not narrowly focused on science alone but on what science meant to the world in his time.

Regarding his sense of humility, Einstein wrote, "My religion consists of a humble admiration of the illimitable superior spirit who reveals himself in the slight details we are able to perceive with our frail and feeble minds." He spoke further and freely about the interplay of science and religion. Speaking at the Princeton Theological Seminary in 1939, Einstein acknowledged that it was "widely held that there was an irreconcilable conflict between knowledge and belief." However, he said, "science can only ascertain what *is,* but not what *should be.*"[43]

Einstein died at the age of seventy-six in April 1955. A *Time* obituary titled "Death of a Genius" suggested that if one were to look beyond the surface of "a self-effacing little man, careless-clad in baggy pants and a blue stocking cap," who for twenty-two years "stepped down from the front porch of a modest frame house at 112 Mercer Street, Princeton, N.J.," one would find "a compound of soaring intellect and wide-ranging imagination that carried Albert Einstein past the confines of man's old scientific certitudes and deeper into the material mysteries of the universe than any man before."[44]

Einstein's respectfulness, accessibility, and humility were traits that built trust at a time in which the public had to agree with the proposals of scientists concerning such "material mysteries," even if they could not independently judge the evidence offered in support. Today, that trust—and scientists in Einstein's mold—are needed more than ever.[45]

PART FOUR

The Age of Harm

Crisis, Change, and Scandal

CHAPTER 13

Eisenhower's Atoms for Peace (and War)

The United States pledges before you . . . to devote
its entire heart and mind to find the way by which
the miraculous inventiveness of man shall not be
dedicated to his death, but consecrated to his life.

—President Dwight D. Eisenhower (1953)[1]

The urgency created by Hitler's (and Heisenberg's) agenda in World War II meant that global conflict governed the deployment of Einstein's famous equation. His revolutionary theory would be used to make nuclear weapons in a very real, dangerous, "hot" war.

That the first nuclear technology was intended for war meant that all nuclear technology would forever bear that association. Even when nuclear experimentation turned to peaceful purposes, the link remained. It is still common to hear questions such as, "Can a conventional nuclear power plant explode like a nuclear bomb?" (It cannot.) Or, "Can nuclear power plants help to create materials that can be used to make nuclear weapons?"

(They can.) Nuclear power technology was born of war, and that will likely never be forgotten.[2]

After nuclear power technology was born of a "hot" war, it was nurtured to serve as a weapon in a brand-new type of global conflict: the Cold War. The federal government was the major proponent of nuclear power, not because it was a low-cost source of electricity, but because it was valuable in the Cold War effort. President Dwight D. Eisenhower promoted nuclear power as an economic weapon in his high-profile speech before the United Nations General Assembly in 1953. In addition, Admiral Hyman Rickover used nuclear power to create a military weapon for the Cold War with the nuclear navy; nuclear reactors would be used to power submarines and other war ships. Somehow, Rickover's choice of technology for a small-scale nuclear reactor on ships later became the technology of choice for large-scale, commercial nuclear power.[3]

The story of nuclear power technology is the story of a rise and fall over five decades. The rise, of course, began by tapping into some of the most important science of the century: Einstein's famous equation and the science and engineering of the Manhattan Project. In this sense, nuclear power could have taken America back to the Age of Franklin, a time of awe and discovery. Unfortunately, as it has turned out, nuclear power fits better into the Age of Harm: a time when the public stopped thinking solely of the benefits of abundant, affordable electricity for all, and instead focused as much, or more, on the harm electricity generation might do. Three major accidents punctuated the Age of Harm: first at Pennsylvania's Three Mile Island nuclear power plant in 1979, then at Soviet Ukraine's Chernobyl plant in 1986, and at Japan's Fukushima plant in 2011.

Regarding the fall of nuclear power, the discussion must be put in perspective: nuclear power still provides about 20 percent of America's electricity. Clearly, a "fall" did not mean an end to nuclear power. Rather, the "fall" has meant an end to *new* nuclear reactors. Until recently, no new nuclear reactor had been licensed in America for decades. However, unlike coal and other fossil fuels, nuclear power plants emit virtually no greenhouse gases such as carbon dioxide. Some now speak of a renaissance for nuclear power, in large part because of concerns about global climate change.[4]

THE SCIENCE OF NUCLEAR POWER (AND WEAPONS)

The basic physics of nuclear power and nuclear weapons is the same. In both technologies a neutron is shot into the nucleus of an atom, causing it to break apart. For the public this was often put in terms of "splitting the atom"; scientists call it nuclear fission. Einstein's landmark equation ($E=mc^2$) predicts that the result of nuclear fission—that is, the result of breaking mass apart—is the release of a massive amount of energy. To get a sense of the scale involved, consider that the energy released with one gram of refined nuclear material that undergoes fission is "2.5 million times the energy released in burning one gram of coal." Along with that massive amount of energy is the release of additional neutrons, which cause additional fissions. The intent is to create a sustainable chain reaction; that is, more and more neutrons hitting more and more nuclei.[5]

While the basic physics is the same, the purpose of achieving the chain reaction is different with nuclear weapons than with nuclear power. This difference in purpose creates an important and fundamental fork in the road between the two technologies. For nuclear weapons the goal is to get lots of neutrons hitting lots of nuclei at the same moment to cause an explosion. *Fast neutrons* are what are needed for nuclear weapons. To get these fast neutrons, highly purified or "enriched" uranium is needed. Ninety-nine percent of the naturally occurring uranium that can serve as the starting point for the fuel for both nuclear weapons and nuclear power is designated as U-238. "Enrichment," for nuclear weapons, means that the U-238 must be converted into U-235, so that the weapon contains over 90 percent U-235. The terms U-235 and U-238 refer to two variations (isotopes) of uranium. The numbers come from the sum of protons and neutrons in the nucleus of each.[6]

For nuclear power, by contrast, "slow neutrons" are needed instead of the "fast neutrons" needed for nuclear weapons. Nuclear power needs a steady source of heat from the fission and that heat is used for the mundane task of making water boil into steam. Once the steam is created, it is back to conventional power generation in the sense that the steam drives a steam turbine, and that turbine is connected to an electric generator that produces electricity. The central point is that the version of uranium needed for power production is usually enriched so that only about 3 percent of the uranium fuel is U-235. It is this fact that leads to a very comforting conclusion by Berkeley physicist Richard Muller in his book *Physics for Future Presidents* (2008): "A nuclear reactor cannot explode like an atomic bomb. The public image is not correct."[7]

The link between nuclear power and nuclear weapons has gained intense attention more recently because of the US-Iran nuclear deal of 2015. Iran claimed to be developing peaceful nuclear power, but the United States and others feared Iran was on the path to nuclear weapons. Part of the deal reflects the discussion of enrichment above. For example, the negotiated agreement includes a provision by which "for 15 years, Iran will keep its level of uranium enrichment at up to 3.67%"; the deal notes that Iran had "enriched uranium to near 20%." Iran also agreed to limit its access to the equipment for enrichment. For example, Iran will reduce the number of one class of centrifuges to 5,060 at Natanz. Iran had nineteen thousand of these and will now store the excess.[8]

An alternative way a power plant can yield bomb-grade material is through "reprocessing" spent fuel. Plutonium is in the "waste" product of a nuclear reactor and, if separated from other reactor waste during reprocessing, bomb-grade material can be made from it. More than any other power technology, nuclear power depends on advanced science both to make it work and to make it safe.

This book is not the place for a full and fair assessment of the US-Iran nuclear deal. However, true to the fundamental premise of the narrative here, there is much more driving the ultimate outcome of the Iranian deal than nuclear science and technology. Culture will also be a central force in the sense that some might be betting on Iranian youth coming of age, coming to power, and seeking to join the world community. Others will disagree.

THE ATOMS FOR PEACE SPEECH

On December 8, 1953, President Dwight D. Eisenhower stood before the General Assembly of the United Nations to deliver what is known as his Atoms for Peace speech. Reading his speech today is somewhat surprising and considerably encouraging, because this famous and victorious American general from World War II proceeded to give what he called his "recital of atomic danger and power." The United States had "set off the world's first atomic explosion" on July 16, 1945, he recalled, and, since then, had set off another forty-two test explosions. The president reported that contemporary bombs were twenty-five times more powerful than the initial bombs and tests, and that the United States' "stockpile of atomic weapons . . . exceeds by many times the explosive equivalent of the total of all bombs and all shells that came from every plane and every gun in every theatre of war in all the years of World War II." Eisenhower warned that America no longer had a monopoly on the "the dread secret and the fearful engines of atomic might." Addressing fears of a nuclear war, he said that even the possibility of a "devastating retaliation" would not guarantee that such a war would never occur. And yet he made it clear that, if attacked, the United States would retaliate even though there is no "victory in such desolation" or in a "civilization destroyed."[9]

With this backdrop, President Eisenhower declared that America preferred "to be constructive, not destructive," and wanted "agreements, not wars, among nations." Here his speech turned more hopeful. The president said that "it is not enough to take this weapon out of the hands of the soldiers. It must be put into the hands of those who will know how to strip its military casing and adapt it to the arts of peace." America could do that right away because it already had the technology; in his words, "peaceful power from atomic energy is no dream of the future."[10]

Eisenhower then called for the creation of an International Atomic Energy Agency (IAEA) under control of the UN to take control of all the "fissionable material" to be shared with the world. The IAEA would promote the use of this material to "serve the peaceful pursuits of mankind." Atomic energy would be used for "the needs of agriculture, medicine, and other peaceful activities." He called attention to the "special purpose" of providing "abundant electrical energy in the power-starved areas of the world." All this would be done "to serve the needs rather than the fears of mankind." In closing, he pledged that America would "devote its entire heart and mind to find the way by which the miraculous inventiveness of man shall not be dedicated to his death, but consecrated to his life."[11]

Writing in 1954, soon after the president's speech, Gordon Dean, the former chairman of the US Atomic Energy Commission, or the AEC, continued along the lines Eisenhower laid out and argued for sharing America's nuclear know-how unrelated to nuclear weapons with other countries. After World War II, secrecy had been built into the 1946 Atomic Energy Act to prevent businesses from sharing commercial knowledge of nuclear technology overseas. However, America was now ready to promote a "bold interchange of information and materials pertaining to the peaceful uses of atomic energy."[12]

Robert McKinney was another contemporary who wrote of the significant benefits of spreading peaceful nuclear technology across the globe. He was the chairman of a nonpartisan citizens' panel which reported to the Joint Committee on Atomic Energy on the impact of the peaceful uses of atomic energy. In 1957 McKinney wrote of the great contributions that nuclear power could make to world peace. His first major example was that it had the potential to greatly reduce the dependence of western Europe and other regions on Middle Eastern oil, and another was that it could spur economic development in the world's undeveloped countries.[13]

Clearly, Atoms for Peace was part of a much broader geopolitical strategy influenced by the end of World War II and the beginning of the Cold War. The promotion of nuclear energy around the globe, and the related sharing of nuclear know-how, was a central part of the Cold War strategy of the United States and the USSR. While the arms race and space race were clear avenues of competition, the push to use nuclear energy knowledge to gain allies is often overlooked. However, its legacy is just as important. John Krige, a professor of science and technology at Georgia Institute of Technology, writes: "For the United States (and the Soviet Union) a demonstration of scientific and technological generosity and prowess on the international stage was intended to win hearts and minds and to confirm the legitimacy or even the superiority of rival politico-economic systems." Furthermore, a key tactic with undeveloped countries was to engender a dependence on American technology as well as to embed surveillance through technological sharing.[14]

JUMP-STARTING AN INDUSTRY

From the start, the Atomic Energy Commission, created by the Atomic Energy Act of 1946, faced two conflicts. The first was the trade-off between keeping nuclear weapons secret and the desire to promote nuclear energy. The second was the tension between *promoting* nuclear energy and the responsibility of *regulating* it for safety. After the president's Atoms for Peace speech the United States hoped to share scientific and commercial knowledge in order to help grow a private export industry in nuclear energy, but protections were put in place to prevent military uses. The most tangible evidence that promotion was winning out was the effort by the AEC to team with private businesses to demonstrate nuclear power technology.

In the early stages, the AEC pushed the development of facilities to serve the dual goals of producing plutonium for nuclear weapons and producing nuclear power, but the AEC's industrial partnerships did not flourish under this mandate. So the AEC decided to build five power plants with private partners. Shippingport was the first nuclear reactor to serve the sole purpose of producing electric power. Key to its location near Pittsburgh, Pennsylvania, was the pledge of a local utility, Duquesne Light Company, to provide the site as well as $5 million in funding. Shippingport started construction in September 1954 and came on line in December 1957. It operated until 1982.[15]

Shippingport was a high-profile project throughout. For the ground-breaking ceremony, President Eisenhower waved a symbolic wand to signal a bulldozer to begin construction. Admiral Rickover supervised the construction of the plant. At its start, Phillip A. Fleger, chairman of the Duquesne Light Company, said it was fitting that the first nuclear plant was built not far from the birthplace of the oil industry and on top of vast coal reserves, signaling that nuclear was the fuel of the future. By the 1960s, when plant owners were accused of covering up a radiation leak, a more unfortunate signal began to be heard.[16]

In 1955 the AEC began a series of efforts to promote nuclear power. The first round resulted in the choice of four demonstration-scale power plants and four different technologies used. Not all efforts succeeded, as should be expected with demonstration projects. Round two of the AEC's demonstrations resulted in two efforts chosen out of seven proposals, but both failed. In round three, the AEC limited applicants to proven technologies. In a fourth stage of demonstration projects the AEC received two more proposals and accepted one: the 575-MW project in Connecticut sponsored by Connecticut Yankee Atomic Power Company, a consortium of utilities. It was the largest power plant built under the AEC Power Reactor Demonstration Program, and the last. These demonstration projects were substantially subsidized by the AEC as well as by private industry trying to jump-start the nuclear power business. Connecticut Yankee enjoyed an AEC subsidy for about 15 percent of its cost plus a Westinghouse subsidy equivalent to about 30 percent of its cost.[17]

With the end of the AEC demonstration program in 1963, private enterprise was expected to more fully advance nuclear power. The Westinghouse technology was winning against General Electric's technology, so to beat the competition, General Electric began to offer fixed-price "turnkey" contracts. That meant General Electric would be fully responsible for building the nuclear power plant, making sure it worked, and then turning the plant over to the utility for the payment of a fixed amount of money.[18]

These turnkey contracts assigned a significant risk to General Electric: they required the company to shoulder the risk of cost overruns in the project. Because of these turnkey projects, the subsidies to nuclear power did not stop when the private sector took the lead in promoting nuclear technology. If the actual cost of a nuclear power plant was $150 million, but the guaranteed turnkey price was $100 million, GE or Westinghouse absorbed the cost overrun, thus subsidizing the electric utility by $50 million. General Electric sold seven turnkey plants, while Westinghouse secured contracts for six. The effect of the implicit subsidy is remarkable. The average subsidy for General Electric's turnkey projects was 50 percent of construction costs, while the average subsidy for Westinghouse was not far behind at 43 percent.[19]

The early history of nuclear power was characterized by strong government promotion and major subsidies both public and private. This "commercialization" did not achieve commercial feasibility as it should have. It did, however, reflect a faith won at Hoover Dam and throughout World War II. There was a deep-seated trust that big government working with big corporations could succeed at big projects.

THE RICKOVER FACTOR

New products that win market share typically do so by offering a better deal in terms of price and performance. Nuclear power introduced its own complications into that basic model. How did nuclear power secure a 20 percent market share in the American electricity business? At least in large part, it was because the federal government set the use of nuclear power as a strategic goal to help win the Cold War. Furthermore, the government, through the AEC, subsidized the technological development of civilian nuclear reactors. Private businesses, specifically Westinghouse and GE, then subsidized commercialization with loss-leader turnkey deals.

A related question is, how did "light water" nuclear power technology secure a 100 percent market share among commercial nuclear power plants? To answer this question, it is crucial to consider the role played by then Captain, later Admiral Hyman G. Rickover, often called the father of the nuclear navy. Rickover was the first to build a nuclear reactor that produced substantial electricity. That reactor was called the Submarine Thermal Reactor (STR) Mark I. It was first operated in May 1953. Rickover then developed the STR Mark II, which was the reactor actually used in the first nuclear-powered submarine, the USS *Nautilus*, which made its first voyage in 1955. His STR Mark II reactor also served as the basis for the first nuclear power plant in America, the Shippingport Plant near Pittsburgh. One historian concluded, "Much of the success achieved at Shippingport is directly attributed to the highly disciplined planning and execution of Admiral Rickover." That is, Rickover is credited with the leap from small-scale nuclear reactors in ships to large-scale commercial electric generation.[20]

Rickover's path to nuclear-powered ships started after World War II, in 1946, when he was assigned to what is now the Oak Ridge National Lab to study the fundamentals of nuclear power. It was clear from the start that the Navy needed nuclear submarines. The submarines of the time could not stay submerged for long periods of time because they had to surface to recharge their batteries with diesel engines. This diminished the primary military advantage of a submarine, which was to sneak up on an enemy. The introduction of nuclear power had the potential to revolutionize naval warfare. There was also the concern that the Navy was in danger of becoming irrelevant in a world of nuclear weapons; the Navy would have to launch nuclear weapons from the sea to compete with the capability on land and in the air. Given all this context, it is not surprising that in 1947 Admiral Chester W. Nimitz, then acting chief of naval operations and former submarine officer, approved the development of a nuclear-powered submarine.[21]

The Navy was in a hurry to develop this capability, so Rickover was in a hurry, too. He set a goal for the production of the first nuclear-powered submarine in 1955, which meant the first test of the STR Mark I had to be in 1953. A colleague wrote of the risks Rickover took with his own career by pushing so hard for rapid progress. One episode involved a planned forty-eight-hour test of the Mark I. After just twenty-four hours, engineers on the project recommended that the test end right then and there due to safety concerns. Rickover said no; he wanted to test it for the time needed to cross the Atlantic Ocean. The test went to forty-eight hours and then beyond, on to sixty hours and then to sixty-five, with Westinghouse's engineers and Rickover's own technical staff protesting. Rickover pushed forward, and, as a result, he was able to predict that the *Nautilus* would be capable of crossing the ocean fully submerged and at full speed.[22]

In addition, with his self-imposed deadline looming, Rickover chose to push the light-water reactor—it uses regular water, rather than "heavy," deuterium-containing water, as both the coolant and the moderator—and Westinghouse was the company with the lead in that technology.[23] However, Rickover did give a parallel try to another technology. The Mark A reactor was developed by General Electric, and it used a sodium-cooled reactor. This competing model was used in the second nuclear submarine, the *Seawolf*, but problems arose, thus supporting Rickover's initial choice of light-water technology.[24]

Rickover's successful effort with a nuclear submarine has been the subject of studies on the nature of scientific and technological development. Looking back on that period, sociologist Scott Frickel wrote in 1996 of the holistic nature of such development; the stars must be aligned in the sense that the full range of forces must work in the same direction. Politics and public policy have to set the strategy for the effort, the science has to be right, the technology has to work, and the costs and performance must be on target. Moreover, there must be champions, the "engineer-sociologists," who are the glue of these networks that bring about change. Rickover was the champion who ensured that all the relevant parties were brought in. Another lesson, however, is that narrowly focused efforts get equally narrow results. Rickover's goal was not to build a nuclear reactor. Rather, it was to build a nuclear reactor *on a submarine*. And the central goal was not necessarily to do this on the cheap; cost mattered, but no one in the Navy was asking for the lowest-cost nuclear reactor.[25]

Rickover's accomplishment was impressive, and its impact was felt well into the future. By the late 1990s, the Navy had built 240 reactors and had safely decommissioned 50 of them. However, today there are those who say that, while the decisions might have been right for the Navy, the choice of light-water technology was not the best for the creation of a civilian nuclear power business. The question of whether America made a wrong turn in picking a preferred technology for nuclear power has not yet been answered in full. Three broad conceptual doubts come to mind. First, it was the nation's and the Navy's *strategic* needs that created the demand for new nuclear reactors. A military need is different from a civilian or commercial need; no one was asking for a technology for civilians that was the best deal for consumers in terms of price and performance. Second, the urgency created by the Cold War set the fast pace for the effort to build a new reactor, but developing fast is not always best for a commercial business. Third, subsidies were always necessary. Subsidies for the light-water reactor came from both the Navy and the AEC, plus from private companies in the form of the turnkey projects by Westinghouse and General Electric. At some point, a technology must demonstrate it can make it on its own.[26]

THE "FALL" OF NUCLEAR POWER

The "fall" of nuclear power took place mainly in comparison to the expectations for the technology. The most-cited indicator is the number of plants canceled: one estimate is that, of the 253 orders placed, 114 of them, or 45 percent, were canceled. The fall is also evidenced by the fact that America stopped building new nuclear plants: among the hundred operating reactors today, the date of the last construction license granted was 1978. Indeed, almost all the operating reactors received their construction licenses in the narrow window between 1964 and 1978.[27]

Any reference to nuclear power's "fall" must be put in proper context. Nuclear power is not going away. The data make this clear: The U.S. Nuclear Regulatory Commission (NRC) reported that the United States had one hundred commercial nuclear power reactors operating at sixty-two different sites across thirty-one states in 2014. The NRC reported that about 20 percent of America's electricity generation in 2016 came from nuclear power. Global data reflect a similar state of affairs; nuclear power is a key resource, and phasing it out would be a complex challenge. The United Nation's IAEA reports that there are 450 operating reactors in thirty countries around the world. In 2012 nuclear power provided 11 percent of total worldwide electric generation.[28]

Rapidly Rising Costs

With this context in mind, what caused the cancelations and the long drought in nuclear licenses in America? The major reasons were financial. The rapidly rising costs of building nuclear power plants have significantly undermined the nuclear power business. Construction costs rose for coal, too, which was nuclear power's main competition for base load service, but the costs to build new coal plants did not rise as steeply as they did for nuclear power; *base load* refers here to the fact that coal and nuclear plants run around the clock because of their low fuel costs. A statistical study to document these cost increases for both new nuclear and new coal-fired power plants was done for plants built in the 1971 to 1978 time frame. The resulting estimate was that the cost of building a new coal plant over those years increased by 68 percent in real terms. But the real cost for new nuclear plants increased even more: by 142 percent. With these increases, nuclear power lost its ability to compete with coal on a cost basis alone. This study estimated that a nuclear power plant cost just under 6 percent more to build than coal in 1971, but it was 52 percent more expensive to build by 1978.[29]

Charles Komanoff, activist, analyst, and author of the study, attempted to sort out the causes of the construction cost increases for both coal and nuclear power. It is telling that he was able to point to specific regulations to explain most of the increase for coal, but could not do the same for nuclear. For coal, he reported that 90 percent of the increase was statistically linked to the "new equipment to reduce the environmental impact of coal plants." Komanoff did not find a similar correlation between the cost change for nuclear power and specific regulations. Instead, he looked for indirect evidence in the form of a "seven-fold" increase in "regulatory guides" issued by the AEC, and then the NRC, during the 1971 to 1978 period. He also cited the fact that the result of such regulatory guides was a doubling of "the amounts of materials, equipment, and labor" going into new nuclear power plants, plus a tripling of the "design engineering effort" needed. Nuclear plant owners built in an "environment of constant change," and there was an inherent reason for this: Komanoff concluded that, as the number of nuclear power plants grew, the probability of an accident at each plant must be reduced; he wrote, "otherwise, nuclear expansion could lead to such a high rate of accidents *per year* that the public's confidence in nuclear power would collapse and plants would be forced to close."[30]

Not only were costs rising but they were unpredictable, too, and cost overruns persisted through the 1960s and 1970s. For all the plants initiated from 1966 to 1977, cost overruns averaged 207 percent. And the costs kept rising rapidly well after the first two decades of nuclear power development. A comparison of construction costs for nuclear plants first operating from 1975 to 1979 to those first operating from 1990 to 1995 shows that construction costs rose threefold.[31]

Steve Cohn, a professor of economics at Knox College in Illinois, argues against thinking of new technologies as coming from out of the blue. Rather, he recommends a "holistic analysis of technical change," and he introduces the concept of an "official technology." An official technology enjoys "strong state support, the promoted image of the coming technology, and capture of 'critical mass' . . . such as scale economies." Cohn concludes that nuclear power expansion in the 1960s and 1970s should be understood in the context of its status as an official technology. Nuclear's success was not a matter of it having known "ex ante technical superiority over available alternatives," nor was nuclear's status as an official technology based on any "prescient anticipation of future energy or environmental crises."[32]

In broad terms, what is most important to see here is that nuclear failed even before its first public incident: the March 28, 1979, accident at the Three Mile Island nuclear plant near Middletown, Pennsylvania. Although many might attribute nuclear's fall to that event, it was in fact rooted in nuclear's failure to be cost competitive with coal-fired power plants. Moreover, the failure to be cost competitive might best be traced to nuclear's designation as an official technology. That designation probably meant that commercialization came too soon. Rickover's success was impressive, but it served a particular purpose—creating a nuclear navy—not large-scale commercialization.

The Three Mile Island Accident

This is not to say that accidents at nuclear power plants and other nuclear facilities did not play a compounding, crucial role in the fall of the nuclear power business in America. They did. And the accident at the Three Mile Island nuclear power plant is the one that many in the United States recall when the issue of safety is debated. There were two nuclear units at Three Mile Island. The first came on line in 1974 and had about 775 MW of generating capacity. The second unit had just been put into service a few months before the accident and it had 880 MW of capacity. It was the second unit that was destroyed in the accident; the first unit continues to operate today.[33]

Looking back at that accident some thirty years later, the World Nuclear Association published a blow-by-blow account. The accident started at 4 AM on March 28, 1979, with "a relatively minor malfunction" in the cooling system. The temperature of the coolant rose and, as designed, the reactor shut down automatically. Another malfunction involved a valve that should have closed, but stayed open. That open valve allowed the loss of reactor coolant water. Operators did not know the valve remained open because there was no instrument to report the actual position of that valve alone. Because cooling water was leaking, another cooling system pumped water into the reactor system. Mistakenly, the operators thought what was happening was that *too much* water was in the system, so they reduced the flow of replacement water. By this time, with the loss of coolant, "the reactor's fuel core was uncovered . . . the fuel rods were damaged and released radioactive material into the cooling water." Another primary concern was the buildup of hydrogen gas caused by a chemical reaction of water and the uncovered fuel rods. It took "an anxious month" for operators to bring the cooling issue under control.[34]

The President's Commission on the Accident at Three Mile Island delivered its report to President Carter on October 30, 1979. The commission characterized the Three Mile Island incident as the country's "worst accident in the history of commercial nuclear power generation," and concluded that it was "human failures," not equipment failures, which primarily caused the accident. Specifically: "The equipment was sufficiently good that, except for human failures, the major accident at Three Mile Island would have been a minor incident."[35]

Emphasizing that President Carter tasked the commission with making recommendations that would "prevent any future nuclear accidents," the commission made it clear that sweeping changes were needed. The commission stated: "To prevent nuclear accidents as serious as Three Mile Island, fundamental changes will be necessary in the organization, procedures, and practices—and above all—in the attitudes of the Nuclear Regulatory Commission and, to the extent that the institutions we investigated are typical, of the nuclear industry."[36]

What about the immediate impact on people living nearby? The evacuation ordered by Pennsylvania governor Richard Thornburgh had a major impact on what people did and on what they thought. There was terrible uncertainty and fear over the four-day period in which the president's commission wrote that the severity of the accident was unknown. Two days after the accident, the governor ordered the evacuation of pregnant women and small children within a five-mile radius. At the same time, federal agencies planned for the worst. The Food and Drug Administration, for example, requested 250,000 bottles of potassium iodide solution from manufacturers, which would block the absorption of radioactive iodine related to cancer.[37]

The US Nuclear Regulation Commission still calls Three Mile Island the most serious US nuclear accident. It concludes, however, that "its small radioactive releases had no detectable health effects on plant workers or the public."[38]

Shoreham: The First "Stillborn" Nuclear Power Plant

The Shoreham Nuclear Power plant built by Long Island Lighting Company (LILCO) suffered the most famous cancelation in the industry, and it may be nuclear power's most expensive failure. It is the most famous because the plant had been fully constructed when it was abandoned by its owners as part of a deal negotiated with New York's governor, Mario Cuomo. The Long Island Power Authority (LIPA), a specially created entity, subsequently bought Shoreham for one dollar and then proceeded to dismantle it. Even though Shoreham produced no electricity, ratepayers still had to pay for a substantial part of the construction costs with a premium on their monthly bills.[39]

In 1965 LILCO had decided to build a 500 MW nuclear power plant on New York's Long Island. LILCO made that proposal at a time when its customers' power needs were growing at 10 percent per year and the AEC was heavily promoting nuclear power. By 1968 needs had grown even more, and LILCO decided to make the plant 52 percent bigger. Early indications of the popular opposition to nuclear power were seen in 1970 when local residents successfully opposed the construction of two other nuclear plants proposed by LILCO. Still, construction of Shoreham started in 1973 with the intent to have it in operation by 1979. LILCO missed that initial on-line date and, of course, 1979 was the year of the Three Mile Island nuclear accident in Pennsylvania, which had a profound effect on the licensing of Shoreham, including protests by local residents fearful of having a power plant open up in their backyard. Despite the opposition, Shoreham forged ahead, and in 1985, a fully built Shoreham nuclear power plant was allowed to do some low-level tests. Then, in 1986, the Chernobyl nuclear power accident happened in Ukraine.[40]

In addition to facing the aftermath of two infamous nuclear accidents, the cost estimates for completing Shoreham skyrocketed and there were significant delays. In 1970 the construction cost estimate was $250 million and commercial operation was expected to begin in 1975. By 1980 the cost estimate had risen to $2.2 billion and the on-line date was pushed out to 1982. By 1989 the cost estimate had grown to $5.5 billion.[41]

The stated cause of the abandonment of Shoreham was the failure to develop an evacuation plan in the event an accident happened. There was a deeper fear of an accident itself. The *New York Times* reported on Nora Bredes, "the primary organizer of the grass-roots campaign that kept the Shoreham nuclear power plant" from opening. In 1987, just eight months after the Chernobyl accident, Ms. Bredes testified at a public hearing on Shoreham. She held up a picture of her two-year-old son and said, "Along with all the other evidence you collect and weigh, you should weigh this." The photo, she said, "argues that Shoreham shouldn't be opened, and it reminds you what you are risking if you allow it to operate." Bredes died in 2011 at age sixty. After his mother's death, Nathan, the son in the photo, told another *New York Times* reporter that the recent Fukushima Daiichi nuclear power plant accident in Japan "was exactly what they were trying to avoid on Long Island."[42]

The Shoreham nuclear power plant was started and completed because it made sense for LILCO. The utility wanted a nuclear power plant because with cost-plus rate making, it would make a big profit on such a big plant. Moreover, the company probably believed it did not face a risk of failure; under traditional cost-plus regulation, it may have presumed that as long as its decision-making process was prudent, it would recover its costs. Surprisingly, LILCO's presumption turned out to be largely right.

Shoreham was abandoned, however, because it was not something the utility's customers wanted. The accidents at Three Mile Island and Chernobyl proved there was danger and without an effective evacuation plan, there would be no way to escape that danger. Well after Shoreham's abandonment, on March 11, 2011, Fukushima only confirmed those fears.

The regulatory bargain is also a major factor in explaining the spectacular failure. Specifically, it was the improper allocation of risk that caused it. No private company in another business would presume that customers would cover its costs no matter what the costs turned out to be; there is no cost-plus pricing in private business as there is under electric utility regulation. No private company would presume it faced no risk of failure. And it is unlikely that any private company would have wanted to locate itself where thousands of its customers opposed it. That is where the contrast with a regulated utility comes in. Shoreham was as much or more a failure of traditional cost-plus regulation as it was a failure of technology.

THE LONG GOODBYE: STORING NUCLEAR WASTE

With nuclear power plants supplying up to 20 percent of America's electricity, substantial radioactive nuclear waste has already accumulated. Business consulting group Navigant reports that there are about 65,000 tons of waste stored at seventy-six sites across America, all but one a nuclear reactor site.[43] The United States faces a relatively more challenging nuclear waste issue because it does not reprocess its spent fuel.[44]

Congress passed the Nuclear Waste Policy Act of 1982, requiring the US Department of Energy to "accept and dispose of spent nuclear fuel and high-level radioactive waste beginning no later than January 31, 1998." The department selected ten possible sites and, in 1987, Congress narrowed that to a single site: Yucca Mountain in Nevada, about ninety miles from Las Vegas. In 2002 Congress and President George W. Bush approved the choice of Yucca Mountain, but they did so "over the State of Nevada's veto." By 2010 the political situation had been reversed and President Obama and the Congress defunded the development of Yucca Mountain.[45]

Federal courts have another opinion on Yucca Mountain. They have found the US Department of Energy to have breached its contract to provide storage, a contract under which it has collected $750 million a year from nuclear utilities. A blue-ribbon panel, created by the secretary of energy in 2010, recommended "an independent waste management organization" and "a consent-based siting policy" among other guidelines. Given all this, centralized nuclear waste storage is stalled politically. The unresolved storage issue only adds to the fear of harm from nuclear power.[46]

FUKUSHIMA

On March 11, 2011, Japan was hit with the largest magnitude earthquake in its recorded history, the Great East Japan earthquake. Forty-one minutes after the earthquake, the first of several massive tsunami waves made landfall. Adding to this tragedy was the severe accident at the Fukushima Daiichi nuclear power plant about 163 miles northeast of Tokyo. The earthquake led to the emergency shutdown of the operating units, and the tsunami led to the total loss of off-site electric power for the plant because transmission lines were downed. In addition, the tsunami destroyed onsite backup power including diesel engines as well as the seawater cooling system.[47]

The facts of what happened are not the core of the dispute; the dispute is concerned with *why* it happened. In the government's report to the International Atomic Energy Agency, Fukushima was distinguished from the other two modern, famous nuclear accidents, Three Mile Island and Chernobyl: Fukushima was preceded by a massive natural disaster. The report then detailed extensive lessons learned such as "strengthen measures against earthquakes and tsunamis," "ensure power supplies," and "ensure reliable cooling function[s]."[48]

A blunter answer came from the Independent Investigation Commission formed by the Japanese legislature, the National Diet of Japan. As an indicator of the seriousness of the accident, the commission noted that it was the "first independent commission chartered by the Diet in the history of Japan's constitutional government." The devastating critique of all those associated with nuclear power—in government, regulatory agencies, and the utility (TEPCO)—began with the chairman's message: The accident was not a "natural disaster," it was a "manmade disaster." And, further, it was a disaster "that could and should have been foreseen and prevented." Even more pointedly, the chairman said, "What must be admitted—very painfully—is that this was a disaster 'Made in Japan.' Its fundamental causes are to be found in the ingrained conventions of Japanese culture: our reflective obedience; our reluctance to question authority . . . and our insularity."[49]

The criticism does not fade after the chairman's introductory message. The main text is just as hard-hitting and it gets more specific. Ultimately, the commission points to an old-fashioned cause: regulatory capture. The government, regulators, and TEPCO were focused on promotion of nuclear power, not nuclear safety. The concern over regulatory capture would seem to be heightened by the designation of nuclear as an official technology. It accentuated the difference between the patriotic promotion of that official technology versus strict regulation for safety.[50]

NUCLEAR POWER'S LEGACY

The scientific pedigree of nuclear power is unmatched: it was born of Einstein's amazing science as well as that from the Manhattan Project. In this sense, nuclear power could have marked a return to the excitement about scientific discovery that predominated in the time of Franklin. Great science, however, is not enough to make a technology successful, and nuclear power, from the start, belonged more to a time preoccupied with the harm the science could cause. One reason is that the first application of this science was to nuclear weapons. The link between nuclear power and nuclear weapons grabs headlines even today, as seen in the heated debate on Iran's nuclear power plan as a path to nuclear weapons. No other electric generation technology has such an explicit link to weapons and the fear of war.

Another reason nuclear power belongs in the Age of Harm is the fear of accidental harm inherent to it. Despite the calming, clear-eyed risk assessment of Richard Muller in *Physics for Future Presidents*, the possibility of other kinds of harm from nuclear power was confirmed in the minds of the general public with the accidents at Three Mile Island in 1979 and Chernobyl in 1986. No historical event better reflects the concern about harm than the fact that the Shoreham Nuclear Power Plant, built at a cost of $5.5 billion, was dismantled before ever starting to generate power due to the fear of another nuclear accident. The tragic accident at the Fukushima nuclear power plant in 2011 in Japan gave further support to the public's fear of accidental harm. Indeed, it raised it to the level of a "black swan risk," as defined by economist and scholar Nassim Nicholas Taleb—the risk of an improbable event with catastrophic consequences.[51] The demonstrated black swan risk of nuclear power sets it apart from other technologies and, ultimately, places it squarely in the Age of Harm.

Given all this, how can the rise of nuclear power be best explained? How did nuclear power lay claim to a 20 percent market share in America? Cohn seems to be nearest to the truth when he argues that nuclear rose because it was designated as an "official technology." Official technologies are promoted to serve broad, overarching goals. For nuclear power, one overarching goal was to serve as an *economic* weapon for the Cold War—it was meant to show the economic prowess of capitalism as compared to communism. Another such goal was to serve as a *military* weapon in the Cold War—small-scale nuclear reactors were designed by Rickover to fuel the nuclear navy.

Given these roles in the Cold War, it seems that few focused on the essential question of whether nuclear power was the best deal for electricity ratepayers in terms of low cost, low risk, and high reliability, which are generally the right qualifications for choosing technologies for America's electricity business. Nuclear power lost to competing technologies because of its rising and unpredictable costs—a turn that began even before the accident at Three Mile Island. Its designation as an official technology aggravated this. There was a rush to commercialize the technology, and little vetting of alternative technologies or management techniques to assure nuclear power was on the right path.

Because of its 20 percent market share in America, nuclear power is best characterized as an industry that has become "too big to go away." It will not go away because of the steep costs involved in shutting down a power source responsible for a large share of electricity generation. Also, America has yet to find an alternative to on-site nuclear waste storage. Finally, there is today a second attempt to designate nuclear power as an official technology because it does not emit greenhouse gases and can have a big impact in this regard because it has been built to a large scale; that, too, may preserve nuclear power as a viable option around the world.

CHAPTER 14

Rachel Carson's
Silent Spring

There was once a town in the heart of America where all
life seemed to live in harmony with its surroundings . . .
Then a strange blight crept over the area . . . No witchcraft,
no enemy action had silenced the rebirth of new life in
this stricken world. The people had done it themselves.

—Rachel Carson (1962)[1]

Rachel Carson did not write about electricity in her best-selling books,
including her most famous work, *Silent Spring*, published in 1962.
Regardless, few people have had a bigger impact on the electricity business than Carson; with that book, she launched the modern environmental movement. That movement took the world much deeper into the Age
of Harm.[2]

Many factors contributed to Carson's impact, but first and foremost
was her writing. Carson had written three popular books about nature
and the sea before writing *Silent Spring*. Her writing is beautiful; however,

it would be a mistake to come to *Silent Spring* expecting only beauty. *Silent Spring* is replete with direct punches to the gut, aimed at those in industry and government who allowed, in her view, the indiscriminate use of deadly pesticides and insecticides.

The other factors that explain her impact include the barrage of case-by-case evidence she launched concerning the damage done to nature and to man and her placing the burden of proof squarely on the shoulders of business. And then there was Carson's very character: it took great personal courage to take on big industry and big government, and it took great conviction to testify before Congress when she was gravely ill, dying from cancer.

What were the major milestones of the Carson-inspired environmental movement? Even before Carson, there was clear and accessible evidence of harm from what are simply called the "episodes." These were extreme air pollution events that sickened and killed large numbers of people in cities as diverse as Donora, Pennsylvania; London, England; and New York City. Next there was the historic Clean Air Act of 1970, which came abruptly and from an unlikely source—the Republican president Richard Nixon. The 1970 act created the command-and-control mechanics for air pollution regulation, and these allowed the federal government to tell businesses, such as large electric power plants, which air pollution emissions to control and which control technology to use. The 1990 amendments to the Clean Air Act, proposed by President George H. W. Bush, created market-based regulations as an alternative to the command-and-control approach. The most famous (or infamous) market-based approach that emerged was cap-and-trade, the centerpiece of proposed 2009 legislation addressing global climate change. Today, command-and-control regulations still predominate, but market-based approaches have taken a foothold, too.

Carson inspired these later milestones primarily by setting a template for environmental advocacy. This template consisted of compelling writing, overwhelming case-by-case evidence, and a burden of proof placed fully on industries to prove they do not cause harm; Carson also gave no quarter to possible offsetting benefits. The electricity business has been and still is a principal target for such advocacy.

THE BOOK READ ROUND THE WORLD

Carson began *Silent Spring* with a section titled "A Fable for Tomorrow." In vivid language, her fable reveals a foreboding for man's future; she opens, "There was once a town in the heart of America where all life seemed to live in harmony with its surroundings." Ominously, Carson continues, "Then a strange blight crept over the area and everything began to change." She concludes, "No witchcraft, no enemy action had silenced the rebirth of new life in this stricken world. The people had done it themselves."[3]

That is the end of the fable and the end of any impression that her book might have a light touch. From that point onward, Carson hits big and hard. She warns that America has experienced an epochal change. Only in the twentieth century had "one species—man—acquired significant power to alter the nature of his world." She writes of "lethal materials" and man-made pollution as a "chain of evil." Strategically, she links modern chemicals to a prominent object of fear of the time—nuclear weapons. In the end, dichlorodiphenyltrichloroethane (DDT) is the chemical Carson attacks most frequently and vehemently; she notes with irony that the discoverer of DDT, Paul Müller, a Swiss chemist, won the Nobel Prize.[4]

Harming Nature

Chapter by chapter Carson takes the reader through all of the elements of nature that have been harmed: surface waters, soil, birds, and a host of other living things. After reviewing a US Fish and Wildlife Service report from 1960, Carson concludes that surface waters are "universally contaminated." Robins were indirectly harmed by an effort to control

Dutch elm disease because they ate earthworms after DDT was sprayed on the trees. Salmon were harmed on the east coast of Canada when DDT was used to combat the spruce budworm. In 1956 in an effort to eradicate the gypsy moth, Carson recounts that "nearly a million acres were sprayed [with DDT] in the states of Pennsylvania, New Jersey, Michigan, and New York." On Long Island citizens led by ornithologist Robert Cushman Murphy asked for a court injunction against the liberal use of DDT, but the spraying went on. The legal case went on, too, all the way to the Supreme Court, which decided not to hear it. Justice William O. Douglas dissented, saying that "the alarms that many experts and responsible officials have raised about the perils of DDT underline the public importance of this case."[5]

Carson passionately questions our passive consensus on who makes the decisions to weigh the costs and benefits of chemical use: "Who has placed in one pan of the scales the leaves that might have been eaten by the beetles and in the other . . . the lifeless remains of the birds that fell before the unselective bludgeon of insecticidal poisons?" Carson's answer is that bureaucrats in government and corporations make that judgment, while millions of people who might differ attend to their daily chores.[6]

Harming Man

For Carson, it is clear that harm to nature was enough of a motive to push back on the use of chemicals. For many readers, however, then and now, harm to man is more compelling, and Carson provides strong evidence on that front as well. She first quotes Dr. Arnold Lehman, an expert pharmacologist at the Food and Drug Administration, who found chlordane to be "one of the most toxic of insecticides." Carson says lawn treatments containing chlordane were used widely. She then relays the story of "one victim who accidentally spilled a 25 per cent industrial solution on the skin developed symptoms of poisoning within 40 minutes and died before medical help could be obtained." Another chemical, endrin, is the "most toxic of all the chlorinated hydrocarbons." Endrin was also used widely as an insecticide. In Venezuela, a little boy and his dog had been taken out of their home while insecticides with endrin were applied, and then the house was thoroughly cleaned before their return. Despite this, one hour after they came back, the dog died, and later that night the small boy "lost consciousness" and was left, in Carson's words, "little more than a vegetable."[7]

With great disdain reflected in her words, Carson recalls how easy it was to get these chemicals and how plainly they were promoted, despite, she said, their known capacity to sicken and to kill us. She states, "So thoroughly has the age of poisons become established that anyone may walk into a store and, without questions being asked, buy substances of far greater death-dealing power than the medicinal drug for which he may be required to sign a 'poison book' in the pharmacy next door."[8]

Carson was alerting readers to the fact that not every technological advance was good, and that some could do grievous harm. This change in mentality would soon be brought to the electricity business. No longer would more electricity be automatically seen as a good thing; the possible environmental harm from electricity generation would get equal or more attention.

The Science of Harm

Carson convincingly shows that a great deal of hard science must be understood to discern what harm actually results from any given chemical. She demonstrates her grasp on that science throughout the book, which increases the credibility of her allegations. At the cellular level, Carson explains, the threat comes in the form of a chemical assault on the production of energy for all the functions of our body. Carson goes on to describe the cellular creation of energy in more detail. She writes that the "transformation of matter into energy in the cell is an ever-flowing process," and each step of that process is "controlled by an enzyme of so specialized a function that it does this one thing and nothing else." She reports that this energy production through oxidation is achieved "in tiny granules within the cell called mitochondria." These are the "Mighty Mitochondria" we learn about in school, "tiny packets of enzymes" that serve as the "'power-houses' in which most of the energy-producing reactions [of the human body] occur." This energy production within the body can be irreversibly disrupted by radiation and chemicals like insecticides and weed killers.[9]

The Author and the Aftermath

Rachel Carson's most impressive trait was her personal courage. She demonstrated this courage by facing up to big industry and big govern-ment. *New York Times* journalist Eliza Griswold provides especially com-pelling evidence on this front; while Carson kept on with her commitment to protect the lives of creatures as diverse as robins and man, she knew she was dying of cancer. Griswold describes Carson's condition a year after publishing *Silent Spring*: "She was 56 and dying of breast cancer. She told almost no one. She'd already survived a radical mastectomy. Her pelvis was so riddled with fractures that it was nearly impossible for her to walk to her seat at the wooden table before the Congressional panel. To hide her baldness, she wore a dark brown wig."[10]

As her biographer Linda Lear has written, Carson was an "improba-ble revolutionary, even an unlikely reformer." She was born in Springdale, Pennsylvania, a town north of Pittsburgh, in 1907. Her mother, Maria, was the "most important influence on her intellect and outlook." Included in that outlook was an interest in and a respect for nature honed by walks through "the woods, wetlands, and river flats" near the Allegheny River. In 1925 Carson entered the Pennsylvania College for Women, now Chatham University in Pittsburgh. While there, she won a fellowship at the presti-gious Marine Biological Laboratory in Woods Hole, Massachusetts. She then earned a master's of arts in zoology from Johns Hopkins University. Her first job was with the Bureau of Fisheries, then in the Department of Commerce. She worked for the next sixteen years in the Fish and Wildlife Service, rising to editor-in-chief of all service publications in 1949.

Somewhat surprisingly, this is where and how Carson's literary career blossomed.[11]

In 1941 she published the first of three books on the natural wonders of the sea, *Under the Sea Wind*. She later won the George Westinghouse Science Writing Award and a Guggenheim Fellowship. Following that, *The Sea Around Us* was published in 1951. Carson gained an international reputation with this second book after it was serialized in the *New Yorker* magazine and won the National Book Award. It also was a selection of the Book of the Month Club, and remained on the *New York Times* best-seller list for almost two years. In 1952 Carson retired from the Fish and Wildlife Service to write full time, and in 1955, her third "sea biography" appeared, *The Edge of the Sea*.[12]

Her success as a writer gave her the skills, reputation, and credibility to speak out, but her years at Fish and Wildlife gave her the experience with nature that would shape each of her books. Lear writes that Carson had been climbing the ladder at "the *one* agency in the government, which by the mid-1950s, had a long standing record of concern about the widespread use of synthetic pesticides." By virtue of her job, Carson was already highly attuned to the implications of the rise in toxic chemicals used in daily life and on the natural environment.[13]

Silent Spring was a sensation: six hundred thousand copies sold soon after publication, and it became an enduring best seller. (As of today, more than two million copies have been sold.) However, it should not be forgotten that *Silent Spring* was an indictment of three powerful groups: the industry that made these synthetic chemicals, the governments who used them, and the scientists who claimed they did no harm. There was pushback, of course. An agricultural lobby spent $25,000 to refute Carson's allegations while another chemical company, Velsicol, threatened Carson's publisher, Houghton Mifflin, with a libel suit.[14]

Carson, however, had even more powerful supporters won over by her book. President John F. Kennedy saw the importance of the political issue and ordered his science adviser, Jerome Wiesner, to investigate Carson's charges. The president's advisory report was issued in May 1963 as Senate hearings began. Equally important, CBS ran a prime-time special titled "The Silent Spring of Rachel Carson" on April 13, 1963. Although very ill, the calm and competent Carson appeared to be winning the debate against the volatile corporate scientist sent to rebut her by the chemical industry. One corporate scientist even foresaw a return to the "dark ages" if the use of pesticides were restricted.[15]

In the Senate hearings in June 1963, Senator Abraham Ribicoff, a Democrat from Connecticut, opened by paraphrasing Abraham Lincoln's remark to Harriet Beecher Stowe, the author of *Uncle Tom's Cabin*. The senator said, "Miss Carson . . . we welcome you here. You are the lady who started all this." "All this" was the modern environmental movement. Carson made environmental concern a topic of policy that the voting public wanted to hear more about. No longer would communities accept projects on the basis of their being "big," "more," or "new," unless attention was paid to potential risks. That new view would be applied to the electricity business and soon meant that environmental concerns became a major factor driving decisions about electric power.[16]

Rachel Carson died on April 14, 1964, not quite a year after the Senate hearing.

THE EPISODES

Major incidents before and after *Silent Spring* showed that air pollution could sicken and kill. The first of these episodes that brought air pollution to the headlines in America occurred in Donora, Pennsylvania, in 1948: six thousand people fell ill and twenty died. Quick on its heels, London experienced a smog in 1952 during which at least four thousand people died of respiratory and heart ailments. Throughout the 1960s there were other internationally reported pollution incidents, like the deadly smog that killed 169 people in New York City in 1966.[17]

All these episodes had common culprits. There were, of course, air pollution emissions from sources ranging from zinc factories to home furnaces. In addition, due to climate conditions, the air at the ground level was colder than the air above it; thus the air at the ground level was trapped and could not disperse. This is called an inversion; the warm air higher up creates a "lid" trapping ground-level air, which was heavily polluted in these incidents. The result is that people are exposed to pollution for extended periods of time until the temperature level allows the air to dissipate. Terrain contributed, too, with riverbed valleys and mountains making matters worse. The episodes revealed the damage that could result from severe levels of air pollution, and they also revealed early on the complexities of understanding what happened, who was to blame, and what should be done.

Donora: "Hell with a Lid On"

Donora is a factory town about thirty miles southeast of Pittsburgh along the Monongahela River. The Zinc Works in Donora was center stage in this episode. It had been built in 1915 and was run by the American Steel & Wire Company, an affiliate of U.S. Steel; steel mills were the major consumer for the zinc produced there. Writing in 1994, policy analyst Lynne Page Snyder notes that air pollution concerns were not new; American Steel & Wire paid its first fine for air pollution soon after it opened in 1918, with more suits following.[18]

The "mysterious, death-dealing smog" started in Donora on Tuesday, October 26, 1948. By Friday, hospitalizations began. Doctors ordered that those with heart or respiratory disorders be evacuated, but the smog and clogged roads had made that impossible. National news soon picked up the story, with Walter Winchell leading the way. Dr. William Rongaus of the local Board of Health walked an ambulance through the darkened town in order to deliver the sick to the hospital and the dead to the morgue. Later he would say that what happened was "just plain murder."[19]

American Steel & Wire retained scientific experts led by Dr. Robert Kehoe at the Kettering Laboratory to defend it from blame for the disaster. Kehoe concluded, "The crux of the problem is the cost of control. Control is possible in most cases now, but the cost is exorbitant." In the

end, Kehoe's recommendation was not to control the emission of pollutants but to adjust the timing of the industrial processes at the Zinc Works
to avoid dangerous weather conditions.[20]

Dr. Clarence Mills, from the College of Medicine at the University of
Cincinnati, conducted some of the early studies on Donora as well, and he
pushed back on the government health agency report for several reasons.
In the end he concluded that Donora was "America's first mass killing
from industrial air pollution." Though a study by the US Public Health
Service report "relieved" the Zinc Works of responsibility, the Donora
episode demonstrated the danger of air pollution and started the debates
on whether and how to mitigate the danger.[21]

"Hell Is a City Much like London"

Sadly, when the poet Percy Bysshe Shelley wrote in 1839 that "Hell is a
city much like London," he presciently gave an apt description of London
in early December 1952. With four thousand deaths reported at the time,
this was the world's worst air pollution episode. Again, it had the typical
interrelated culprits: temperature inversion and polluted air trapped at the
surface. London had high emissions levels because the city's homes used
a large amount of coal to fuel home furnaces. The effects, however, were
dire, even for a city noted for its fog and smog: visibility was so poor that an
indoor performance of *La Traviata* had to be cut short because the audience
could not see the stage, and nurses at hospitals could not see "from one end
of their wards to another." Undertakers reported a shortage of caskets.[22]

The London episode reveals the challenge of estimating the number
of deaths caused by a single episode. The general impression held for many
years was that the spike in deaths in the first months of 1953 was caused by
an influenza epidemic. However, in 2001, researchers started to discount
the impact of the influenza. Michelle Bell, a professor at Johns Hopkins
at the time, and Devra Lee Davis, then a visiting professor at Carnegie
Mellon, concluded that a more accurate estimate of the death toll was not
4,000 but in fact "about 12,000 excess deaths occurred from December

1952 through February 1953 because of acute and persisting effects of the 1952 London smog."[23]

Another look at unique physical evidence from the London episode helped to better define how air pollution kills. In 2003 researchers used autopsy tissue samples from the 1952 London episode to support the more recent view that among the most dangerous air pollutants for human health is inhalable (or fine) particulate matter. The smallest form is $PM_{2.5}$, which is particulate matter in tiny particles no more than 2.5 microns in width; to give a sense of just how tiny they are, there are 25,000 microns in an inch. Noting that "massive mortality events such as that in London in 1952 no longer occur," the authors conclude that these autopsy samples provide unique evidence of pollution's effects. It is unique evidence that $PM_{2.5}$ can be an invisible killer. As it turns out, today, concerns about $PM_{2.5}$ are a major issue in environmental regulations for the electricity business.[24]

New York City suffered an episode only a year later, and then again in Thanksgiving week of 1966. There was a temperature inversion at the time, as was the case with the episodes elsewhere, and air pollution concentrations spiked. Sulfur dioxide averaged forty-six parts per hundred million in the worst three days, which was three times the standard set by regulators. In all, a total of 169 deaths were attributed to the Thanksgiving episode.[25]

THE 1970 CLEAN AIR ACT

With the backdrop of Rachel Carson's *Silent Spring* and the mounting number of episodes (and resulting deaths), the federal government became more directly and substantially involved in environmental regulation during the 1960s. Congress passed the Clean Air Act of 1963, which gave the federal government added "authority to enforce existing state laws" and to promote adding state laws on air pollution. The federal government also regulated interstate pollution. In 1965 Congress added the Motor Vehicle Air Pollution Control Act, which controlled emissions from new vehicles.[26]

More legislation quickly followed. In 1967 Congress added the Air Quality Act, which allowed the federal government to define "air quality control areas" and "recommend pollution control techniques." With few measurable results through these laws, however, the president and Congress ultimately took aggressive action to give the federal government "primary responsibility" for regulation of pollution. President Nixon created the Environmental Protection Agency (EPA) in 1970, and Congress passed the Clean Air Act that same year.[27]

Looking back, MIT professor Judith A. Layzer found in her 2006 book, *The Environmental Case*, that "the surge in environmental policymaking in the early 1970s was not a response to a sudden deterioration in the condition of the nation's air and water." Rather, it was the result of "redefining, or reframing" the issue, which gained support and popularity from the public for environmental regulation; that public support pressured politicians to take action. Layzer pointed to Carson's *Silent Spring* and Paul Ehrlich's *The Population Bomb* (1968) as two major works that shaped public opinion and popularized environmentalism. Events such as the oil spill off the California coast at Santa Barbara, the Cuyahoga River in Ohio catching fire, and increasing concern about mercury in seafood also contributed to the environmental movement. Finally, a wealthier and better-educated population contributed to the demand for regulation. This trend toward environmental awareness culminated with the inaugural Earth Day on April 22, 1970.[28]

The political backstory to President Nixon's proposal for the creation of the EPA and the 1970 Clean Air Act was his expectation that he would face Senator Edmund Muskie of Maine in the presidential race of 1972. Indeed, the senator submitted a rival act, which, among other things, proposed a strict 90 percent cut in auto emissions from new cars by 1975. The Senate passed the Clean Air Act and the president signed it on December 31, 1970. In *Environmental Law and Policy*, their book from 2007, James Salzman and Barton H. Thompson Jr., law professors at Duke and Stanford, respectively, at the time, call the 1970 Clean Air Act "historic," a "massive law" that "stand[s] apart." They write that the act "boasted *uniform, national standards* covering a wide range of pollutants and sources." The authors conclude that the 1970 act was the "*first* truly comprehensive national pollution law." Similarly, Helen Ingram sees the 1970 Clean Air Act as a big leap forward, not just an "evolution" in policy. The goals for air pollution got more demanding as the legislative process went on. And those goals were not limited to what could be achieved with current technology or with only limited cost increases.[29]

The Clean Air Act of 1970 imposed several layers of regulation. It established both primary and secondary National Ambient Air Quality Standards (NAAQS). Primary standards were to protect public health. Secondary standards were aimed at protecting visibility and the ecology. NAAQS were set for six "criteria pollutants": smog (or ozone, O_3), nitrogen oxides (NOx), carbon monoxide (CO), sulfur dioxide (SO_2), lead, and soot or "fine particles." The act required the states to meet the NAAQS through State Implementation Plans by 1975. If the EPA did not approve a state plan, then a federal implementation plan would be imposed on that state. It also set New Source Performance Standards for stationary sources of emissions like power plants and other manufacturing plants; the federal regulations would set the rules for *new* sources, meaning the states were left with control only over older facilities.[30]

Despite the sudden support for environmental regulations, it was not all clear sailing for the act. By 1975 not one State Implementation Plan had been approved. The 1977 Clean Air Act codified postponement by extending compliance deadlines to 1982 for most places and to 1987 for others, such as California. In addition, the plans now had to consider the effect on jobs and broader economic impacts. Still, just about every nook and cranny of this regulatory structure would affect the electricity business. Power plants were a clear and immediate focus of concern because they emitted air pollutants and were accessible politically.[31]

THE 1990 CLEAN AIR ACT

Looking back at the 1990 Clean Air Act, Richard Schmalensee and Robert Stavins, professors at MIT and Harvard, respectively, published the intriguingly titled "The SO_2 Allowance Trading System: The Ironic

History of a Grand Policy Experiment." Title IV of the 1990 Clean Air Act established an acid rain program, which the authors characterize as "a grand experiment in market-based environmental policy." This was the first large-scale cap-and-trade program and was seen as "quite novel." Acid rain occurs because of air pollution emissions—sulfur dioxide and nitric oxides—reacting in the atmosphere to form sulfuric and nitric acids. This acid rain then damages forests, lakes, and rivers, and concern over its negative effects heightened in the 1980s.[32]

The acid rain program was implemented in two phases and aimed precisely at the electricity business. Phase I (1995–99) focused on the 263 coal-fired power plants with the highest sulfur dioxide emissions. Phase II, beginning in 2000, would regulate a much broader range of power plants—3,200 in all. The goal of the acid rain program was to ultimately cut sulfur dioxide emissions by 50 percent as compared to emissions in 1980. Again, the means to this end was a novel cap-and-trade program. The cap set the total allowed sulfur dioxide emissions, and it would be steadily reduced over time to achieve the desired 50 percent reduction. Sulfur dioxide allowances—each giving the right to emit one ton of sulfur dioxide—were allocated to existing sources of such pollution. As to "trading," the hope was that those power plants that could reduce emissions cheaply would do so and then sell their excess allowances to a power plant for which emission reduction was expensive. The goal was to minimize the overall cost of emissions reduction for the industry.[33]

The results were good in terms of achieving the planned emission reduction. Schmalensee and Stavins report that emissions fell by 36 percent from 1990 through 2004 despite coal-fired power generation increasing 25 percent. And they conclude further that the cap-and-trade program reduced emissions at a lower cost than would have been incurred with command-and-control regulation.[34]

However, and importantly, if "results" are defined not strictly in terms of emission reductions, but rather in terms of the reduction in acid rain and reductions in acidic lakes, a different story emerges. Indeed, it is an ironic one; perhaps, in the end, the acid rain program did the right thing for the wrong reason. The program did not solve the acid rain problem. However, say the authors, it mitigated a new problem that had emerged over time to become an urgent concern: the emission of $PM_{2.5}$, which can get deep inside the lungs, as already discussed here relating to the London episode.[35]

A second irony was that the lower cost of reducing emissions was caused, in part, by an unexpected source: the unrelated deregulation of railroad rates. Most of the polluting coal-fired power plants were located east of the Mississippi River. The cheapest low-sulfur coal, however, came from mines in the West; specifically, in the Powder River Basin in Montana and Wyoming. The expensive part of using Powder River Basin coal lay in the transportation from the mine to the power plant. Deregulation of railroads cut that cost by 50 percent.[36]

A third irony was that sulfur dioxide emissions were reduced with other programs that, in effect, sidestepped the cap-and-trade program. These other programs undermined the demand for allowances so that, by 2012, the price of an allowance had fallen to near zero. As Schmalensee and Stavins conclude, "When the government creates a market, it can also destroy it."[37]

Looking at the acid rain program now, it is hard to call it an unequivocal success. The stated goals for acid rain mitigation were not achieved. The mechanics of cap-and-trade were successfully demonstrated, and the fact that it opened the door broadly to other sources of cost cutting (lower rail rates) is a plus. In the end, however, the cap-and-trade mechanism was undermined by another parallel regulation.

MEASURING SUCCESS

More than fifty years have passed since *Silent Spring* inspired the modern environmental movement. It would be hard to identify a subsequent book with the same impact in this field. It is not hard, though, to find studies that go well beyond the case evidence Carson used to sound the alarm. Today, researchers use far more rigorous methods to assess the links between death or illness and air pollution. However, their published studies are, most of the time, addressed to a scholarly audience. Rigorous analysis is essential, but it has also put a greater distance between the person on the street and the case for air pollution controls. Moreover, there is still great uncertainty in estimating the health benefits of cutting emissions.

Sorting out the causes of illness and death is a challenge. There is no official designation at the hospital or morgue that an illness or death was caused by air pollution. In a rigorous study of $PM_{2.5}$ emissions published in the *Journal of the American Medical Association*, for example, the authors are cautious but clear about their results. They are careful to say their studies find an "association" rather than a clear cause and effect. However, they do say that their study is the "strongest evidence to date that long-term exposure" to $PM_{2.5}$ is an "important risk factor for cardiopulmonary mortality." And the researchers emphasize their efforts to control for other factors affecting human health when they state, "The associations between fine particulate air pollution and lung cancer mortality, as well as cardiopulmonary mortality, are observed even after controlling for cigarette smoking, BMI [body mass index], diet, occupational exposure, [and] other individual risk factors." Another uncertainty lies in whether other public investments might have more direct and substantial health benefits than the investments made to cut emissions.[38]

RACHEL CARSON'S LEGACY

Rachel Carson served as the herald for the Age of Harm for the electricity business. Her work spoke to an age when Americans stopped thinking solely of the benefits of abundant, affordable electricity for all and, instead, focused much more on the harm electricity generation might be doing to the physical environment or, more compellingly, to human health. Carson created a template for environmental activism that included persuasive and moving writing, a barrage of case evidence, the use of hard science to explain how the harm might have occurred, and a burden of proof placed squarely on business. More importantly, that burden of proof often implied a standard of zero risk of harm to human health—a standard to be met at any price. Critics claim that the template also included a serious deficiency: Carson seldom acknowledged the possible benefits of the pesticides she railed against, and surely never considered that the benefits might outweigh the costs.

Carson's template had huge implications for the electricity business as time went on and attention turned to air pollution. The well-publicized episodes in Donora, London, and New York City gave clear and accessible physical evidence that air pollution can kill and sicken. Even though environmental regulators and activists have had to shift from physical evidence to less compelling statistical studies to prove harm, tightening of emission limits has become all but inevitable. As an example, between 1980 and 2014 national concentrations for sulfur dioxide were cut 80 percent. Still, regulations have been proposed, and are pending in the courts, that would cut those remaining emissions significantly for electric power. It is interesting, too, that the template still works decades after Carson first used it. Former vice president Al Gore's award-winning 2006 documentary, *An Inconvenient Truth*, followed in Carson's tracks as it warned of the dangers of global climate change.[39]

The debate has moved from whether environmental regulations should be tightened to how to tighten them. Command-and-control regulation was the core of the effort established in the 1970 Clean Air Act. Complex layers of regulation and constant litigation are the main legacy of the 1970 Act, but so is a significant reduction in air pollution emissions. The 1990 Clean Air Act offered an alternative to command-and-control mechanics called market-based regulations. Where the market kicks in is in determining how that government-created demand for reduction will be supplied; this makes sense since markets are good at minimizing the cost of supply, but much still depends on an administrative determination of demand.

In a very short time environmental regulations became a powerful factor driving the electricity business. As will be seen even more clearly in later chapters, by 2010, environmental regulation evolved from being a powerful factor to being the dominant factor in the electricity business—most notably in terms of dictating the types of power plants to build. Concern about air pollution has become so deeply embedded in American culture and law that there is no electricity policy per se; one might say that there is only environmental policy. No longer do regulations ask "how do we minimize the cost of electricity." Rather, they seek to minimize the harm electricity generation might cause.

CHAPTER 15

California's Electricity Crisis (and Enron's Greed)

> We emphasize that the trading strategies—while bearing
> Enron's name—were not limited to Enron but appear to
> have been widely engaged in by numerous parties.
> —Federal Energy Regulatory Commission Staff (March 2003)[1]

*In 1994 California embarked on an effort to change the way the electricity busi-*ness had been regulated for almost ninety years. The old way had been originally defined by Samuel Insull when he formed Commonwealth Edison in Chicago in 1907. By 1996 California governor Pete Wilson claimed, "We've pulled the plug on another outdated monopoly and replaced it with the promise of a new era of competition." It was not the only state to change the rules around this time, but, as is California's wont, it was the first and the most provocative. Sadly, soon after the structure California had built started to operate, it collapsed. This collapse and its repercussions became known as the California electricity crisis.[2]

As does any collapse of a major industry, this one harmed a great number of people, and the harm came in many forms. Prices increased dramatically in the short-term or "spot" electricity market that California had created. A constant threat of rotating blackouts of electric service loomed over consumers across the state for months at a time. Pacific Gas and Electric (PG&E), the largest electric utility in the state, went bankrupt. The state government had to take over electricity purchases because the electric utilities were not creditworthy. And California governor Gray Davis became the first governor in California history to be recalled and removed from office—the second ever in the United States to suffer this fate.[3]

Two assertions are often made about the causes behind this crisis: first, that Enron caused the California electricity crisis, and second, that Enron's dirty deals in the deregulated electricity business in California caused Enron's collapse. Regarding the first assertion, Enron did not cause the California electricity crisis. The crisis had three major causes: bad market design, bad market conditions, and bad market behavior. Enron was certainly part of the story with respect to bad market behavior, but Enron did not act alone. However, a smoking gun exists for Enron's role in the form of a memo that lays out in considerable detail the schemes Enron used to manipulate California's electricity markets.

As to the second assertion, Enron's dubious deals in a deregulated electricity business were not the proximate cause of Enron's collapse. The major causes were the far-too-aggressive accounting and financial moves that the Enron board and outside auditors failed to, or refused to, see.

Of course, Enron's rise and fall is a big story no matter what its link to the electricity business. Enron's bankruptcy in December 2001 was, at the time, the largest bankruptcy in US history, involving $65.5 billion of assets. That bankruptcy marked a fall from the highest heights: Enron was ranked seventh on the Fortune 500 list in 2001 with over $100 billion in revenue, up from $40 billion just a year earlier. Enron's fall is a big story, too, because few, if any, saw it coming. In their widely read 2003 book on the fall of Enron, *The Smartest Guys in the Room*, Bethany McLean and Peter Elkind cleverly sum up Enron's reputation in its heyday with the chapter title "Everybody Loves Enron."[4]

CAUSES OF THE DISASTER

Bad Market Design

Since Samuel Insull's day, the electricity business had been run as a collection of local monopoly franchises. That is, the United States was essentially carved up into areas in which a single electric utility was given the exclusive right to supply all the electricity. That utility would be "vertically integrated"; that is, it would control all phases of the business in that franchise area: generation, transmission, and distribution of electricity. Having been granted that monopoly, the local utility took on the responsibility to assure sufficient and reasonably priced electricity was available to all ratepayers. That utility took on what was characterized as an "obligation to serve" with cost-plus rates, reflecting actual costs plus a fair return on investment.

With California Assembly Bill 1890 (AB1890), all that would change. In most of its features, the deregulation prescribed in the bill followed a standard template. The electricity business would be "unbundled" into its three parts: generation, transmission, and distribution. Generation was the segment that would become competitive; that is, individual customers—homes and businesses of all sorts—would be free to choose their generation supplier. This is termed *retail choice* or *direct access*. If a customer chose not to buy electricity from a particular supplier, the local utility would procure electricity on that customer's behalf from a newly created competitive marketplace. In California, this competitive market would be managed by a new organization, called the California Power Exchange or simply the PX.[5]

To ensure there would be competitors other than the incumbent utilities, the law required that the two biggest utilities divest at least half of the fossil-fueled power plants that they owned; in the end, the utilities actually divested 100 percent of these plants for reasons ranging from promoting competition to taking advantage of the expected favorable market for sale of these assets. Notably, the natural gas–fired power plants were divested to five independent generators. Another important feature was "open access" to the electric transmission system. Any generator would have access to the long-distance transmission lines owned by the regulated utilities to get its electricity to market. To ensure that the access was fair or "non-discriminatory," a brand-new institution called the California Independent System Operator, or California ISO, would take operational control of the transmission system from the utilities.[6]

California went well beyond the template, however, with one feature that became the primary cause of the crisis: the state-mandated deregulation was based on the presumption that market prices would always stay below $65. The law required the utilities to sell power at a fixed price of about $65 per MWh, but to buy power in the newly created competitive market, the PX, the utilities had to pay whatever the market price might be. These prices varied freely, hour by hour, day by day. At the start, these hourly or "spot" prices were well below the limit, but no one could guarantee that would continue. When the spot price went above $65, the utilities would be selling at a loss, with obvious consequences.[7]

Paul Joskow, an economics professor at MIT and a highly respected expert on the electricity business, concluded in his paper "California's Energy Crisis" for the *Oxford Review of Economic Policy* that California was "an accident waiting to happen." He was right. In both 1998 and 1999 the weighted average price across all hours in the real-time and day-ahead markets was low, at about $30 per MWh. By May 2000 the average price ticked up to $50. By June the price was up to $132 per MWh, which was a fivefold increase from June 1999. Then, in December 2000, the PX price increased to $386—almost thirteen times higher than the price in December 1999. In January 2001 the PX had to close its doors; it was simply not getting paid for the power transacted in its market, a circumstance that the bill had not anticipated. Transactions shifted to the California ISO's market. In February, March, April, and May of 2001 the prices per MWh in the California ISO market were $363, $314, $370, and $275, respectively. Clearly, the utilities were losing a lot of money selling at \$65 per MWh and buying at these very high real-time prices.[8]

Bad Market Conditions

Joskow identifies five interrelated reasons for the price spike. (Another well-respected entity, the US Congressional Budget Office [CBO], also addressed the motives, methods, and results of California's restructuring plan. CBO's diagnosis of the California electricity crisis in its 2001 report overlaps significantly with Joskow's.) The first was an increase in natural gas prices. These prices were up nationwide, but California's were up five times higher than the rest of the country's. Natural gas–fired power is usually the most expensive, so it is the last to be called upon and, therefore, sets the spot market price.[9]

The second reason was the increase in the demand for electricity due to economic growth and hot weather. Joskow points out that it was not that the peak demand in one or a few hours increased but rather that the average demand across all hours of the month increased. In May, June, July, and August of 2000, the average demand was up 10.8 percent, 12.7 percent, 2.2 percent, and 7.1 percent, respectively, as compared to average demand in those same months in the previous year.[10]

The third reason for the price spikes was that imports of electricity into California from outside the state were lower in 2000 than in 1999. In May, June, July, and August of 2000, as compared to the same months in the previous year, net imports into California had decreased by 27 percent, 41 percent, 67 percent, and 75 percent, respectively. This decline in imports reflected a decline in the availability of hydroelectric power across the West.[11]

The fourth reason cited by Joskow was a tightening of the requirements for an air pollution program in Southern California called RECLAIM. The goal was to reduce nitrogen oxide emissions from both electric power plants and oil refineries. It was one of the first attempts at a cap-and-trade program for pollution control. Under this regulatory scheme, a power plant must have a permit or an "allowance" to emit pollutants. In 2000 the price for nitrogen oxide emission allowances increased almost tenfold. Because it was a cost incurred by natural gas–fired power plants, that increase, in turn, increased the spot market price for electricity in California.[12]

Joskow's fifth reason was the exercise of market power. He points out that this was not the classic antitrust situation in which suppliers join together in a scheme to fix a high market price. It was not the exercise of market power through *collusion*. Rather this was *unilateral* market power, with each supplier acting on its own. That is, a supplier raises its bid price, betting it can earn more overall by receiving a high price for the MWh it does sell, despite the loss of revenue from selling fewer MWh. Joskow finds "at least a third of the wholesale price can be attributed to market power during June, July, August, and September 2000, *after* accounting for changes in fundamental supply and demand conditions."[13]

Another reason, however, got the greatest attention from the press and politicians. It was the reduction of supply due to increased outages of existing power plants. These power plants were simply not available anymore to generate power. Joskow reports that between November 2000 and May 2001, as much as 35 percent of the generating capacity within the California ISO area was out of service. The generators themselves attributed the increase to the fact that these plants had broken down because they had overworked in the summer months, or that they had to be off line so new nitrogen oxide emission controls could be installed. In sharp contrast, Joskow writes, "California government officials argued that the plants had been withdrawn from service at least partially for strategic reasons."[14]

Soon the biggest problem facing California was the creditworthiness of the major utilities: because of the fixed price of $65 per MWh, they could not pay their bills at the PX or to the California ISO. PG&E and another large electricity supplier, Southern California Edison (SCE), were technically insolvent. In April 2001 PG&E, the largest utility in California, actually went into bankruptcy. Ultimately, to address the lack of a creditworthy buyer, California ordered a state agency, the California Department of Water Resources, to buy power under long-term contracts to serve consumers. Finally, the fever broke in late summer 2001, and market prices fell back to normal levels seen before the crisis.[15]

Bad Market Behavior

In March 2003 the staff of the Federal Energy Regulatory Commission (FERC) published the results of its investigation into the California crisis. The discussion so far had reflected a balanced attribution of the crisis to bad market design, bad market conditions, and bad market behavior. The FERC assessment marked a big change because it narrowed in on only bad market behavior and it named companies at fault, including Enron, the most infamous of all. The FERC found "significant market manipulation."[16]

Another important change with this FERC report was the greatly sharpened focus on natural gas prices in the diagnosis of the causes of the California electricity crisis. The FERC report starts by stating that electricity and natural gas prices are "inextricably linked" in California, and goes on to say that the natural gas spot price "rose to extraordinary levels, facilitating the unprecedented price increase in the electricity market." Turning to the bad market behavior that caused natural gas prices to rise, FERC concludes that the price increases "appear to stem, at least in part, from efforts to manipulate price indices compiled by trade publications." The link between these published price indices and actual natural gas prices is straightforward: it is common to see a contract to buy natural gas that sets the price equal to the index prices. The reasoning is that the index—an average or the midpoint of prices reported by buyers and sellers to trade publications—would be an objective reflection of true market prices. According to FERC, the manipulation of natural gas price indices included reports of false data. The report unequivocally states that "false reporting became epidemic." Only with honest information can a market offer buyers and sellers an honest chance to win or lose, which is the essential ethic of free markets.[17]

FERC soon zoomed in on one company for its "high-volume, rapid-fire trading strategy"—a strategy that FERC characterized as "churning." Churning is pushing up the market price by buying a large volume of natural gas with no intention of actually using that gas. A company can profit from churning in several ways. One is by increasing the electricity price in California by artificially increasing the natural gas price and then profiting on spot electricity sales. A second is buying the natural gas it actually needs to supply its customers at low prices in the morning, but then driving up the index price—the price it charges those customers—by artificially buying and selling natural gas in the afternoon. A third takes advantage of a feature in Enron's trading platform that also allows a supplier to artificially create a gap between the prices for buying and selling. FERC says the company "often bought and sold many times its needs in quick bursts, which significantly increased the price of gas" at one point in that market.[18]

Importantly, FERC concluded that the company's churning did not violate the rules under which it sold natural gas because, in essence, there were no rules to cover such tactics. Still, churning fails to comply with the implicit, underlying ethic of free markets.[19]

Addressing the real-world consequences of churning, FERC notes that had churning by specific companies increased the price for all natural gas bought at the California-Arizona border, the cost increase for the eight-month period was $1.15 billion. The possible impact on the California electricity prices was not straightforward to calculate; making some simplifying assumptions, FERC estimates that the churning could have increased what was paid for spot electricity by $1.6 billion. If these are accurate estimates of the impact of churning, they constitute a significant harm.[20]

The other way FERC found that market participants manipulated the natural gas price index is by knowingly submitting false information to the trade press. FERC mentions that five companies had already admitted to manipulating natural gas prices, including "fabricating trades, inflating the volume of trades, omitting trades, and adjusting the price of trades." The report goes on to say that it was "particularly troubling" to find that "everyone knew that everyone else was manipulating the indices by reporting false prices and volumes."[21]

FERC conceded that market conditions had also driven natural gas prices up but quickly added that manipulation "contributed significantly" to the price increase, too. In the interests of fairness, however, the report said that it "cannot calculate the portion" of the price increases "due to manipulation." Besides, market conditions were pushing natural gas prices up; FERC pointed to the fact that the use of natural gas for electricity generation in California had increased by 44 percent in the May to October time frame in the year 2000, as compared to the same period in 1999. In the western United States as a whole, that demand had gone up by a comparable amount—about 46 percent. These are substantial increases in demand that increased natural gas prices, but also increased both the opportunity for bad behavior and the magnitude of the harm it caused.[22]

ENRON'S GREED

A Brazen Trade

Today, Enron has become the poster child for bad market behavior. It showed its true colors early on in its effect on the new California electricity markets. In an article titled "Brazen Trade Marks New Path of Enron Probe for Regulators," the *Wall Street Journal* highlighted one proposed transaction by an Enron energy trader, Timothy Belden, on May 24, 1999. On that day, a series of offers proposed that Enron sell 2,900 MW into the PX market. The trouble was that he proposed the 2,900 MW be sent across a part of the transmission system with the capability of transmitting just 15 MW. His aim was, in fact, to cause transmission "congestion" that Enron would then be paid to relieve, and also to drive up the price for power sold into the market by cutting off lower-cost bids.[23]

The PX found the proposal to be "pretty interesting" and even "odd," so much so that the PX marked it for further investigation. The *Wall Street Journal* reported that a state investigation showed that "California electricity customers were overcharged by $4.6 million to $7 million that day as a result of Mr. Belden's scheme." Still, Enron would be fined only $25,000 for the incident. Fast-forward to October 2002, when Belden faced far more serious consequences for this and other trading schemes. It was then that Belden signed a plea agreement with the federal government in which he pled guilty to one count of "conspiracy to commit wire fraud." With that plea Belden faced up to five years in prison, a fine of $250,000, and a restitution payment of at least $2.1 million for his schemes in the California markets.[24]

A Smoking Gun

On May 6, 2002, in the context of the FERC investigation of possible manipulation of electricity and natural gas prices in California, counsel for Enron handed three memos to FERC staff. Although the memos were marked clearly with the stamp "Attorney/Client privilege"—the legal equivalent of a No Trespassing sign and barbed wire—Enron counsel waived all privilege. Two of the memos are titled "Traders' Strategies in the California Wholesale Power Markets/ISO Sanctions," and they appear to be near duplicates. The memos are dated December 6, 2000, and December 8, 2000. The purpose of the memos appeared to be for Enron to get ready for the investigations into the bad behavior that contributed to the California electricity crisis.[25]

The memos describe strategies actually used by Enron in California. One was called "Inc-ing" or "Fat Boy"; the California market rules have to be well understood to make Fat Boy workable. Enron (or any market participant) would have to come into the market with what is termed a *balanced schedule*, which means the electricity Enron intended to generate would need to be equal to the electricity it intended to provide to its customers: generation must equal customer need or what insiders call *load*. What Enron did was to essentially lie about how much customer load it had to serve. The memo gives an example: Say that Enron, a day ahead, claimed it had generation and customer load of 1,000 MW, when it knew the actual load would be only 500 MW. The next day Enron delivered the 1,000 MW of generation, then found that its load was just 500 MW, leaving 500 MW of excess generation. Enron would be paid for the excess generation in the real-time market, so it would artificially overstate its load only on days in which it believed the real-time price would be high. The payoff to overstating load a day ahead lay in the high-priced sales Enron would achieve in the real-time market the following day. How would Enron know that the real-time price the next day would be higher? Well, it would be likely if, while Enron was *overstating* its load, other market participants were actually *understating* theirs. And, in fact, the big California utilities were doing just that.[26]

This is intricate manipulation. Again, implementing Fat Boy took a deep understanding of the rules that were to be gamed as well as the insight that false information submitted by the utilities was necessary to make the strategy profitable. Consistent with the view that Enron was not alone, the memos suggest that Enron implemented the Fat Boy scheme on behalf of others. The memos name other companies that joined in, including government-owned power companies in both the United States and Canada.[27]

The Enron memos describe several other schemes. One outlines its "Export of California Power" scheme as follows: "As a result of the price caps in the PX and ISO (currently $250), Enron has been able to take advantage of arbitrage opportunities by buying energy at the PX for export outside California. For example, yesterday (December 5, 2000), prices at Mid-C [a major trading hub in the Northwest] peaked at $1200, while California was capped at $250. Thus, traders could buy power at $250 and sell it for $1200."[28]

There were further clever names and blunt admissions. "Death Star" was one scheme. Enron says that the net effect of Death Star was that "Enron gets paid for moving energy to relieve congestion without actually moving any energy or relieving any congestion." "Get Shorty" came next. The upshot of this one was that Enron sold something that it did not have to sell—essentially a specific power plant sitting idle, ready to quickly provide more power if customer needs soared. Enron itself recognized that this was explicitly against the rules. "Ricochet" was another scheme out-lined in the memos. With this strategy, "Enron buys energy from the PX in the Day Of market, and schedules it for export. The energy is sent out of California to another party, which charges a small fee per MW, and then Enron buys it back to sell the energy to the ISO real-time market" at a higher price.[29]

Pointing to the smoking gun memos, FERC investigators noted that Enron exhibited "great eagerness to experiment," to "game the system," or to "simply provide false information." FERC added, "We emphasize that the trading strategies—while bearing Enron's name—were not lim-ited to Enron but appear to have been widely engaged in by numerous parties. Indeed, it would appear to Staff that the majority of public utility entities and some nonpublic utilities engaged in at least some of the trad-ing strategies at some time during the 2-year review period." The staff goes on to list thirty-seven entities that "should be required to show cause why their behaviors did not constitute gaming."[30]

ACCOUNTING FOR SCARCITY

Beyond the increase in demand there was, in fact, an actual physical scar-city of natural gas in California at the time. In August of 2000 a major natural gas pipeline into Southern California ruptured. It was the El Paso Natural Gas Company line, which was one of three major interstate pipelines; it alone accounted for 65 percent of the transport capacity into Southern California. El Paso lost about one-third of its capacity for about two weeks in August, and then, under orders from the US Department of Transportation, it had to further reduce its capacity by about 8 percent for a prolonged period into the winter of 2000/2001.[31]

What happens to prices for any commodity—natural gas, electric-ity, copper, corn, and so on—when there is a physical shortage of either the commodity or the means to transport it? Let's assume that in the calm of the market equilibrium, the quantity of supply and demand are equal at one hundred units. Let us assume further that a shortage cuts that supply abruptly to sixty units. The central question is, who gets the sixty units?

There are two ways to allocate the supply. For convenience, we can call one way the "policymaker's way" and the other the "economist's way." With the policymaker's way, a government agency steps in to determine which of the original customers for the one hundred units actually gets a share of what they want; that is, some share of the sixty units that can be produced and delivered with the shortage. The government agency presumes the sixty units will be sold at the old market price. The economist's way is to let the market price do the allocating. This sounds coldhearted, but the way it works is that the price keeps rising until a critical mass of people say, "Enough, it is not worth it to me at that high of a price." It is necessary that customers who would have bought forty units drop out in this way. The crucial distinction is that the price in a market suffering a shortage will reflect the *willingness of customer to pay, not the cost of supply.* In a shortage, if the economist's approach is followed, the price needed to allocate the shortfall is well above the old market price.

Both the policymaker's and the economist's ways have merit. The policymaker's approach emphasizes fairness. The premise is that everyone is in this crisis together and should pay the same price, which should not necessarily increase because of the crisis. However, the economist's approach has advantages, too. If prices rise to clear the market, the people who get the sixty units are, in theory, the ones who value them most; that is, the ones willing to pay the most. Moreover, the economist's approach has in it the seeds of a solution to the crisis. Not only does the higher price drive customers to reduce their demand but it also drives suppliers to find ways to increase their supply. Reducing demand or increasing supply are ultimately the only ways to actually resolve a shortage.

A deliberate, balanced weighing of the economist's approach and the policymaker's approach can readily give way in a crisis to a full-throated politicization. The tale of how that happened in California is briefly outlined in this chapter, but a thorough version can be found in Arthur J.

O'Donnell's 2003 book, *Soul of the Grid: A Cultural Biography of the California Independent System Operator.*

GOVERNOR DAVIS IS RECALLED

In 2001, with California well into its electricity crisis, an interesting exchange appeared on the op-ed page of the *Sacramento Bee*, between two men deeply involved in the original California deregulation legislation and its implementation. The first was former governor Pete Wilson, who had signed AB1890. The other was David Freeman, the man asked to jump-start the implementation of AB1890 and the former general manager of the Los Angeles Department of Water and Power.

Governor Wilson's opinion was printed first, on June 3. He wrote, "State Government cannot ignore the law of supply and demand," and the "plain and simple reason" that California would "suffer hundreds of hours of blackouts" in the coming summer was that the state "will not have a sufficient supply of electrical power to meet demand." He faulted his successor, Governor Davis, for not acting decisively once the crisis hit. Wilson argued that Davis should have used his emergency powers to suspend the rate cap to allow utilities to secure long-term contracts, and speed the permitting process for new power plants. Had enough new power plants been built, according to Wilson, the state would have avoided the blackouts as well as the bankruptcy of Pacific Gas and Electric.[32]

A week later, David Freeman responded. He started by saying it was "embarrassing" for Governor Wilson to "absolve himself of blame" for the crisis with "his current who-me? campaign." He said that Governor Wilson could not "wash his hands" of the deregulation plan that had become California's "Frankenstein." The fatal flaws of the Wilson effort, wrote Freeman, included forcing the utilities to divest their fossil-fuel plants, which were bought by "out-of-state energy companies—several in Texas" who demanded "excessive prices." Another fatal flaw in the Wilson effort was to give control to the "toothless" FERC, which failed to impose needed price caps. He applauded Governor Davis' energy conservation efforts, which achieved an "11 percent reduction in electricity use in May," and for finalizing "nearly 50 long-term power supply contracts." Freeman argued that it was Governor Davis who successfully got power plants to be proposed and to start construction under "fast-track" process rules. Governor Davis also had created a "California public power authority" that could build power plants if the private sector did not.[33]

Freeman closed by calling Governor Wilson a "free-market ideologue" who pushed a deregulation plan that he "acknowledged was screwed up." Then he asked Governor Wilson to help Governor Davis by calling on President George W. Bush "to support the governor's effort to bring sanity to the wholesale power market by cracking down on the gouging out-of-state generators."[34]

These op-eds offer a succinct example of the conflicting views held by those on the ground during the crisis. They reveal the philosophical divide between those who attribute the crisis to bad market design and bad market conditions (Wilson) and those who see only bad market behavior (Freeman). So, too, they reflect the divide between the economist's approach to resolving a shortage (Wilson's call for an end to the rate caps) and the policymaker's approach (Freeman's focus on conservation).

In October 2003, just over two years after this op-ed exchange, Governor Davis lost a recall election and was replaced as governor by Arnold Schwarzenegger. It was the first-ever recall election of a governor in California; although several other governors had been threatened with a recall, proponents had not previously acquired the requisite number of signatures on the petition. At the time it was only the second gubernatorial recall in United States history.

ENRON'S DIRTY DEALS

There is widespread perception that the financial collapse of Enron was caused by Enron's dirty deals during the California electricity crisis. This is inaccurate for three reasons. First, Enron's bankruptcy was triggered by the significant restatement—meaning, revision and publication—of its financial accounts. Those restatements, in turn, were triggered by the unraveling of long-term financial deals. The bankruptcy, therefore, had little if anything to do with Enron's attempts to manipulate electricity markets in California or elsewhere. Second, the faulty accounting treatments of these long-term deals could have been and were readily applied to any business—there was nothing peculiar to the electricity business. Third, in the trial of Enron's two top executives, the argument made by the winning prosecutors was that the collapse of Enron was "not about accounting, it is about lies and choices." That is, it was about those executives covering up the true financial condition of the firm.[35]

The restatement of Enron's accounts deserves further elaboration here, as it had far-reaching effects on the rest of the American economy. On October 16, 2001, Enron announced large reductions in Enron's financial stature related to transactions between Enron and an investment vehicle called LJM2, which had been created by Andrew Fastow, Enron's chief financial officer. The acronym stood for Lea Jeffrey Michael, the names of Fastow's wife and children, and the number symbolized that this was the second in the series of entities created by Fastow for Enron. On October 22 storm clouds began to gather when the US Securities and Exchange Commission started to ask questions about LJM2. Two days later Enron announced that Fastow would take a leave of absence and be replaced. The board of directors then announced its own investigation of the "related-party transactions" like JLM2, to be led by William C. Powers Jr., the dean of the University of Texas School of Law.[36]

The Powers Report used plain language, but it did not mince words when presenting its findings. It started by naming Enron employees who enriched themselves at the expense of the company. The report then went after the accounting schemes, which "were used by Enron Management to enter into transactions that it could not, or would not, do with unrelated commercial entities." Had there been unrelated commercial entities on the two sides of these transactions, that would have been evidence that the transaction did something real to benefit both partners. With Enron on both sides, the transactions were used only to artificially make Enron financial statements appear healthy.[37]

As an example of faulty accounting, the Powers Report pointed to the "Chewco" transaction, Fastow's use of a *special-purpose entity* (SPE) run by Enron personnel to keep an investment off its balance sheet. SPEs are common in the electricity business as well as in other businesses. When deployed appropriately they are used to allocate and isolate risk, but the Powers Report found they were not used to that end. Moreover, the Powers Report stated that to legally use an SPE, the transaction must meet two straightforward standards. The first is that "an owner independent of the company [independent of Enron] must make a substantive equity invest-ment of at least 3% of the SPE's assets, and that 3% must remain at risk throughout the transaction." The second is that "the independent owner must exercise control of the SPE." The Powers Report makes it clear that Enron's SPEs did not meet these two basic requirements.[38]

The US Senate Permanent Subcommittee on Investigations, of the Committee on Governmental Affairs, issued its own report on July 8, 2002. The subcommittee framed it as an expansion of the Powers Report, and the title of the report says a lot about the focus of the Senate's investiga-tion: it is called "The Role of the Board of Directors in Enron's Collapse." The Senate panel started by documenting the extensive experience and credentials of the members of the board. Then the panel concluded, in no uncertain terms, that the Enron board failed to do its job. Specifically, the Senate report stated the board allowed "Enron to engage in high risk accounting, inappropriate conflict of interest transactions, extensive undisclosed off-the-books activities, and excessive executive compensa-tion." They noted that Enron's accounting schemes were so aggressive that even the company's independent auditor, Arthur Andersen LLP, debated whether keeping Enron as a client was prudent.[39]

The Senate makes it clear that the collapse was not about missing the 3 percent equity requirement for an SPE. It was about transactions that fundamentally had no true value, certainly not in terms of shield-ing Enron shareholders and employees from risk. And it showed that the standard protections against financial manipulation did not, in practice,

protect anyone. Specifically, having an independent board and an independent auditor did not protect against Enron's financial collapse.[40]

ARTHUR ANDERSEN'S COLLAPSE

Enron's auditor, Arthur Andersen LLP, was one of the largest professional
service companies in the world with eighty-five thousand employees and
$9.3 billion in revenue. Its ninety-year run ended when it was convicted of
obstruction of justice on June 2002; its downfall was directly related to its
shredding documents relevant to the investigation of Enron's bankruptcy
and collapse. Describing the extent of the shredding, the *Chicago Tribune*
reported that Andersen's "housecleaning would go on virtually around
the clock."[41]

The *Chicago Tribune* further noted that Enron came at the end of a
"decade of sliding standards and audit debacles" at Andersen. The article points to the conflict of interest between Andersen auditing the books
and, at the same time, doing other consulting work for the client, as was
the norm with Andersen and Enron. In hindsight, both the lucrative consulting fees and the closeness of Andersen and Enron employees contributed to lax accounting standards when Andersen was asked to approve
special-purpose entities. Even worse, it seemed that Enron had the power
to push out those at Andersen who did not support the accounting schemes
Enron wanted to use.[42]

Andersen argued all along that their conviction was the result of "over-
zealous prosecutors and a dishonest client." And, indeed, in June 2005,
the *New York Times* reported that the US Supreme Court, in "a pointed
and unanimous opinion . . . overturned Arthur Andersen's conviction for
shredding Enron accounting documents as that company was collapsing
in one of the nation's biggest corporate scandals." The *Times* went on to
report that the reason for the reversal was that "the court held that the
trial judge's instructions to the jury failed to require the necessary proof
that Andersen knew its actions were wrong."[43]

THE TRIAL: "ABOUT LIES AND CHOICES"

In his opening statement in the trial of Enron's top two executives, the federal prosecutor, John Hueston, said, "This is a simple case. It is not about accounting, it is about lies and choices." That set the tone. The prosecutor's case would *not* be a painful lesson in accounting and finance; the jurors would *not* be asked to go through a primer on SPEs and the like. Rather it would be about bad guys doing bad things to unwitting investors.[44]

Hueston told the jury he would take them "inside the doors of what was once the seventh largest corporation in this country, Enron." And once inside those doors the jury would find that the two people at the very top, Ken Lay and Jeff Skilling, lied repeatedly. Hueston said, "In the year before Enron declared bankruptcy, two men at the helm of the company told lie after lie about the true financial condition of Enron, lies that propped up the value of their own stock holdings and lies that deprived the common investors of information that they needed to make fully informed decisions about their own Enron stock."[45]

Hueston briefly mentioned the LJM deals to the jury, but mainly to make the point that these were not deals with people independent of Enron. Hueston cleverly came up with a simple image for the asset sales in special-purpose entities. He likened one kind of transaction to the purchase of a truck for $25,000 that gets put on the books. He went on to explain, "The truck gets driven around. It gets into an accident, blows out a couple of tires and starts to rust . . . It's now worth about $5,000." Hueston explained that central to the "magic" was the fact that Fastow, who ran the deals, never wanted to see the true value of the assets. He was not a true independent buyer who would put the assets through a legitimate test to verify what their value really was.[46]

Hueston, asking for guilty verdicts on all counts against Lay and Skilling, closed with the same point he made at the beginning: "Ladies and gentlemen, lies and choices. It's a simple case."[47]

Separate counsel represented Lay and Skilling, with Mike Ramsey speaking for Ken Lay. He had a clear explanation for Enron's bankruptcy.

In its wholesale division, Enron did transactions worth "literally millions of dollars a day" and did so "with the trust of a phone call." "When that trust was eroded," Ramsey said, "that machine froze up." With the uncertainty surrounding Enron in the fall of 2001, Enron's trading partners said, "We're not going to trade with you on a phone call anymore. You're going to have to post full collateral." No company in America, opined Ramsey, "could have withstood that call for cash." Enron was no exception.[48]

In defending Ken Lay against charges of criminal behavior, Ramsey said it was all about Fastow and his inner circle; only the Fastow gang had known about the accounting gimmicks. Ramsey asserted, "He hid it. He hid it from the board of directors. He hid it from Ken. He hid it from Jeff. He hid it from everybody . . . He was a crook. He was a liar. He was deceitful." Even though Fastow really only stole the equivalent of "coffee money," Ramsey argued, it was enough to cause trading partners to "stampede."[49]

Responding to the prosecutor's central point that Lay and Skilling deceived everyone because of their "choices and lies," Ramsey pointed out that a lot of other people were involved and "70 percent of them are very, very sophisticated people." He added that the auditor, Andersen, "actually existed and officed in the Enron building."[50]

In the wake of its financial collapse, Enron's two highest-ranking corporate officers were convicted on multiple criminal counts. Ken Lay, Enron's founder, was found guilty on six counts of fraud and conspiracy and four counts of bank fraud. While awaiting sentencing, Lay, age sixty-four, died of coronary artery disease just about forty days after his conviction. Jeff Skilling, who resigned as Enron's chief executive officer just before Enron plunged into bankruptcy, was convicted on eighteen counts of fraud and conspiracy and on one count of insider trading; he was acquitted on nine other counts of insider trading. Skilling was sentenced to twenty-four years; his sentence was later reduced by ten years.

The *Houston Chronicle* reported that thirty-two people and one company had federal charges brought against them. Skilling received the longest sentence by far. Andrew Fastow, the chief financial officer said to have created many of the financial schemes that led to Enron's bankruptcy, ultimately received a six-year sentence in a plea deal that covered both him and his wife and required him to testify against Lay and Skilling.[51]

THE LEGACY OF THE CRISIS AND THE COLLAPSE

The primary legacy of the California crisis and the collapse of Enron is a prejudice against efforts to create a competitive electricity business—in particular, efforts to deregulate at least the generation sector of that business. That prejudice persists today. It is unfairly placed, since it seems clear that bad behavior by Enron and others was not the only cause of the California crisis. The wide-angle lens of history shows that there were three major causes; bad market design and bad market conditions should be taken into account, too.

Yes, bad market behavior played a role. And that should remind the country of the famous line attributed to Herbert Hoover: "The trouble with capitalism is capitalists; they're too damn greedy." Hoover's quote, sadly, is still an apt way to characterize the behavior of some in the crisis. What kind of businessperson deliberately "churns" sales and purchases and knowingly provides false information to manipulate price indices that, in turn, are used widely as the prices charged in natural gas and electricity supply contracts? If businesses of any sort want consumers to trust markets—to trust capitalism—the businesses themselves must be trustworthy.[52]

A consequence of this bad behavior is that, today, both federal and state regulators focus on tighter market rules and more aggressive enforcement to catch the bad guys. While enforcement has merit, this focus diverts attention from ensuring good market conditions prevail—essentially, making sure enough of the right kind of power plants are in place and able to operate. To this end, regulators must balance the interests of power plant development against the possible impediments of more and more detailed market rules.

Another lesson of the California crisis is that presumptions can do great harm. The most serious mistake in market design—forcing the utilities to sell to their customers at a *fixed* price and to buy from their suppliers at a *volatile* price—was based on the presumption that the electricity price in the spot market would always be below the price cap of sixty-five dollars. Clearly, someone should have tested the system under the presumption that the opposite might happen—just as it did in reality.

The taint of the crisis, however, did not derail deregulation. Restructuring of the electricity business to allow for competition continued apace in large parts of the United States. Indeed—perhaps amazingly—California is one of those parts. Why didn't the California electricity crisis derail deregulation elsewhere? One reason may be that the evidence identified the crisis as inherent to a particular time and place. The most persuasive evidence is that similar crises did not occur elsewhere, even though markets were restructured elsewhere.

Also important is the fact that Enron's financial collapse was not solely because of its participation in deregulated markets in California or elsewhere. Enron's collapse was driven by accounting gimmicks that could have been and were readily used on commodity businesses other than electricity. The open space of a partially deregulated electricity business was not more fertile ground for abuse than any other commodity market.

Another lesson of Enron's collapse is that the protections put in place did not protect against the far-reaching fallout of the crisis. Auditing by one of the world's largest accounting firms—Arthur Andersen—did not protect the public from the consequences of Enron's far-too-aggressive accounting and financial moves. An experienced board of directors gave no protection, either.

More important, though, is the fact that the typical protections did not protect the bad guys. The defense attorney in the Enron case said that nothing Enron did should be judged to be illegal because everything it did was checked by very smart outside lawyers, accountants, and investors. The jury must have tossed out that view from the start. The verdicts sent a clear warning to those who rely on procedural protections rather than on creating value for customers. Looking back to Samuel Insull's acquittals only makes this warning even clearer. The acquittals in Insull's trials were based on a much simpler accounting of good offsetting bad. Insull's good was that he made electricity available and affordable to all. The jury must have valued this highly enough to offset the bad of the murky accounting he used for his holding companies. It is the absence of this kind of public good that, in the end, led to many of the Enron convictions.

CHAPTER 16

Competition's Steady Rise

> The literature of public regulation is so vast that it must
> touch on everything, but it touches seldom and lightly on
> the most basic question one can ask about regulation: Does
> it make a difference in the behavior of an industry?
> —George Stigler and Claire Friedland (1962)[1]

The 2003 electric blackout followed hard on the heels of the California crisis and Enron's collapse. It shut off power to fifty million people, bringing the number of deregulation-related disasters to three. Why was deregulation—better termed *competitive reform*—not killed off by these three significant, negative events? Did people conclude that what happened in California and to Enron was the exception, not the rule? Did they discern that the 2003 blackout was due to long-standing problems with trees, tools, and training, not competitive reform? Or was there a compelling logic to competitive reform that gave it political, economic, technological, or cultural momentum? What's clear is that, for whatever reason, competitive reform did have momentum. It marched forward nationwide for decades before the California crisis, Enron's collapse, and the blackout and, although stalled in some ways, the march continues today. This chapter tells the story of the steady rise of competition in America's electricity business.[2]

THE END OF ECONOMIES OF SCALE

A belief in and regulation of economies of scale produced the golden age of regulated, monopoly electric utilities. Fittingly, the end of economies of scale ushered in an era marking the end of those monopolies.

The FERC writes that the story began with the Federal Power Act of 1935, which was "enacted in an age of mostly self-sufficient, vertically integrated electric utilities." Into the 1960s, the FERC reports that price increases were not necessary because of economies of scale, technological change, and tame input prices. Given this, there was "no pressure" to restructure the electricity business to allow competition.[3]

Major change came in the late 1960s and 1970s when electricity consumers faced rising electricity bills in an era of inflation and high interest rates. FERC noted that nuclear power cost overruns also pushed rates up, as did the fact that utilities overestimated demand, leaving excess capacity to be paid for and thus driving rates up even more. FERC reported that "between 1970 and 1985, average residential electricity prices more than tripled in nominal terms, and increased by 25% after adjusting for general inflation." For industrial customers it was even worse: a quadrupling in nominal terms and an 86 percent increase in inflation-adjusted or "real" terms.[4]

By FERC's reckoning the 1970s marked the end of economies of scale. In the 1960s bigger meant cheaper. Therefore, big, vertically integrated utilities were able to build big and gain a competitive price advantage over smaller entities. In the following decade bigger utilities hit the ceiling of economies of scale, and bigger began to mean higher operating costs and longer downtime. As FERC put it, "Bigger was no longer better."[5]

CREATING COMPETITORS

The Public Utility Regulatory Policies Act of 1978 (PURPA) was the first major milestone for competitive reform. Interestingly, reform was not its intent. PURPA came out of the upheaval caused by the Arab oil embargo in the early 1970s. The embargo put Americans into long gasoline lines because of shortages, and it threw Americans out of work because of the recession it caused. Suddenly, energy policy became important, and its central motive was to cut US dependence on foreign oil and thereby make the United States less vulnerable to another embargo. Thus, geopolitics drove energy policy. PURPA was a small part of President Carter's National Energy Act and, as such, a small part of the effort to reduce dependence on foreign oil. It was meant to motivate the use of designated technologies to produce electricity that either used oil and natural gas more efficiently or those that did not use oil at all—solar, wind, and waste power, for example.[6]

Power plants using any of the designated technologies were called *qualifying facilities* (QFs). They included cogenerators: often big industrial plants that *co*generated in the sense that they could use the same fuel to make both steam needed for industrial processes as well as electricity used at the industrial site or sold back to a local utility. The incentive for QF investment was that the local utility was required to buy that electricity at "avoided cost"—the cost the utility would incur had it produced the electricity itself.

The controversy over setting these avoided-cost prices can be readily imagined: the local utility would want that number to be low, while the QFs would want it to be high, and any price estimate would be based on a lot of assumptions that could be challenged. In this context, it is important to discern another motive for PURPA: the search for an alternative to utility monopolies building nuclear power plants with the attendant cost overruns and poor performance. Sometimes that motive would be revealed in the form of high avoided costs based on the high cost of nuclear power. More important was the fact that many states would willingly endorse long-term contracts for power sales from QFs based on forecasted avoided costs that would be locked in for ten or twenty years. Those power purchase contracts became an essential element for financing a QF. They literally could be "taken to the bank."

PURPA's success in attracting new entrants made it the first major step to opening the electricity business to competition, but this opening was inadvertent. Soon after developing QFs became popular in the early 1980s, states would find that far more QF power capacity was being offered than was needed, sometimes at a ratio of ten to one. So, rather than offering to pay full avoided cost, the states asked that the QFs compete on price by having them bid a discount off avoided costs. This is what brought forth the most basic element of any competitive market: competition on price.

PURPA also ushered in a complete change in the preferred technology. Under traditional cost-plus-profit regulation, the preferred technologies were those with the highest capital costs because the larger the capital cost, the higher the basis on which profit would be calculated. Coal and nuclear power plants were built with this model in mind. With PURPA, all of that changed. The preferred technology now became one with a relatively low capital cost: a natural gas–fired combined-cycle power plant. With combined-cycle technology, electricity can be produced twice with the same fuel, which fit well with PURPA's goal of promoting more fuel-efficient technology. In the first phase (or cycle), electricity is produced by burning natural gas in a combustion turbine—akin to a jet engine. That combustion turbine is hooked into an electric generator, thus creating the first round of electricity. The second phase, or second cycle, would take the exhaust heat from the turbine, which came out at a very high temperature, and use it to make steam from water in what is called a waste heat boiler. Then that steam would be sent through a steam turbine, also connected to an electric generator, to make electricity a second time.

Natural gas–fired combined-cycle technology mitigated many of the problems of the times: the 1973–74 Arab oil embargo, nuclear power cost overruns and poor performance, and an increasing concern with air pollution. Combined-cycle technology used natural gas (or oil) far more efficiently than other technologies, making it appealing to policymakers eager to do damage control following the Arab oil embargo. To address the concerns about the soaring costs of nuclear power, combined-cycle project developers were quite willing to offer a fixed price or a fixed-formula price. By offering a fixed price, these new entrants took on the risk of capital cost overruns; that is, the price simply stayed fixed and would go no higher if the developer suffered higher costs. A fixed price also meant the supplier took on the risk of poor performance and only got paid if the combined-cycle plant reliably delivered electricity.

To be fair about the risk of capital cost overruns, it was a lot easier for a combined-cycle plant than a nuclear power plant to take responsibility for this risk. Just one-third of the total cost of electricity from a combined-cycle plant can be traced to capital cost while that share for a nuclear power plant is much greater. The other two-thirds were attributed to the cost of the natural gas used in the facility. While the capital cost portion of the price might be fixed, the combined-cycle facility typically offered a price to cover fuel costs that would go up and down with the market price of natural gas. That "indexed" or "floating" price to cover fuel costs was seen as a good thing because it reflected the up-to-the-minute market price of natural gas; however, it also meant that a rapid rise in natural gas market prices would mean a rapid rise in electricity prices—as evidenced by the price spikes during the California crisis.

From an environmental perspective, combined-cycle facilities have far fewer air pollution emissions than coal-fired power plants. They emit virtually no sulfur dioxide nor the particulates emitted with coal. And, while combined-cycle plants do emit nitrogen oxides, equipment manufacturers—including Westinghouse, General Electric, and Siemens—were able to make steady reductions in such emissions during the late 1980s and into the 1990s.

PURPA was crucial to starting competitive reform because it created competitors—the QFs. There can be no competition without competitors. Moreover, those competitors offered solutions to the energy supply and environmental concerns of the time. And these new competitors became an interest group that could push for even more competitive reforms.[7]

OPENING ACCESS

The next frontier for these newly created competitors was to secure the transmission service needed to deliver the electricity they generated. Their competitors—the vertically integrated utilities—controlled the transmission system, both in terms of how it was operated (for example, who got to use it) and whether it would be expanded to accommodate new generators. Vertically integrated utilities were the norm; they could be found across the country—Southern California Edison, Middle South Utilities, New England Power, and so on. The primary complaint against them was that vertically integrated utilities used their control of the transmission system to favor their own electric generation when competing against generation by the new competitors. To put that complaint in perspective, consider the case of a municipal utility that agreed to buy electricity under a long-term contract with a new competitor. If the local utility wanted to serve that municipality instead, it effectively could deny transmission service to the new competitor by claiming that the transmission system was fully booked and that the cost to expand the system made the competitor's power too expensive. There was no law or regulation at the time that required the utility to give competitors equal access to transmission service.

It was clear from the start that, to enable the newly created competitors to compete, some form of open-access transmission policy would have to be put in place. In the late 1980s, FERC began a campaign to create open access on a case-by-case basis. If, for example, two vertically integrated utilities came to the FERC seeking approval for a merger—an approval required under the Federal Power Act—FERC would condition its approval of the merger with a requirement that the utilities give at least a rudimentary form of transmission open access to all parties, including new competitors. FERC also used utility requests for so-called market-based rates to promote open access. If FERC granted a request for market-based rates, a utility would be allowed to charge a price other than the cost-based price under traditional regulation—but it had to offer open access first.

FERC's case-by-case push for open-access transmission helped build the momentum behind competitive reform. Congress contributed to that momentum during the same time frame with the Energy Policy Act of 1992 (EPAct 1992). This widened the range of technologies that new entrants could use to compete. Instead of being limited to the QF technologies under PURPA, a new category of competitors, called exempt wholesale generators (EWGs) was created and they could build a full array of stand-alone power plants. Now, for example, a new entrant could build a stand-alone natural gas–fired combined-cycle power plant and not be limited to the small size or unique design required for a cogeneration plant.[8]

By 1996 FERC was ready to codify the open-access transmission requirements that it had been establishing case by case. It formalized that with its Order 888. The choice of the numerical designation 888 showed, in an insider's way, that this order was a big deal; FERC's address for its brand-new headquarters building in Washington, DC, was 888 First Street, N.E. Before FERC announced its draft order in 1995, it reported that "38 public utilities had filed wholesale open access transmission tariffs." Presumably, this was the progress made under its case-by-case campaign. In the first year after the draft of Order 888 had been announced, the tally had grown substantially: FERC reported, "106 of the approximately 166 public utilities that own, control, or operate transmission facilities used in interstate commerce have filed some form of wholesale open access tariff." This evidence showed that even the promise of a commission order was faster than a case-by-case approach.[9]

The purpose of Order 888 was to ensure that public utilities "cannot use monopoly power over transmission to unduly discriminate against others."[10] Preventing undue discrimination was, and is, the "legal and policy cornerstone" of Order 888. To do that, Order 888 required all public utilities that owned or controlled transmission lines to file an open-access transmission tariff. And, importantly, those public utilities would have to take transmission service under the same tariff and abide by the same rules.[11]

As always, states' rights were a central concern. FERC noted that the Order "clarifie[d] Federal/state jurisdiction over transmission in interstate commerce," and claimed to provide "deference to certain state recommendations." Later in the order, FERC added that while "jurisdictional boundaries may shift," the order would not "change fundamental state regulatory authorities." Specifically, states would continue "to regulate the vast majority of generation asset costs" as well as "the siting of generation and transmission facilities."[12]

Four years later, FERC was ready to take the next step to promote competitive reform in the electricity business by taking open access to the next level. This was achieved with Order 2000. FERC took this major next step because the open-access transmission tariffs from individual utilities were not enough to end discrimination in granting transmission service. For example, FERC pointed to the inconsistencies in determining how much transmission capability was available. The remedy in Order 2000 was to take operational control of the transmission system from individual utilities and place it in the hands of a new, independent entity: a regional transmission organization (RTO). [13]

RTOs are a remarkable step forward in competitive reform. They break down the walls separating the old monopolies and create something akin to a large free-trade zone for electricity. Everyone plays by the same rules in that free-trade zone and everyone has a say in what those rules should be. In addition to transparent rules for gaining transmission access, RTOs also create a transparent market for real-time or "spot" electricity sales. Importantly, that spot market creates essential liquidity in the sense that a supplier always has the opportunity to sell into the RTO's transparent spot electricity market, rather than having to negotiate with a utility for a bilateral sales contract.

Still, RTOs do not mean deregulation. They bring another layer of governance to the regional market with an independent board of directors. That board takes votes and FERC gives considerable weight to those votes, but FERC has the ultimate say.

FUELING REFORM

Natural gas plays a central, supporting role in regulatory reform of the electricity business in two ways. First, regulatory reform of the natural gas business was a template for reform in the electricity business. Second, from the start, and continuing to today, natural gas is the fuel of choice for the new competitors that made reform happen in the electricity business.

Natural gas regulation began and evolved at essentially the same pace and in the same way as electricity regulation. Very early on, when gas was mostly manufactured from coal and was therefore a local affair, municipal governments regulated the industry. As sales of that manufactured gas crossed municipal boundaries, the states took over the regulatory role. Then, when natural gas supply and transport crossed state lines, the states were told by the US Supreme Court that they could not regulate interstate sales of natural gas—the same ruling the Supreme Court gave regarding interstate electricity sales. And as with the electricity business, it was the FTC's report that substantiated concerns about concentrated ownership and the potential for unnecessarily high natural gas prices.

The Natural Gas Act of 1938 began the regulation of interstate sales and transport of natural gas. It did so by requiring just and reasonable rates for interstate natural gas sales including sales for resale. The act also required that all new natural gas pipelines win approval from the Federal Power Commission—the predecessor of the FERC.

However, it took several court cases to get the federal government more fully into the regulation of natural gas prices at the source—or, as was commonly said, at the "wellhead." The Supreme Court ruled in *Phillips Petroleum Co. v. Wisconsin* that all natural gas sold into interstate pipelines would be regulated. Three failed efforts to regulate natural gas prices at the wellhead followed, and these failures led ultimately to the deregulation of those prices. In the first, the federal government attempted to regulate price supplier by supplier; the sheer number of suppliers led to failure. In the second, the federal government attempted price regulation by region, but this, too, failed because cost-plus regulation did not work for natural gas supply. Not every well drilled will produce usable natural gas, which changes the game entirely. If the cost-plus rate is based only on the cost of successful wells, the added cost of failed exploration would be ignored and no one would explore for natural gas. The third attempt was to set a nationwide price for natural gas—at forty-two cents per million cubic feet. While that price was double the previous regulated price, it still was well below the price a free market would yield. And for that reason it led to continued shortages—an artificial gap between demand and supply.[14]

The Natural Gas Policy Act of 1978 was Congress' response to the shortages. This act was part of the same National Energy Act that included PURPA. The Natural Gas Policy Act set in motion the forces that would lead to the full deregulation of natural gas prices at the wellhead. To start, it took the tactic of setting prices by vintage. That is, prices for new natural gas supply were higher than prices for natural gas developed before the act. Having been starved for natural gas in the era of shortages, pipelines rushed to sign long-term, fixed-quantity, fixed-price contracts with suppliers—called take-or-pay contracts. At first, prices increased after the act, but those high prices decreased demand and increased supply so that natural gas prices then fell. Over time the spot market price fell well below the high prices in the take-or-pay contracts. With this take-or-pay overhang, customers clamored for the right to buy their natural gas on their own so they could access lower spot prices.[15]

In response to this clamor, FERC issued Order 436 in 1985. This important order allowed the major interstate pipelines to voluntarily offer unbundled natural gas transport service; that is, interstate transport would be offered separately from the supply of the natural gas commodity as well as independent of other related services such as gas storage. Order 436 became known as the "Open Access Order." The change brought on by Order 436 was immediate and important. FERC reported that, in 1984, only 8 percent of the throughput of natural gas in pipelines was shipped with unbundled pipeline service. After Order 436, however, that share rose to 79 percent by the early 1990s. Congress then passed the Natural

Gas Wellhead Decontrol Act in 1989; this legislation ultimately would fully deregulate wellhead prices.[16]

In 1992 FERC's Order 636 addressed the remaining obstacle to natural gas competition. The obstacle was that pipeline service taken under open-access rules was not equivalent to that embedded in bundled service. Order 636 fully unbundled pipeline transport from commodity sales and also required comparable service from transmission-line owners to all gas suppliers.[17]

The data overwhelmingly showed that natural gas had become the fuel of choice for new competitors. By 1995 electric utilities used natural gas for just 10.3 percent of their electricity generation. By sharp contrast, the new entrants—termed *nonutility producers*—used natural gas for 56.2 percent of their electric generation. As for other fuels, electric utilities used coal for 55.2 percent of their generation that year, and nuclear for 22 percent. Again, in sharp contrast, the second-ranked fuel source for nonutilities was renewables, which accounted for 22.9 percent of generation, and the third was coal, which accounted for 15.4 percent.[18]

These statistics give powerful, real-world evidence that the method of price regulation drives fuel choice for electric suppliers. Regulations are not neutral because cost-plus rates give big incentives to build huge coal and nuclear power plants. Market-based rates give incentive for smaller-scale, versatile natural gas–fueled power plants; combined with significant tax incentives, they also provide incentives for nonrenewable fuels. New regulations mean new fuels, and that understanding is essential when designing policies for the future.

WRITING DOWN THE UNWRITTEN

Competitive reform is, as we have seen, different from hands-off deregulation. What changed with competitive reform, and what remained the same? The short answer for the latter question is that with or without competitive reform, a complicated electric system must be operated at the knife's edge; the amount of power supplied must equal exactly the amount of power demanded moment to moment. As long as we use a transmission grid to connect distant power plants to serve dispersed customers—as long as we use *system* power—this will be true and rules will be needed. When the system was operated by one company, the local monopoly, few people knew the rules. Decisions about whether and how to run each of the power plants and how full to allow transmission lines to get were made in secure control rooms deep inside each monopoly's headquarters. There were rules, of course, but not rules that were frequently subject to public scrutiny.

What changed with competitive reform is that companies that owned power plants rightfully wanted to know how the decisions affecting their power plants would be made. Specifically, they wanted to know that the system would be run in a way that gave their power plants a fair chance to win or lose in the marketplace. Because of all these new players, the rules that were once unwritten now had to be written out, subjected to a thorough public vetting, and approved by regulators. The written rules for a competitively reformed electric system go into excruciating detail, but each detail has a purpose. The overall purpose is the same as it was with the monopoly control: to reliably keep the lights on and to do so at the lowest reasonable cost to the customers. The hoped-for payoff to opening the door to new competitors, of course, is that they will come up with new ideas that increase reliability and decrease costs.[19]

THE CHICAGO SCHOOL

Does Regulation Make a Difference?

A broad, analytic assessment of the effectiveness of regulation also contributed to the momentum of deregulation in America. It is at least somewhat ironic that the intellectual roots of deregulation would be planted in Chicago, the place where the roots of cost-plus regulation were planted several decades earlier by Samuel Insull. However, many of the central writings for the reassessment nevertheless appeared in the University of Chicago's *Journal of Law and Economics*.

One of the earliest and most important articles was written in 1962 by George Stigler and Claire Friedland of the University of Chicago; it was titled "What Can Regulators Regulate? The Case of Electricity." The authors start with a provocative, straightforward claim that the most basic question about regulation was seldom asked: Did regulation make a difference? Specifically, they asked: "The literature of public regulation is so vast that it must touch on everything, but it touches seldom and lightly on the most basic question one can ask about regulation: Does it make a difference in the behavior of an industry?" After extensive analysis, their conclusion was no, it does not make a difference in the case of the electricity business. Part of the problem, they wrote, was that observers assess regulation too often on what it *does* rather than what it *achieves*. Specifically, does it affect prices paid by ratepayers?[20]

Stigler and Friedland asserted that the price of electricity can only be affected by regulation if regulation affects supply and demand. The authors proceeded to conduct three quantitative tests of the effect of regulation by comparing results with and without—or, more precisely, before and after—regulation across the states. In the end, they wrote, "We conclude that no effect of regulation can be found in the average level of rates."[21]

An Unrelenting Reassessment

That article was the first in what would become an unrelenting series of studies questioning regulation, especially as applied to large utilities. In his notable 1968 article, "Why Regulate Utilities?" Harold Demsetz of the University of Chicago undermined the presumption that regulation was necessary because utilities were natural monopolies. He argued, in short, that a large number of *bidders* for the original monopoly franchise could lead to a competitive price even if the bidding resulted in a single *supplier*.[22]

In 1971 Stigler wrote another, equally influential article in the *Bell Journal of Economics and Management Science* titled "The Theory of Economic Regulation." This time, his scope went far beyond regulation of the electricity business to include a wide range of government interventions—money subsidies, tariffs, price controls, and the like. One of his insights was to write of the supply of and demand for regulation. In this second article, Stigler undermined the presumption that regulation was designed to benefit the public interest, arguing, "as a rule, regulation is acquired by the industry and is designed and operated primarily for its benefit."[23]

In 1974 Richard Posner, professor of law at the University of Chicago, picked up where Stigler left off. Posner focused on the challenge of finding a theory to explain "economic regulation"—more broadly, the pattern of "government intervention in the market." The options were a "public interest" theory, in which regulation comes from a public demand to correct "inefficient or inequitable market practices," or a "capture" theory, in which "interest groups" call for regulation that benefits them economically. Posner cited Stigler as the source of a new, improved, and more elegantly stated capture theory—the "economic theory of regulation."[24]

Writing in 1993, George Priest, a professor of law and economics at Yale, revisited the decades-long debate triggered by Stigler and Posner. While he offered reasons to challenge the work, Priest found that the debate "has exerted an extraordinary influence on public policy." The real-world experience in the electricity industry supports his conclusion. Over the ensuing twenty years public policy opened the door wide for competitive reform. It started by creating competitors in the 1970s and, at the time of his writing, by expanding the range of technologies these competitors could use while giving them fair access to the transmission grid. Extraordinary influence indeed.[25]

THE BLACKOUT OF 2003

On August 14, 2003, a large portion of the US Northeast and Midwest as well as Ontario, Canada, suffered a major electric power blackout. The blackout affected fifty million people. Electric customers in eight states (Ohio, Michigan, Pennsylvania, New York, Vermont, Massachusetts,

Connecticut, and New Jersey) were caught up in it, many of them left without light, refrigeration, air conditioning, or any of the many other benefits of electricity taken for granted. It took up to four days to restore power in the United States and outages persisted in Ontario for more than a week. The cost of the blackout in the United States was estimated to be in the $4 billion to $10 billion range.[26]

The 2003 blackout is important to the narrative here because some attributed it to the deregulation of the American electricity business. President George W. Bush and Canadian prime minister Jean Chrétien called for an investigation by a joint US-Canadian task force; their report was issued in April 2004, and it ultimately concluded that "restructuring was not the cause of the August 14, 2003 blackout." Still, even after this conclusion was made public, opinions varied.[27]

In a follow-up report, *The Relationship between Competitive Power Markets and Grid Reliability*, the task force called on ten diverse industry leaders and technical experts to write issue papers. As expected, the range of opinions was diverse. The first issue paper was written by three authors who added to their bylines the designation "Power Engineers Supporting Truth." Their conclusion was simply and boldly stated: "Deregulation and the concomitant restructuring of the electric power industry in the US have had a devastating effect on the reliability of North American power systems, and constitute the ultimate root cause of the August 14, 2003 blackout."[28]

In contrast, the North American Electric Reliability Council (NERC), the long-standing industry group running the then-voluntary program for grid reliability, was much more optimistic about the future. Still, NERC acknowledged challenges and the need for change. For example, it noted that "construction of new transmission has been inhibited by the uncertainty associated with financing and cost recovery as well as local resistance to siting and building new transmission facilities." Most importantly, NERC concluded that the old voluntary system could not ensure reliability: "NERC rules must be mandatory for all users of the bulk electric system."[29]

What were the actual causes of the blackout? As analysis around it grew, a catchphrase emerged; it became popular to say that the causes were "the three *t*'s: trees, training, and tools." For example, in its investigations of the 2003 blackout NERC concluded that the utility "did not adequately manage tree growth in its transmission rights-of-way." NERC pointed to "training deficiency," which meant there was not enough "situational awareness" of the emergency. And it cited the failure of the utility to "control computers and alarm systems" and the regional transmission organization's "incomplete tool set." While the task force acquitted deregulation, its conclusions were troubling nonetheless. The neat phrase "trees, training, and tools" suggests that the blackout was caused by fundamental problems.[30]

THE ENERGY POLICY ACT OF 2005

The Energy Policy Act of 2005 (EPAct 2005) was Congress' response to a pile-up of crises. It was meant to address the 2003 blackout as well as the California electricity crisis and the collapse of Enron (related to the manipulation of electricity markets). The FERC certainly saw the law as a major milestone. It wrote, "The Energy Policy Act of 2005 (EPAct) is the first major energy law enacted in more than a decade, and makes the most significant changes in Commission authority since the New Deal's Federal Power Act of 1935 and the Natural Gas Act of 1938." In addition, it wrote that, with the act, "Congress signaled a strong vote of confidence in the Commission."[31]

That confidence was reflected in new responsibilities for FERC to oversee "the reliability of the nation's electricity transmission grid" in the wake of the 2003 blackout. As FERC wrote, "Today's rulemaking heralds a historic new turn in the Commission's regulatory responsibilities." FERC emphasized that the problem had been that reliability standards were voluntary in the past, but going forward would be mandatory and enforceable: "Every major regional blackout since the 1960s was caused in part by violations of voluntary, unenforceable reliability standards." The 1965 blackout was the largest to date, affecting thirty million people with outages lasting up to thirteen hours, and it led to the creation of a voluntary NERC. Proposals for mandatory federal standards came in the wake of the New York City blackout in 1977 and the major blackouts in the western United States in 1996, but the proposals were unheeded. FERC said that "with these important new rules from EPAct 2005, compliance with reliability standards is not optional, but a legal requirement subject to substantial civil penalties."[32]

To define and enforce reliability standards, FERC established a new institutional framework. Overseeing all of this would be a newly formed Electric Reliability Organization (ERO). In July 2006, FERC certified the North American Electric Reliability Corp. as the ERO. FERC also defined a regional aspect to the new institutional framework. In part to allow regional flexibility, NERC would work with eight regional entities. FERC had to approve all reliability standards proposed through the institutional framework, and began by reviewing 102 reliability standards proposed by the ERO.[33]

FERC also gained enhanced powers to address the kind of bad behavior seen in the California electricity crisis. For example, FERC made it clear that government-owned utilities were now included under its enforcement. The new law, wrote FERC, covered any manipulation with respect to commodities or transport. Notably, the act increased FERC's enforcement authority by "increasing the maximum civil penalty under these statutes to $1 million per day per violation." Finally, as if to note a new era, FERC explained that the act also repealed the seventy-year-old Holding Company Act and also allowed "termination of the PURPA mandatory purchase obligation."[34]

EPAct 2005 sent two distinctly different messages. To keep the lights on—to assure reliability—more federal law and regulation was needed. However, to keep electricity prices low and to encourage innovation, old-school federal regulation of generators should be scrapped in favor of using competitive markets governed by the threat of huge penalties for bad behavior. It was not immediately clear how these two messages—one calling for more regulation and the other less—could be reconciled and put into action.

THE LEGACY OF COMPETITIVE REFORM

Einstein wrote that new theories gained currency when old theories failed to work. Something akin to that happened in the electricity business. After decades of regulated monopolies delivering lower rates, in the 1970s and 1980s rates went up. Bigger was no longer better because the monopoly utilities had no ready solutions for the problems of the day: cost overruns and poor performance at nuclear power plants, the Arab oil embargo, and growing concerns over air pollution.

Apart from these burgeoning challenges, three widely publicized, negative events could have derailed the deregulation of America's electricity business: the California electricity crisis, the collapse of Enron, or the 2003 blackout. Taken together, it would seem that derailment should have been a sure bet, but it was not. The major reason is that substantial momentum had been built up in the decades preceding these events. That momentum had at least three sources: the US Congress, FERC, and the University of Chicago. In the nearly three decades between 1978 and 2005 Congress passed three laws that promoted competitive reform. The Public Utility Regulatory Policies Act of 1978 was the first; its promotion of competition was in large part inadvertent. The next two—the Energy Policy Acts of 1992 and 2005—explicitly promoted competition by easing the regulation of new competitors and by facilitating their access to electric transmission systems. What is most interesting is the fact that the last of these three—the Energy Policy Act of 2005—promoted competitive reform at the same moment it was addressing what some would say were the three crises of competitive reform: California, Enron, and the 2003 blackout.

In this same period, often spurred by the congressional legislation, FERC issued rulings that added to the momentum toward open-access electric transmission. The start was its case-by-case effort to require open access as a condition of approval for transactions such as utility mergers. Then open access was codified by FERC's Order 888 in 1996 and taken to a new, higher level by its Order 2000 in the year 2000. Congress and FERC teamed up to literally fuel competitive reform with the Natural Gas Policy Act of 1978 and the subsequent Orders 436 and 636.

Finally, the University of Chicago added intellectual fuel to the fire. They did so by publishing, in the 1960s and 1970s, an unrelenting and thoughtful attack on the effectiveness of economic regulation of all sorts, including specifically the economic regulation of the electricity business.

The broadest legacy of this momentum is that, today, over 60 percent of Americans get their electricity through six region-wide markets created to achieve competitive reform. That is, they live in the footprint of one of the five regional transmission organizations created and overseen by FERC, plus one exclusive to the state of Texas.[35]

In a period of about twenty-five years (1980 to 2005), America dismantled the monopoly regulation that had prevailed for decades. The electricity business was permanently altered, and together, government and business created momentum for competitive reform. Still, at the start of the new millennium, more change was to come.

PART FIVE

The Age of Uncompromising Belief

CHAPTER 17

President Obama's Clean Power Plan

> But I am convinced that no challenge poses a greater
> threat to our future and future generations than a changing
> climate. And that's what brings us here today.
> —President Obama (2015)[1]

The Nobel Peace Prize for 2007 was awarded jointly to former vice president Al
Gore and the United Nations' Intergovernmental Panel on Climate
Change (IPCC). The Norwegian Nobel Committee said that the award
was made for "their efforts to build up and disseminate greater knowledge about man-made climate change, and to lay the foundations for the
measures that are needed to counteract such change." Of Vice President
Gore in particular, they said, "[He] is probably the single individual who
has done the most to create greater worldwide understanding of the measures that need to be adopted." To complete an awards hat trick of sorts,
along with his Nobel Prize, Vice President Gore's documentary on climate change, *An Inconvenient Truth* (2006), won two Oscars in that same

year, for best documentary and best original song, as well as a Grammy for the audiobook version.[2]

The Nobel and acclaimed film together reveal the high-profile status conferred on the issue of global climate change in the first decade of the twenty-first century. Global climate change or, to use the narrower term, global warming, was a subject for scientific research for decades, and although well understood in its simplest form, it garnered little or no attention outside a small circle of experts. Indeed, until well into the 1970s, the focus of climate science was on the chance of another ice age—global cooling, not global warming.

The basic science has been well understood for a long time: the sun shines on Earth, and some of that sunlight and heat is reflected back into space. Some heat, however, is trapped by elements in our atmosphere, and that is a good thing since Earth would be too cold for humans if this greenhouse-like trapping of heat did not occur. However, Earth can have too little or too much of a good thing: too much of a greenhouse effect leads to higher average global temperatures and too little would result in average temperatures falling. Carbon dioxide is one of the more important elements in the atmosphere that traps the heat and also is one of the most abundant "greenhouse gases."

The direct link to the electricity business is clear: when fossil fuels like coal, oil, and natural gas are burned to generate electricity, the power plants emit carbon dioxide and other greenhouse gases. Indeed, around 30 percent of all the greenhouse gases emitted each year in the United States can be traced to producing electricity. If America was to decide that carbon dioxide emissions had to be cut significantly—as Vice President Gore and the IPCC recommend—the electricity business would have to change fundamentally.[3]

Major legislation almost made its way through Congress that would have required an 83 percent reduction in greenhouse gas emissions from 2005 levels and, in so doing, would have touched every aspect of American life. The Waxman-Markey bill passed in the US House of Representatives in 2009 but was never brought to a vote in the Senate, for reasons explained later in the chapter. With the legislative route blocked, environmentalists simply shifted to an alternate route through the courts to gain the authority to regulate greenhouse gas emissions. The result was President Obama's Clean Power Plan, first introduced in August 2015.[4]

THE SLOW RISE OF CLIMATE SCIENCE AND POLITICS

All the Right Questions, Early On

The science of global climate change was first formalized early in the 1800s. Between 1824 and 1827 French scientist and mathematician Joseph Fourier conducted a series of mathematical studies on Earth's atmosphere; it was Fourier who first suggested the gases that trapped heat on Earth had a greenhouse effect. John Tyndall, a British scientist, also studied the absorption of heat by gases in the atmosphere. Ironically, considering today's context, Tyndall's studies were aimed at trying to understand ice ages and looked favorably upon the absorptive powers of greenhouse gases. In 1896 a Swedish scientist, Svante Arrhenius, worked on a paper that calculated the impact of carbon dioxide on global warming. Arrhenius concluded that a reduction of carbon in the atmosphere by 55–62 percent of then-current levels would lead to a decrease in temperature of 4–5 degrees Celsius in inhabited regions (7.2–9 degrees Fahrenheit). He, too, focused on ice ages. Why the emphasis on ice ages at the time? Ice ages were perceived as a proximate threat because they caused poor harvests, which in turn would cause famines.[5]

By 1910, presaging an important element of the scientific debate today, some scientists were saying that Arrhenius' calculations were wrong. They believed Earth already contained the maximum amount of greenhouse gases possible. In addition, they said carbon dioxide would not build up in the atmosphere because it would be absorbed by the oceans. This early scientific debate asked all the major questions that have persisted into today's theories on global climate change.[6]

Can Man Change the Global Climate?

One of the most persistent and difficult questions is whether man can affect global climate in the first place. In the 1920s a Russian geochemist, Vladimir Vernadsky, concluded that "Mankind taken as a whole is becoming a powerful geological force." In 1938 Guy Stewart Callendar, a British scientist, came to the same conclusion, because an increase in carbon dioxide caused by man could, he said, increase temperature. He added, however, that there was no reason for alarm since the increase could even be beneficial for agriculture in colder regions. Gilbert Plass, at the US Office of Naval Research, in a 1955 article titled "The Carbon Dioxide Theory of Climate Change," concluded that human "activities" increased global temperature by 1.1 degree Celsius (1.98 degrees Fahrenheit) every hundred years.[7]

Other scientists pursued further lines of inquiry, most notably about whether carbon dioxide emitted by man through fossil fuel combustion would be absorbed by the oceans. In 1957 Hans Seuss and Roger Revelle published an influential paper while at the Scripps Institution of Oceanography at the University of California, San Diego. The two found that from the beginning of the Industrial Revolution in the middle of the nineteenth century to about 1950, the amount of CO_2 "added to the atmosphere through consumption of fossil fuels" per decade increased almost twelvefold. Revelle would later serve as a member of the committee that produced "Restoring the Quality of Our Environment," a report to President Johnson's Science Advisory Committee in 1965. The report echoes the Revelle-Seuss article in its warning that "man is unwittingly conducting a vast geophysical experiment."[8]

Keeling's Curve

Charles David Keeling is one of the most respected authorities on the greenhouse effect and worked at the Scripps Institution for most of his career. His primary contribution to the science of global climate change was to develop a precise way to measure carbon dioxide concentrations in the atmosphere. This was a substantial feat because it enabled scientists to test central hypotheses, such as the Revelle-Seuss question on whether the carbon emissions are all or only partly absorbed by the oceans. Keeling's measurements at the Mauna Loa Observatory in Hawaii are the basis for the Keeling Curve, which began tracking the carbon dioxide concentration in the atmosphere in the 1950s and continues on through today. The concentration he found was 310 parts per million (ppm) in 1958. By 2005 that had increased to 378 ppm, and it has since crossed over 400 ppm. The recorded rise in the concentration of carbon dioxide is at the heart of the concern over rising global temperatures and climate change.[9]

In 1963 Keeling, Plass, and other experts issued a report through the Conservation Foundation that gave one of the first blunt warnings about climate change. It found that a doubling of carbon dioxide concentration could increase average global temperatures by more than 3.8 degrees Fahrenheit. In 1966 came a recommendation by NASA for an extended research plan and an expansion of the research budget from $5 million to $30 million.[10]

Keeling was recognized for his contribution by those on both sides of the aisle. In 1997 Vice President Gore gave Keeling a special-achievement award at a White House ceremony. In 2002 President George W. Bush selected Keeling for the National Medal of Science, America's highest lifetime achievement award for science.

Hard "Core" Evidence

Most people would prefer to have physical evidence of global warming rather than output from computer models. Some of the best physical evidence comparing today's carbon dioxide concentrations and temperatures to the past comes from cores drilled in the layers of ice in both Greenland and Antarctica, and cores drilled in seabeds and lakebeds. This "core" evidence allows scientists to compare carbon dioxide concentrations today to those hundreds of thousands of years ago; to detect the number, intensity, and pace of temperature changes in history; and to judge which theory might best explain the variation in climate over time. Each layer of ice was formed at a specific point in time, so the cores yield something akin to a stacked set of time capsules. Some of the best ice core data came from the cores drilled by a Soviet-French team at Vostok Station in Antarctica, which provided data from four hundred thousand years ago. A 1969 report on the ice cores from Greenland and Antarctica said that scientists had found variations in average global temperature of up to 10 degrees Celsius (18 degrees Fahrenheit). This hard scientific evidence of temperature change was among the first to reveal increasing carbon dioxide concentrations.[11]

A Staccato Call for Action

What came next from some in this small club of scientists was a staccato awareness campaign: a series of small-scale, distinct calls to action that together told the world that the scientists believed danger lay ahead and something must be done. In 1971, a conference was held in Stockholm with the title "Study of Man's Impact on Climate." Further concern was driven in the mid-1970s by Reid Bryson at the University of Wisconsin, who concluded that climate change could happen in hundreds of years, not thousands. In 1976 a congressional committee held hearings and heard scientists warn of the effect of carbon dioxide emissions. In 1977 the *New York Times* ran a front-page story entitled "Scientists Fear Heavy Use of Coal May Bring Adverse Shift in Climate." In 1978 Congress passed the National Climate Act. Spencer Weart, at the Center for the History of Physics, concluded, however, that "nobody of consequence proposed to regulate CO_2 emissions or make any other significant policy changes to deal directly with greenhouse gases."[12]

The 1970s was not a time of consensus on global warming. It is often pointed out that the concern was "alarmism" over global cooling at the time. The proffered evidence of this is reflected in headlines throughout the decade. In 1970, for example, the *Washington Post* wrote, "Colder Winters Held Dawn of New Ice Age—Scientists See Ice Age in the Future." In 1974 *Time Magazine* asked, "Another Ice Age?" In 1979 the *Christian Science Monitor* worried, "New Ice Age Almost Upon Us?"[13]

Still, the staccato beat went on, and the focus of concern shifted from the possibility of an ice age to global warming. In 1977 the National Academy of Sciences concluded that the correct concern was warming—not cooling. Jim Hansen, a scientist at NASA's Goddard Institute for Space Studies, concluded that, while temperatures had been cooling up through the 1970s, in the 1980s the reverse was occurring. Hansen said he was convinced that clear evidence of warming should be seen by the year 2000. In time, Hansen would become one of the major scientific voices raising alarm about global warming.[14]

The Montreal Protocol

The public debate and policy surrounding chlorofluorocarbons (CFCs) helped make global climate concerns more credible to the general population. The mechanics of the CFC policy also may have set a template for how to address global climate change through an international agreement. CFCs are man-made chemicals that were thought to be safe, in part, because they occurred in small amounts and were chemically stable. They were commonly used in aerosol sprays for bug sprays and paint. Their stability—they didn't react with plants or animals—turned out to be a problem, because it meant they stayed in the atmosphere for long periods of time, long enough to drift into the stratosphere where they were activated by the sun's ultraviolet light. Once activated, they catalyzed a reaction that destroyed the layer of ozone; the ozone layer actually protects Earth from ultraviolet light, which can cause skin cancer. In 1987, forty-six governments signed the Montreal Protocol pledging to control the chemicals that destroyed the protective ozone layer, and by 2009 these treaties were the first to have been universally ratified by all 196 nations.[15]

Importantly, the Montreal Protocol gave evidence to the fact that, though man was harming the delicate balance of Earth's atmosphere, nations could come together to mitigate that harm.

THE UNITED NATIONS' INTERGOVERNMENTAL PANEL ON CLIMATE CHANGE

Reports in the 1990s

In 1988 the issue of global climate change was taken to another level when the United Nations formed the Intergovernmental Panel on Climate Change (IPCC). The IPCC was intended to combine science, government, and politics, in order to deal with the issue of man-made global climate change. The IPCC's first report was issued in 1990. One major conclusion was that, if nothing changed—if the IPCC's business-as-usual scenario prevailed—that would "result in a likely increase in global mean temperature of about 1°C above the present value by 2025 and 3°C before the end of the next century," in 2100. The IPCC openly acknowledged "uncertainties in our predictions" and wrote that the uncertainties were due to "our incomplete understanding" of clouds, oceans, and polar ice sheets. And, further, the IPCC in 1990 did not rule out that the warming "could be largely due to . . . natural variability" in climate. [16]

In 1992, at the Earth Summit in Rio de Janeiro, one hundred nations signed the "Convention on Climate Change" as well as agreements on biological diversity, forest principles, and Agenda 21, a "plan for achieving sustainable development in the 21st century." Everyone looked to 1995 for the next IPCC report. Ultimately, that report concluded that "the balance of evidence suggests a discernible human influence on global climate." [17]

The next big event for the IPCC was the UN Conference on Climate Change held in Kyoto, Japan, in 1997. There, the United States proposed cuts in carbon dioxide emissions that would take its emissions back to 1990 levels, but Europeans wanted deeper cuts, and China wanted an exemption, saying it and other developing countries should not pay for a problem caused by developed countries' industrialization. This changed the issue into a political battle between rich countries that had already achieved high levels of industrialization and poorer nations that were looking to catch up through rapid industrialization. Vice President Gore flew to Kyoto to help finalize a deal, which became known as the Kyoto Protocol. Under the Kyoto Protocol, the developed countries pledged to cut average emissions to 5 percent below 1990 levels by 2008–2012, but the developing countries were exempt. They had only to report their emissions and come up with "mitigation" plans. Vice President Gore's deal did not get a warm reception back home: The US Senate voted ninety-five to zero on a resolution saying that they would not accept Kyoto if the developing countries were exempt.[18]

The 2014 Report

IPCC research and reports are looked upon by policymakers as the definitive source of information on the cause and impact of global climate change. Even the US EPA relied on IPCC evidence when making its "Endangerment Finding"—the foundation of its authority today to regulate greenhouse gas emissions under existing law. *Climate Change 2014: Mitigation of Climate Change* includes a *Summary for Policymakers*, with boldly stated conclusions. Overall, the conclusions suggest that the negative impacts of global climate change are already being seen, that human activity is a substantial cause of those impacts, and that to prevent even more serious impacts would require significant reductions in emissions.[19]

Specifically, the IPCC states, "Warming of the climate system is unequivocal, and since the 1950s, many of the observed changes are unprecedented over decades to millennia. The atmosphere and ocean have warmed, the amounts of snow and ice have diminished, and sea level has risen." Regarding any question about human activity being the cause, the IPCC highlights that "human influence on the climate system is clear . . . anthropogenic greenhouse gas emissions have increased since the pre-industrial era, driven largely by economic and population growth, and are now higher than ever. This has led to atmospheric concentrations of carbon dioxide, methane and nitrous oxide that are unprecedented in at least the last 800,000 years." It is an unambiguous conclusion.[20]

Linking global warming to a human cause is central because it implies there is a human cure: cutting carbon dioxide emissions from human activity. To specify the precise human cause, the IPCC stands firm with the oft-stated claim that greenhouse gas emissions that keep carbon dioxide concentrations at or below 450 parts per million in the year 2100 would, in turn, keep temperature increases at about 2 degrees Celsius (3.6 degrees Fahrenheit). That is a threshold that the IPCC does not want to cross. To avoid crossing this threshold, the IPCC reports that "anthropogenic" greenhouse gas emissions—those related to human activity—would have to be cut by 40 percent to 70 percent by the year 2050 as compared to emissions in 2010. In addition, even more aggressive emission control would be required by 2100—by that year, emissions would have to be near zero or below.[21]

These bold conclusions are backed up by arguments and evidence, but those cannot be found in the summary; all of that is in the remainder of the report—and it is a massive report. The *Economist* called it "a behemoth of an undertaking. It runs to thousands of pages, involved hundreds of scientists and was exhaustively checked and triple-checked by hundreds of other boffins and government officials to whom they report." What is found throughout the summary, however, is an honest but awkward attempt to give the reader a sense of the confidence the IPCC has in each of the conclusions.[22]

The IPCC reports are not without their critics and controversy. For example, in January 2010 the chair and vice chair of the IPCC acknowledged a mistake in its fourth assessment in 2007. They first say the broad conclusion about stress on water resources, including accelerating loss of glaciers, is robust. However, they express regret about "poorly substantiated estimates" of the pace at which the Himalayan glaciers would recede. The 2007 report concludes that the Himalayan glacier could disappear in 2035, but that is unfounded.[23]

In the wake of this criticism, the secretary-general of the United Nations, Ban Ki-Moon, asked in 2010 for a review of the IPCC process by the InterAcademy Council. The council concluded that the IPCC process needed to be updated since the issue of global climate change had become ever more complex and "increased transparency" was required. The recommended remedies included better "governance and management" and improvements in the "review process." Since the council is an international organization of science academies, perhaps it would have been inspiring to remind all of us of the great commitment to the scientific method and personal integrity of the scientists who have gone before us: Franklin, Faraday, Maxwell, Tesla, and Einstein.[24]

AN INCONVENIENT TRUTH

In 2006 former vice president Gore brought mainstream attention to what he saw as an inconvenient truth. And that view was ominous: as he argued, man's emissions of carbon dioxide and other greenhouse gases since the Industrial Revolution in the 1700s caused global warming, and the consequences of that warming could be disastrous. The inconvenience of this truth stemmed from the fact that, to avert disaster, America and the entire world would have to fundamentally change the way energy is produced and used in homes, cars, and factories. The film, *An Inconvenient Truth*, is an onslaught of visual and verbal evidence that leads to his unequivocal conclusion that global warming is already upon us, and drastic action must be taken now. In this sense Gore is fully faithful to the template set by Rachel Carson with her indelible *Silent Spring*.

This was not Gore's first involvement with environmental activism. In 1992 Gore published a book titled *Earth in the Balance: Ecology and the Human Spirit*. The book was a full-throated call for enviro-centric thinking that placed environmental concerns at the center of the world agenda.

Taking a different view, *Cool It* is a 2010 documentary film and book by Bjorn Lomborg that responds to Gore's *An Inconvenient Truth*. Lomborg does not attempt to refute the claim that global warming is happening and that it is caused in part by man. Rather, Lomborg's most distinctive contribution to the debate on global climate change is to tee up the choice between prevention and adaptation. He offers human adaptability to problems as an alternative to preventing global climate change.[25]

THE WAXMAN-MARKEY BILL

Those who wanted the American government to take action to reduce greenhouse gas emissions were tremendously encouraged when the House of Representatives passed the American Clean Energy and Security Act of 2009. This act is more often referred to as the Waxman-Markey bill because its sponsors were Representatives Henry Waxman, Democrat from California, and Ed Markey, Democrat from Massachusetts. The vote in the House was quite close: 219 for versus 212 against.[26]

The centerpiece of the bill was a cap-and-trade policy that was first used nationwide as part of acid rain legislation in 1990. The cap meant that carbon dioxide and other greenhouse gas emissions would have a limit or "cap" put on them across the nation. That cap would be reduced over the years so as to reduce emissions to lower and lower levels. By 2020 the cap would require emissions to be 17 percent below 2005 emission levels and 83 percent below that level by 2050. The 83 percent reduction apparently was motivated by the aforementioned IPCC goal of keeping the increase in global average temperature, as compared to pre–Industrial Revolution levels, to about 2 degrees Celsius (3.6 degrees Fahrenheit).[27]

The "trade" part of the deal was meant to signal that a market-based approach to regulation would be used, rather than a command-and-control approach. As with the acid rain legislation, to start there would be a requirement to hold permits to emit, and each permit would allow the holder to send one ton of carbon dioxide emissions into the air; for this reason, individual permits are called allowances. The hope was that the ability to sell or "trade" allowances at a profit would keep the cost of meeting the cap as low as possible. As an additional step to keep the costs of compliance as low as possible, the Waxman-Markey bill allowed "offsets." An offset was created when a source of greenhouse gas emissions that was not regulated by the bill agreed to cut its emissions anyway. Suffice it to say that the bill had further features that made it somewhat challenging to understand and complex to implement.

The Waxman-Markey bill, as passed by the House, was sent to the Senate on July 6, 2009, but no vote was ever taken. Bryan Walsh wrote a postmortem titled "Why the Climate Bill Died" for *Time* magazine. The bottom line is that the public was not demanding change. The lack of such a demand was reflected in the fact that the White House wanted to sell the bill with a green-jobs argument, not as a full-out attack on global climate change. Unable to win the necessary sixty votes in the Senate, proponents simply went to another place to fight for what they wanted—the courts.[28]

EPA'S ENDANGERMENT FINDING

For both those disappointed and those overjoyed that the Senate did not take action, there was a plan B to contend with. Waxman-Markey may have been dead, but greenhouse gas regulation was not. There was another route to regulating greenhouse gases through the court system. For proponents of greenhouse gas regulation, the prize was to get the courts to conclude that the US EPA already had the right to regulate under the existing Clean Air Act—that is, there was no need for Congress to pass a new law.

In 2003 President George W. Bush's administration ruled that the EPA lacked the authority to regulate greenhouse gas emissions under the Clean Air Act. Twelve states and other environmental groups appealed the ruling, and in April 2007 the Supreme Court remanded to the EPA the question of whether greenhouse gases could in fact be regulated under the Clean Air Act. Specifically, the EPA was ordered to determine whether emissions of greenhouse gases from new motor vehicles "cause, or contribute to, air pollution which may reasonably be anticipated to endanger public health or welfare" or explain why "scientific uncertainty is so profound that it precludes EPA from making a reasoned judgment."[29]

These questions took time to address—about two and a half years, to be precise. In December 2009, under President Obama's administration, the EPA found that eleven greenhouse gases taken in combination endanger both the public health and the public welfare of current and future generations. In this "endangerment finding" the EPA listed and explained the harm caused by increased concentrations of greenhouse gases including harm from "heat waves," "extreme weather events," "wildfires, flooding, drought," and "widespread melting of snow and ice." The EPA also gave weight to "the fact that . . . children, the elderly, and the poor, are most vulnerable."[30]

The endangerment finding was an essential first step toward regulating these emissions. Because of it, the EPA expanded coverage not just to cars but to any major stationary source or major emitting facility. That is how the EPA made the great leap from regulating greenhouse gas emissions for new cars to also regulating them for power plants.[31]

The EPA also found that "warming of the climate system is unequivocal" as evidenced by "increases in global average air and ocean temperatures, widespread melting of snow and ice, and rising global average sea level." The EPA reported that "eight of the 10 warmest years on record have occurred since 2001." Looking to the future, using IPCC model output, the EPA noted, "By the end of the 21st century, projected average global warming (compared to average temperature around 1990) varies significantly depending on the emission scenario and climate sensitivity assumptions, ranging from 1.8 to 4.0°C (3.2 to 7.2°F)." Most of these scenarios exceed the IPCC's threshold of 3.6 degrees Fahrenheit (2 degrees Celsius) for change in average temperatures and, therefore, most would trigger significant global climate change impacts.[32]

In the same month—December 2009—as the publication of the EPA's endangerment finding, the fifteenth session under the UN Framework Convention on Climate Change was held in Copenhagen, Denmark. On the last day of the meeting, President Obama negotiated a deal with leaders of some developing nations. Those leaders included Premier Wen Jiabao of China, Prime Minister Manmohan Singh of India, President Luiz Inácio Lula of Brazil, and President Jacob Zuma of South Africa. However, in its assessment of the deal, the Pew Center concluded that it "charts no clear path toward a treaty with binding commitments." In its cover story in November 2010, the *Economist* reported, "In the wake of the Copenhagen Summit, there is a growing acceptance that the effort to avert serious climate change has run out of steam." The article then proceeded to discuss "how to live with climate change," and espoused a shift from prevention to adaptation to the new climate—in other words, the *Cool It* model, rather than the great change inspired by *An Inconvenient Truth*.[33]

THE PARIS ACCORD

In December 2015 the UN held its twenty-first climate change conference since 1995. The progress represented by the conference's major outcome—the Paris Accord—depends a lot on what the person issuing the verdict expected before the meeting began. Two op-eds on December 16, 2015, from two widely respected writers make the point. The headline for the syndicated columnist George F. Will read: "The Paris agreement is another false 'turning point' on the climate." The opposing view came from syndicated columnist Thomas L. Friedman, whose headline read: "Paris Climate Accord Is a Big, Big Deal."[34]

Notably, both writers had low expectations for the conference. Will wrote that any agreement that got almost two hundred countries to sign on must "be primarily aspirational, exhorting voluntary compliance with inconsequential expectations." Friedman also thought expectations should be low, and in the end he concluded that those low expectations had been met "beautifully."[35]

The two writers came at the Paris Accord from fundamentally different perspectives. Will wrote from the vantage point of the Age of Franklin and Edison. That is, he focused on the good fossil fuels had done, citing a source that said that they were needed for "sustained economic growth, a necessary prerequisite for scientific and technological dynamism." Will concluded even more broadly that the "environmental toll from burning coal (it emits carbon dioxide, radioactivity and mercury) has been slight relative to the environmental and other blessings from burning it."[36]

In sharp contrast, Friedman wrote from the perspective of the Age of Harm and focused on the need to prevent the worst of the harm caused by burning fossil fuels. And yet he had great hope from the final accord. He thought that the "willingness" of so many countries "to offer plans to steadily and verifiably reduce their carbon emissions" meant there was a "chance" that the world would keep carbon concentrations below the level that would trigger temperature increases greater than 3.6 degrees Fahrenheit. He also hoped that all would use the Paris Accord to get back to "constructive engagement" that would lead to a phasing out of all fossil fuels by roughly 2050. And interestingly, he hoped that the means to that end was the market—specifically, putting a price on carbon.[37]

The *Wall Street Journal* opined that the claimed success at Paris "is rooted in a conceit and a bribe." The conceit was that the Paris Accord would actually achieve anything when it was "voluntary with no enforcement mechanism"; China and India, the *Journal* pointed out, would prioritize bringing millions out of poverty over reducing CO_2 emissions. The bribe identified by the journal was the $100 billion per year in promised climate aid to the developing world.[38]

The breadth of factors and beliefs motivating opinions on global climate change cannot be overstated. President Obama said there was no greater challenge than addressing climate change and equated it to recovery from America's financial collapse and ending war in Iraq and Afghanistan.[39]

Even Pope Francis felt compelled to weigh in through his encyclical letter *On the Care of Our Common Home* (or *Laudato si*), given on May 24, 2015. He found that the advances on climate change have been "regrettably few" and that fossil fuels need "to be progressively replaced without delay." The pope placed these comments in a much broader context and presented them in a deeply prayerful structure and tone. He spoke of a full range of environmental concerns, including biodiversity. He linked environmental concern to poverty, too, writing that "the same mindset which stands in the way of making radical decisions to reverse the trend of global warming also stands in the way of achieving the goal of eliminating poverty."[40]

PRESIDENT OBAMA'S CLEAN POWER PLAN

What would President Obama do with the authority to address global climate change given to the EPA by the Supreme Court? And, given the failure by the IPCC to create a global policy, would the president take action unilaterally? One indication came on August 3, 2015, when the EPA published a draft of its Clean Power Plan (CPP). Calling it "a historic and important step," the EPA went on to say, "These are the first-ever national standards that address carbon pollution from power plants." Separately, but on the same day, the EPA had issued new source-performance standards for *new* large-base-load power plants. Those standards meant that conventional coal-fired power plants were a thing of the past; given the state of technology, the EPA's rules meant any new coal-fired power plant would have to use new technologies, such as high-efficiency coal-fired power plus some carbon capture and sequestration.[41]

The draft CPP runs 1,500 pages; it is a document that few have actually read but many support or oppose ferociously. Under the CPP, electric power suppliers would have to cut carbon dioxide emissions from *existing* power plants by 32 percent of 2005 levels by 2030. As required by the Clean Air Act for existing facilities, the effort to reach the 32 percent cut had to be a joint federal and state effort. Essentially, the federal government would determine how much each state must reduce emissions, and the states would propose a plan to achieve that reduction. To that end, the EPA kicked off the process. It issued both an emissions rate in terms of pounds of carbon dioxide per MWh and total emissions reduction in terms of tons of carbon dioxide emitted for the state as a whole.[42]

The EPA explained its rationale for the emissions rate and total emissions in terms of three "building blocks." First was the reduction in carbon dioxide emissions achieved "by improving the heat rate at affected coal-fired power plants." Second was the reduction achieved by substituting electricity from "lower-emitting existing natural gas plants" for that from existing coal-fired plants. Third was the reduction achieved by substituting electricity from "new zero-emitting renewable energy" sources (like wind and solar) for generation by existing coal-fired plants.[43]

Also addressed was the need for the CPP. The EPA tends to point to other credible entities who support aggressive action to address global climate change. In its own endangerment finding it relied heavily on the IPCC evidence. Here, it points to the endangerment finding but also to other reports. For example, it cites the 2014 report of the US Global Change Research Program and summarizes the bottom line of that report by stating "climate change driven by human emissions of [greenhouse gases] is already happening now and it is happening in the U.S."[44]

Physical evidence is preferred when it can be found. One example of such evidence is a chart offered by NASA. The chart plots CO_2 concentration over hundreds of thousands of years. This physical evidence comes from ice cores. It shows that CO_2 concentrations were historically in the 200–300 ppm range until climbing to today's 400 ppm. These findings send a clear message that the world is now in uncharted waters.

A legal justification of the CPP is also provided. The legal arguments are quite detailed and expansive. As evidence of the legal challenges, on February 9, 2016, the US Supreme Court stayed implementation of the CPP pending disposition of the petition for review in the US Court of Appeals for the DC Circuit as well as the Supreme Court's own review if granted. The *Wall Street Journal* called the stay an "unexpected move," and the *New York Times* called it an "extraordinary decision."[45]

Bruce Nilles, a Sierra Club lawyer, said, "It is unprecedented for the Supreme Court to stay a rule at this point in litigation. They do this in death-penalty cases." Nilles' comment suggests the court's rationale for the stay. If the court does not stop EPA, it may proceed to implement a plan—the CPP—that ultimately was found to be unconstitutional. As in a death penalty case, it would have been too late for justice.[46]

THE LEGACY OF A HEATED DEBATE

Global climate change policy now dominates the electricity business. Through the Clean Power Plan for existing power plants and the new source-performance standards for new power plants—presuming that they weather the court challenges as well as those from the new president—EPA will drive all major resource decisions in the electricity business. In particular, new source standards and the CPP will define whether any *new* coal power plants will be built and whether and when *existing* coal-fired power plants will be retired. In addition, the CPP greatly influences the nature and extent of the additions of new renewable power plants; substituting renewables would need to be a major source of the planned 32 percent reduction in emissions from 2005 levels. And the CPP will be one more incentive for natural gas–fired combined-cycle power plants—already the technology of choice.

It is disappointing that this dominance was achieved without a public vote on how to address global climate change concerns, let alone a wider debate regarding the aggressive approach embedded in the CPP and new source-performance standards. Without a vote by Congress, there will be no debate on critical issues such as the crucial policy choice between prevention and adaptation. Moreover, the danger of pursuing aggressive prevention unilaterally is that the American people could end up paying twice to address global climate change. First, Americans would pay for a unilateral effort to cut America's emissions in an effort to *prevent* global climate change. Second, Americans would pay to *adapt* to global climate change because, inevitably, a US-only policy could never prevent climate change; that is, if other nations, China in particular, do not do their part, the consequences of global climate change would be realized anyway. And if forces other than carbon emissions meant global climate change occurred anyway, Americans would have to pay for adaptation regardless of whether others did their part or not.

The question of what to do about global climate change deserves an open debate and a vote by the US Congress. This is not a decision for the courts alone. What is needed most is an openness to respectful, effective compromise; something that is nowhere in sight in the Age of Uncompromising Belief.

On March 28, 2017, President Trump signed an Executive Order requiring, among several other items, a review of the CPP that could lead to its withdrawal or replacement. No matter the outcome of the President's review, it is likely that global climate change will be a contested issue for the electricity business well into the future.

CHAPTER 18

George Mitchell's Shale Gas Revolution

Few businesspeople have done as much to
change the world as George Mitchell.
—*Economist* (2013)[1]

George Mitchell is not a household name, but he should be. What he achieved
could affect the availability and cost of energy for decades to come in
households across America and possibly around the world. Through
painstaking trial and error, over a twenty-year span, Mitchell developed
an innovative way to tap into natural gas trapped in shale rock formations
deep under the earth's surface and to produce that natural gas for customers at a historically low price. Combining his innovation with Devon
Energy's horizontal-drilling technology and advances in seismic mapping,
Mitchell unleashed what is now called the shale gas revolution.

It deserves to be known as a revolution because it created an abundant supply of one of the most versatile forms of energy. As recently as
2007 experts warned that North American natural gas production would

decline to the point that the United States would eventually have to import natural gas from overseas, much as it imports oil. Since Mitchell's revolution, the new view is that Americans have enough natural gas under their feet to sustain current levels of consumption for years to come. Indeed, some expect that the United States will export natural gas, not import it.[2]

The link between natural gas and electricity is direct and important. Natural gas is the fuel of choice for electricity production, both because of the opportunities created by opening the electricity business to competition and the challenges presented by global climate change. As the fuel of choice, natural gas often sets the market price for electricity, and natural gas prices have fallen to historical lows in and around the 2010–15 time frame. Many expect the shale gas revolution to keep natural gas prices low and thereby keep electricity prices low, too.[3]

THE PATH TO REVOLUTION

It is crucial to see George Mitchell in the context of his everyday business. In 1953 Mitchell and his partners drilled for natural gas in North Texas, and the well turned out to be one of the most productive wells ever. In 1954 they won a contract to supply natural gas to Chicago through the Natural Gas Pipeline Company of America. This sale was critical to Mitchell's career; indeed, it was the need to sustain supplies for his Chicago contract that led Mitchell to look for new gas resources on his home turf. And that, in turn, helped incite the shale gas revolution.

Mitchell Energy and Development Corporation was established in 1971 and made its initial public offering of stock on the American Stock Exchange the following year. In the late 1970s, faced with his continuing need to fill the Chicago contract, Mitchell pushed his colleagues to find more gas resources on the leases he already had. A geologist with the company wrote a paper suggesting that the Barnett Shale in Denton County, Texas, might be such a source.[4] Despite the protests of his own engineers against using unconventional drilling techniques, Mitchell began experimenting with hydraulic fracturing in the shale.

The International Energy Agency succinctly describes hydraulic fracturing—often called hydrofracking, or simply fracking—as the process of pumping fluid into rock formations at high pressure to release trapped gas. Typically, that fluid is 99 percent water and sand, where the sand serves as a *proppant*—a material that will keep the fractures open after they are made. Small amounts of chemicals are used as part of the fluid, called *slick water*, to serve several purposes like keeping the proppant suspended and minimizing friction, so the fluid can enter the well more easily.[5]

By Mitchell's own reckoning, he spent $7–8 million on well experiments in shale, and it took him over twenty years to get it all figured out. According to his colleague Dan Steward, the author of a history of the Barnett Shale, Mitchell's first well was drilled in 1981, and "from 1982 through 1986, 41 wells were deepened." By 1989 Steward said that enough wells had been drilled and enough seismic data had been collected to know the basic structure of the Barnett formation. From an initial area of 15 square miles, Mitchell then expanded the area to cover 115 square miles in the late 1990s. Still, a lot of trial and error had to be completed to find a commercially viable fracturing fluid. By 1998 Mitchell finished testing slick water, which cut the treatment cost by 80 percent as well as increasing the gas flow. Denton is now considered the center of the Barnett Shale.[6]

THE MEASURE OF A REVOLUTION

Abundance

The consensus of energy forecasts in the early 2000s warned that America would have to turn to imports of natural gas from overseas to sustain current levels of consumption. Indeed, energy expert and author Daniel Yergin himself expressed this view in a *Wall Street Journal* op-ed in January 2007. He warned that "North American supply has flattened out. Yet large amounts of new natural-gas-fired electric power generation have been added over the last decade, which means that demand will increase."

Yergin concluded that the result would be "growing imports of liquefied natural gas—LNG—rising from 3% of our current demand to more than 25% by 2020."[7]

Yergin's op-ed reflected the consensus view. Accordingly, from 2000 to 2009, investors financed a dramatic increase in the capacity to import LNG. To transport natural gas from countries overseas, the natural gas must first be made into a liquid. Then, once it reaches its destination port, it must be reconverted into a gaseous form to be delivered inland through pipelines. In 2000 America had the capacity to convert about 2.3 billion cubic feet per day of LNG into a gaseous form—that is, to import it into the United States. By 2009 that capacity had increased almost tenfold to 22.7 billion cubic feet per day.[8]

Then, suddenly, the view was reversed from shortage to abundance. In November 2009, Yergin coauthored another op-ed in the *Wall Street Journal* titled "America's Natural Gas Revolution." He wrote that the new-found abundance of shale gas "transforms the debate over generating electricity." Yergin suggested that, in the face of global climate concerns, the lower prices of natural gas would make it the fuel of choice for generating electricity. What caused this complete reversal in view, from shortage to abundance? The primary reason was the new accessibility of shale gas, but, more specifically, the cause was the melding of technological innovations by Mitchell and Devon Energy: hydraulic fracturing and horizontal drilling, plus three-dimensional seismic imaging. The shale gas revolution was, at its origins, a technological revolution.[9]

In 2000 shale gas contributed just 1 percent of all natural gas produced in the continental United States. By 2009 shale's contribution was 14 percent, and by 2014 it was 48 percent. The optimism over shale gas can also be found in the ever-changing estimates of how much natural gas still lies beneath the surface in America. For two decades, from about 1984 to 2004, America's proved reserves of natural gas were estimated to be at or below two hundred trillion cubic feet. From 2004 to 2014, those estimates doubled, reaching almost four hundred trillion cubic feet. It is this dramatic change in the United States' energy outlook that justifies using the term *revolution*.[10]

Low-Cost Shale Gas

In the well-publicized price spike in 2008, natural gas prices got as high as $13.31 per million British thermal units (MMBtu).[11] By 2012 the spot price for natural gas had dropped as low as $1.82 per MMBtu—down 86 percent. Prices were not expected to stay in the two-dollar range, because some of the factors for the steep decline were attributed to short-term circumstances. For example, some leases for shale required the lessee to use it or lose it—that is, drill for shale gas or lose the lease. Another factor pushing drilling at the time was that many natural gas wells also produce oil-like liquids—ethane, propane, and butane. With oil prices relatively high at that time, the value of these liquids was high as well, so drilling remained profitable. However, despite a significant drop in oil prices, natural gas prices were still low, averaging just $2.62 per MMBtu in 2015.[12]

Much of the time, we make use of forecasts by assuming a smooth path of prices going forward. But history warns against relying on such patterns because there are substantial risks. The nature of risk can change over time, too. In the past, with conventional sources of natural gas, the risk was primarily in finding the resource. With shale gas, the risk of not finding gas is lower, but the risk around how much it will cost to produce the gas is much higher. This is because not all shale gas sources are the same. An MIT study from 2010 shows this in terms of the likely production costs for wells drilled in different places, or *plays*. The cost of production varies within plays, too. To illustrate these points, for the Barnett Shale play in Texas, MIT presented estimates of production costs ranging widely from about $4.27 per MMBtu to $11.46 per MMBtu. For the Marcellus Shale play in Pennsylvania, MIT presented estimates ranging from $2.88 per MMBtu to $6.31 per MMBtu.[13]

Low-Cost Electricity

Given its research, MIT concluded that the bulk of the shale gas in America can be produced in the $4–8 per MMBtu range. Importantly, prices at these levels make natural gas–fired electricity the technology to beat. Although single-point estimates of costs fail to give a sense of the inherent uncertainty of the industry, MIT estimates the cost of electricity generated with natural gas–fired combined-cycle technology to be about $56 per MWh. This is much lower than MIT estimates for nuclear power plants, at $88 per MWh, and those for coal with carbon emission controls, at $92 per MWh. As for electricity generated with renewable fuel, if the expense of making each technology able to produce electricity around-the-clock is added, rather than allowing intermittent generation, MIT estimated that wind with backup would cost $100 per MWh and solar would be at $193 per MWh. In other words, fully reliable, around-the-clock renewables were not competitive on a cost basis alone.[14]

A similarly aimed study, published in *Daedalus* in 2012 by Adam Looney and Michael Greenstone of the Hamilton Project at the Brookings Institution, came to a similar conclusion: looking at new power plants, new natural gas–fired generation is now the cheapest way to generate around-the-clock power. Even adding estimates of "social costs"—including the cost of harm from carbon dioxide emissions—natural gas–fired combined-cycle power cost $65 per MWh, still the lowest-cost technology, while coal was almost double at $115 per MWh. New nuclear was in the range of $82–105 per MWh, with or without social costs because it has no carbon emissions. New around-the-clock wind cost $97 per MWh, while around-the-clock solar photovoltaics (solar panels) cost $132 per MWh.[15]

By the 2010–12 time frame, the consensus was that natural gas–fueled electricity was the low-cost option when taking into account both environmental performance and reliability.

Credible Doubters (and Optimists)

The US government's Energy Information Administration (EIA) provides widely accessible projections of energy production, use, and prices—including natural gas prices. The EIA runs alternative scenarios to reveal risk, as it should, but its "reference" scenario in 2014 reflects assumptions that natural gas reserves are "abundant," while production costs are expected to increase as "producers move into areas where the recovery of natural gas is more difficult and expensive." Underlying this price projection is the expectation of a 56 percent increase in total natural gas production through 2040. Shale gas production drives this growth.[16]

One credible doubter of this scenario is David Hughes, a geoscientist who spent thirty-two years with the Geological Survey of Canada. He is highly critical of the EIA projections, especially the shale gas reserves and production estimates. Hughes conducted an independent analysis of seven shale gas plays that he sees accounting for 88 percent of future shale gas production through 2040. His conclusion is that "shale gas production from the top seven plays will underperform the EIA's reference case forecast by 39% from 2014 to 2040." Indeed, he expects shale gas from these seven plays will peak as early as 2020, and production in 2040 will be just one-third of that forecast by EIA.[17]

In sharp contrast to Hughes, some experts are optimistic that the technological advances with shale gas production will continue, and they give specific reasons for their optimism. Mark P. Mills at the Manhattan Institute is one of them. Mills' bottom line is that the shale gas revolution has created massive amounts of data on the performance of thousands of wells. He argues that "big-data analytics" of that information will lead to dramatic productivity gains and cost reductions going forward. Another theme is that, while big-data analysis will allow best practices to be discovered and used widely, that same analysis will allow the producers to tailor their approach to drilling and production to the circumstances at a particular site. As a broad policy guide Mills suggests that, rather than looking for alternatives to oil and natural gas, the focus should be on developing new technology to keep oil and gas abundant and priced low.[18]

The debate might best be characterized as man's innovation versus geology's limitations. So far, in the early years of the revolution, man has been winning. New wells are becoming more and more productive because technological advance continues. Still, it must be remembered that a fundamental change in a forecast, such as the reversal of fortunes brought on by the shale gas revolution, *reveals* risk; it does not *remove* risk. Uncertainty remains.

ADDRESSING ENVIRONMENTAL CONCERNS

An Evenhanded View

In the Age of Uncompromising Belief, it is hard to find evenhanded analyses of the environmental impacts of shale gas. One rare example that considers both sides of the issue is a 2012 report by the International Energy Agency (IEA) titled "Golden Rules for a Golden Age of Gas." IEA starts with the basic point that the world will realize this golden age of gas if and only if the unconventional sources of natural gas "can be developed profitably and in an environmentally acceptable manner." The agency readily acknowledges that unconventional sources may have "a larger environmental footprint than conventional gas development" but concludes that "the technologies and know-how exist for unconventional gas to be produced in a way that satisfactorily meets these challenges." The IEA lists "golden rules" that are meant to win popular support for shale gas development or, as IEA puts it, to "earn the industry a 'social license to operate.'"[19]

The golden rules are principles for all involved to follow as environmental and social impacts are addressed. The principles reflect common sense that can lead to common or best practices. For example, "Watch where you drill" should lead shale gas producers to avoid drilling where geologic factors make an area vulnerable to earthquakes when storing wastewater. Similarly, "Isolate wells and prevent leaks" should lead to robust rules for well design and construction and to choosing a drilling depth that keeps impacts away from drinking water reserves. IEA estimates that following these principles would increase shale gas costs by only 7 percent.[20]

Can You Light Your Water on Fire?

Other advocacy-driven assessments of the environmental impact of shale gas development have gotten much more attention than the IEA study. Josh Fox's film *Gasland* is one such effort. It was nominated for an Academy Award for Best Documentary in 2011. *Gasland* is self-described as "part expose, part mystery, part bluegrass banjo meltdown," and it is noted that the film drew Fox into "uncovering a trail of secrets, lies and contamination." Also on the packaging is a quote from Yoko Ono, which reads, "Josh Fox's film *Gasland* will save the American water, air, and soil from being abused by corporate greed."[21]

The subtitle of *Gasland*, "Can You Light Your Water on Fire?", comes from the most dramatic scene in the film: A Colorado homeowner lights the water coming out of the kitchen sink by holding a cigarette lighter to the faucet. The cause of contaminated, flammable water is implied to be nearby drilling and hydraulic fracturing for shale gas.[22]

The EPA's Acquittal (for Now)

The possibility of danger to drinking water is a matter of serious concern. A front-page article in the Sunday *New York Times* in February 2011 provides a good example of another alarm sounded about the possible environmental impacts on water supply from drilling for and producing shale gas. The headline read "Regulation Lax as Gas Wells' Tainted Water Hits Rivers." Inside, the section headlines kept up the alarmist tone: "Overwhelmed, Underprepared," "Little Testing for Radioactivity," and "Plant Operators in the Dark." The reporter got right to his point, which linked wastewater from shale gas wells to the danger of contaminated drinking water. He wrote, "With hydrofracking, a well can produce over a million gallons of wastewater that is often laced with highly corrosive salts, carcinogens like benzene and radioactive elements like radium." His specific concern was that the wastewater "is sometimes hauled to sewage plants not designed to treat it and then discharged into rivers that supply drinking water."[23]

Reports like this led Congress to ask the EPA to study the issue of drinking water contamination. The EPA mapped a comprehensive scope for the study including every one of the five stages of the water cycle for shale gas production. Emphasizing that it was just a draft and not "agency policy," the EPA issued a draft report for public review in June 2015. The report first acknowledged that there were indeed ways or "mechanisms" by which harm could be done: "We conclude there are above and below ground mechanisms by which hydraulic fracturing activities have the potential to impact drinking water resources." For example, the EPA cited "spills of hydraulic fracturing fluids and chemicals" and "inadequate treatment and discharge of hydraulic fracturing wastewater."[24]

However, as its major finding—one that surprised many on both sides of the issue—the EPA wrote, "We did not find evidence that these mechanisms have led to widespread, systemic impacts on drinking water resources in the United States." It was not that there were no problems or that there was zero risk of environmental harm from shale gas production. Rather, it was that, having reviewed data for thousands of wells, there was no evidence of widespread systemic harm. The EPA stated, "The number of identified cases [of harm], however, was small compared to the number of hydraulically fractured wells."[25]

This acquittal was given grudgingly and just for the time being. The EPA pointed out potential limitations of the study, including "insufficient pre- and post-fracking data" and "the paucity of long-term systematic studies." For some, the EPA finding made no difference to their view. In the wake of the EPA study, for example, businesses asked Governor Andrew Cuomo of New York to lift the state's ban on fracking. The governor is, thus far, refusing to lift it based on broader concerns than the impact on drinking water.[26]

The EPA issued its final report in December 2016, and it came with a change in tone. The report was offered as a tool for others to identify factors that "influence the frequency and severity" of the impacts of hydraulic fracturing. And it emphasized the "data gaps and uncertainties" that made EPA "unable to form conclusions" about these impacts. This new tone undermined somewhat the acquittal in the draft and thus restored some of the uncertainty about the future of fracking.[27]

"Is Shale Gas Good for Climate Change?"

Today, environmentalists look at all energy developments through the lens of global climate change. For them, the question about shale gas might be summed up nicely in the article "Is Shale Gas Good for Climate Change?" It was written in 2012 by Daniel P. Schrag, director of the Center for the Environment at Harvard as well as a member of President Obama's Council of Advisors on Science and Technology.[28]

Schrag readily acknowledges the economic and environmental benefits of the surge in shale gas. Regarding economic benefits, he says, "There are enormous benefits in having cheap, abundant natural gas for the United States in terms of the competitiveness of US industry and economic growth in general." As to environmental benefits, he notes that natural gas use is beginning to displace coal use; he concludes that from 2007 to 2011, natural gas use had increased by 15 percent and coal use had decreased by 10 percent. He goes on to say that displacement of coal by natural gas reduces emissions of carbon dioxide, sulfur, and mercury. But while Schrag points out the environmental benefits when natural gas is used in power plants instead of coal, he by no means ignores possible environmental concerns at the well site, including the potential for groundwater contamination and methane emissions.[29]

Interestingly, for Schrag the answer to the question he poses in the title of his article is not found in an assessment of short-term emissions by power plants. Instead, it lies mostly in politics. Schrag conveys this perspective when he says, "By leveraging the financial self-interest of the natural gas industry to broaden political support for anti-coal policies, environmental groups can simultaneously use a grassroots campaign to pressure existing coal-fired power plants to shut down." For Schrag, "The success of this [anti-coal] strategy will determine whether shale gas is indeed good for climate change." One other dimension affects the question of whether shale helps or hurts the cause of climate change: What if its low price and abundance slow the move to low-or-no-carbon technologies like wind and solar?[30]

Still More

Further regulations stemming from the concern over environmental impacts have been proposed. For example, President Obama called for more control of methane emissions from natural gas production. The goal is to cut emissions of this potent greenhouse gas by 40–45 percent by 2025.[31]

Earthquakes are another matter. There does seem to be a clear link between seismic activity and wastewater injection, which can be part of the hydraulic fracturing process. Wastewater injection, done by shale gas producers and others, can cause seismic activity, which in turn can cause damage. The surge in earthquakes in Oklahoma is valid evidence. According to the United States Geological Survey, from 1978 to 1999, earthquakes of magnitude 3.0 or higher averaged about 1.6 per year. The same low level persisted to 2008. By 2013, however, the number of such earthquakes was 109, and in 2015 the number was 890.[32]

A ROLE FOR GOVERNMENT?

Because energy innovations can have far-reaching effects and set precedents for the future, several key questions should be asked about a potentially revolutionary technology like shale gas production. One such question is whether government research and development efforts also helped launch the shale gas revolution. The issue was teed up at the highest levels when President Obama, in his 2012 State of the Union address, first praised the innovation reflected in the shale gas revolution and then implied that the federal government had a lot to do with it. The president said: "And by the way, it was public research dollars, over the course of thirty years, that helped develop the technologies to extract all this natural gas out of shale rock—reminding us that government support is critical in helping businesses get new energy ideas off the ground."[33]

To get a clear picture of the motives for and methods of the revolution, it is helpful to look back to the 1970s. Energy markets in the 1970s were buffeted by at least four major concerns: the oil price spikes due primarily to the Arab oil embargo, coal mine safety, new air pollution regulations, and a natural gas shortage. Research into new technologies became a major focus of government efforts. A result of this focus was that the Energy Research and Development Administration (ERDA) was created in 1974. On ERDA's agenda was the exploration of "mysterious unconventional natural-gas reservoirs." With the oil embargo fresh in their minds, researchers focused on the disappointing fact that just 32 percent of the oil under the surface could be extracted with the exploration and drilling methods the industry used at the time. As early as 1973, Congress gave $1.3 million to ERDA predecessors to study new methods—including the fracturing of underground rock formations.[34]

A 2012 report on the history of government energy research titled *A Century of Innovation* cites examples of ERDA and its predecessors conducting research on the early stages and elements of fracturing; the report was prepared by Sherie Mershon and Tim Palucka of the National Energy Technology Laboratory. While the authors conclude that many of ERDA's efforts led to "dismal failures," the research went on. Ultimately, the report found, ERDA signed "nine new cost-sharing contracts in which representatives of government and industry would further investigate massive hydraulic fracturing and chemical explosive fracturing."[35]

These investigations could not proceed fast enough. American natural gas production fell, and there were natural gas shortages in 1976 and 1977. In response, ERDA focused its attention on "four seldom-explored geological settings that might yield natural gas in large quantities." The eastern shale gas project is one oft-cited example of government interest in shale gas development. Still, for full commercialization of any technology, the private sector must be convinced the technology will work and be profitable. To their credit, the authors of this report on the history of government innovation make this very point and give ample credit to Mitchell's role in the revolution.[36]

LEGACY OF A REVOLUTION

The shale gas revolution has been a game changer for both the natural gas and electricity businesses; the new abundance of shale gas has driven prices down significantly for both of these commodities (and oil, too) and promises abundance for decades to come. More broadly, the revolution has changed America's perspective on its own energy resources, from shortages to abundance, and from the fear over imports of natural gas to excitement over exports. The revolution may also help to revitalize the US economy by creating jobs directly in shale gas and oil production and indirectly as consumers who spend less on energy have more money to spend on other goods and services. Only new technologies that actually lower the cost of energy can assure net job creation like this over the long run.

The shale gas revolution can be transformative for combating global climate change, too. In George Mitchell's words, "Natural gas is the perfect bridge fuel on the way to a less carbon-dependent economy." Using natural gas, rather than coal, to produce electricity can cut carbon dioxide emissions by at least half. Flexible natural gas–fired combined-cycle power plants are able to further optimize energy production by accommodating the intermittent electricity generation by renewable fuels, like wind and solar.[37]

These benefits cannot continue, however, if the revolution does not continue. The revolution is at risk both belowground and aboveground. Belowground there is the threat that improvements in the technology stop; constantly improving technology will be needed as the resources become more and more difficult to tap. Aboveground, meanwhile, new or tightened environmental regulations could potentially make production more expensive.

There are legitimate environmental concerns with shale gas production that must be fairly balanced against the economic and environmental benefits. The underlying danger is that the United States is in no mood to adopt such a balanced view. In one of his columns in the *New York Times*, David Brooks writes: "The country is more divided and more clogged by special interests. Now we groan to absorb even the most wondrous gifts." The gift being, in this case, George Mitchell's innovation. Brooks worries that the United States "is polarized between 'drill, baby, drill' conservatives, who seem suspicious of most regulation, and environmentalists, who seem to regard fossil fuels as morally corrupt and imagine we can switch to wind and solar overnight." This is an era of uncompromising belief on both sides.[38]

The shale gas revolution has demonstrated an essential truth about the nature of innovation. George Mitchell's skill and persistence were central drivers in the revolution. When he died on July 26, 2013, at the age of ninety-four, the *Economist* called him the "Father of Fracking" and said, "Few businesspeople have done as much to change the world as George Mitchell." However, the cliché has it right: success happens at the point where skill and opportunity collide. Looking through the wide-angle lens of history, it is important to see that the opportunity for Mitchell was created by the deregulation of the natural gas and the electricity businesses in the 1970s through the 1990s. Without the deregulation of the natural gas business there would have been no opportunity for Mitchell to earn a profit and, therefore, no incentive for innovation. Without competitive reform in the electricity business there would not have been the full commercialization of the natural gas–fired combined-cycle plants that became an important customer for shale gas and a central means of mitigating global climate change concerns. People matter, but policy matters, too.[39]

Mitchell was not alone in bringing about the transformed energy landscape. It has been said that Aubrey McClendon "wasn't the inventor of fracking, but he was its chief apostle." McClendon was the cofounder and chief executive of Chesapeake Energy, a major shale gas producer. He has a much more mixed reputation than Mitchell, as he was ousted as the chief executive of Chesapeake in 2013 and was indicted in 2016 for "conspiring to rig the price of oil and gas leases." The day after his indictment he was killed when his car went into a bridge embankment. As it was with Insull, the public will have to balance the allegations of wrongdoing with his important contribution to the shale gas revolution.[40]

With the echo of warnings about America becoming dependent on overseas imports still audible, Reuters reported the first exports of American liquefied natural gas on February 24, 2016. Mitchell's and Devon's technologies have changed everything.[41]

CHAPTER 19

Elon Musk's Vision (and History's Lessons)

That men do not learn very much from the lessons of history is
the most important of all the lessons that history has to teach.
—Aldous Huxley (1959)[1]

Understanding the lessons of history is an essential first step in anticipating
the future. Aldous Huxley, however, warns, "That men do not learn very
much from the lessons of history is the most important of all the lessons
that history has to teach." The question, then, should be: What are the
lessons of history and is anyone paying attention to them? One good
test of Huxley's assertion today is to check whether history's lessons are
reflected, intentionally or not, in the vision of the charismatic and sometimes controversial business leader Elon Musk. Two of the three corporations he created are squarely in the electricity business, and both are
squarely focused on revolution. Musk has become a celebrity because of
his ambitious goals, his narrow escapes from failure, and, more important, his successes.[2]

ELON MUSK'S VISION

Musk had two early successes with start-up ventures. The first was Zip2, which started out as a business directory tied to maps—what Musk describes as "a really advanced blogging system." When Zip2 was sold to Compaq, Musk and his brother, Kimbal, walked away with an estimated $37 million. His second successful venture was into disruptive technology for internet banking, which later morphed through a merger into PayPal. Although Musk was pushed out after a leadership coup, he still walked away with an estimated $250 million when PayPal was sold to eBay. What's most impressive is that Musk then put most of that money at risk through three other potentially revolutionary ventures: SpaceX, Tesla Motors, and SolarCity.[3]

SpaceX builds and launches rockets into space to place satellites in orbit for both public and private customers and to resupply the international space station. Reflecting what would become a core business strategy, Musk builds his rockets from scratch in America, and he fully intends to be the lowest-cost supplier. After three failures, and six years of constant work, SpaceX had its first successful launch in September 2008, becoming the first private company in space to orbit Earth using a liquid-fuel rocket. Elon Musk said, "We [SpaceX] want to be the Southwest Airlines of space launches . . . They offer flights for a heck of a lot less, have a strong safety record, and are on time." The ultimate cost savings for SpaceX were to come from the development of a *reusable* rocket. However, low-cost rockets are only an interim goal. The ultimate purpose of SpaceX is to colonize Mars. Yes, that is the intent—to create a backup planet of sorts if all does not go well on Earth. It is easy to see why Musk is known for setting ambitious goals.[4]

Tesla Motors manufactures electric vehicles. Musk's motives here are more down to earth, but no less ambitious: sustainably mitigating global climate change. As with SpaceX rockets, Tesla developed its cars from scratch and started by building them in America; in addition, Tesla benefited from some of the innovative manufacturing practices developed by SpaceX. Lithium-ion batteries power the Tesla cars; ultimately, Musk plans to build his own batteries at full capacity in 2020 at what he calls his Gigafactory in Nevada. With Tesla and those specially built batteries, Musk hopes to achieve economies of scale in manufacturing, thereby cutting the costs of batteries dramatically.[5]

SolarCity is a venture created by Musk's cousins Peter and Lyndon Rive, and Musk is a major investor and leader for the company. SolarCity started as an installer and innovative financier for rooftop solar photovoltaics, or solar panels, for homes and commercial buildings. At first, SolarCity purchased its solar devices, taking advantage of the worldwide glut and the resulting depressed prices. However, like its sister ventures, SolarCity now intends to build its own solar devices in America. On April 30, 2015, Musk announced another device closely linked to SolarCity (as well as to Tesla): an electricity storage device for homes called Powerwall, based on lithium-ion batteries. A larger version for utilities, Powerpack, is also available.[6]

Musk's importance for the electricity business is manifested in two ways, through two of his companies. One is a demand-side revolution: he will create new customers for electricity with his battery-driven Tesla cars, which can be recharged from the existing centralized power system—the electric "grid"—or from solar and other devices separate from the grid. The second is a supply-side revolution: he will create an alternative to the traditional electricity grid. Musk made it clear that ultimately a homeowner or owner of a commercial building may cut its tie to the grid by combining rooftop solar and storage. The combined devices would serve electric needs when the sun shines and then use any excess electricity generated at that time to store power in the Powerwall device; the stored power would then be used to serve electricity needs when the sun does not shine. The rooftop solar combined with Powerwall storage creates an entirely new system of what can be termed *personal power*, and someday it may compete with the grid.

Musk is not without his critics, and success is not guaranteed for any of the three ambitious ventures. In his 2015 biography of Musk, Ashlee Vance cites some of the rhetoric from critics. One calls the Tesla car "nothing but an utterly derivative overhyped toy for showoffs." (Tesla has since announced its Model 3 at a starting price of $35,000, as compared to about $75,000 and up for Tesla's Model S.) Many others point to the subsidies supporting the manufacture and sale of products by Tesla and SolarCity, implying that those subsidies invalidate Musk's success. However, others put Musk in a larger context relative to other current inventors and draw more flattering conclusions. Peter Thiel, one of the founders of PayPal, bemoans the decline of true innovation, saying, "We wanted flying cars, instead we got 140 characters." No one could say that Musk did not go big in this regard with his focus on rockets, electric cars, and solar energy, and by risking all his wealth in the process.[7]

Musk is best characterized as a hero without guarantees. He is a hero—even an old-fashioned American hero—because of what he is trying to do and how he does it. He is trying to revolutionize space transport with reusable rockets, road transport with sleek and powerful electric vehicles, and home heating and cooling with electricity generated by rooftop solar panels backed up by storage. Moreover, all his devices are built from scratch and (for now) built in America. And he did it as the classic entrepreneur; he started with nothing, made his early fortune, and then bet that fortune on three new ventures.

It is impossible not to admire him for these heroics. However, his success is not guaranteed. There are substantial risks made plain by the day-to-day press coverage Musk receives. Here are four examples. First, standard financial risk confronts him. In 2016 he proposed merging Tesla and SolarCity. Some saw danger here in that he was combining two companies that both needed huge cash investment. Some worried that a SolarCity bankruptcy would bring Tesla down.[8]

Second, if the ultimate success for electric vehicles depends on driverless vehicles, the widely publicized death of a person in a driver-assisted Tesla vehicle in Florida in 2016 reflects that risk. The broader risk is whether consumers want to give up car ownership and hands-on driving as required for the world according to Uber, Lyft, and others.[9]

Third, Musk has bet heavily on lithium-ion batteries with his Gigafactory. The risk of fire with those batteries was given significant attention because of the 2016 Samsung recall of some of its smartphones as well as the headlines in 2013 about the new Boeing 787 Dreamliner being grounded for fire problems with lithium-ion batteries. (Tesla has a cooling system to address this risk.) Fourth, in September 2016, a SpaceX rocket exploded on the launchpad.[10]

Still, in the face of these risks, Musk marches on with a new master plan issued in June 2016 that sums up his goals as follows: "Create stunning solar roofs with seamlessly integrated battery storage. Expand the electric vehicle product line to address all major segments. Develop a self-driving capability that is 10X safer than manual via massive fleet learning. Enable your car to make money for you when you aren't using it." And all the while Musk keeps up with his ambitious goal to colonize Mars—he believes people will reach Mars within ten years.[11]

With the rapid-fire technological developments, and hyperbolic media attention, it is hard sometimes to see the enduring legacy Musk is building. It is possible that Musk is leading a revolution in the electricity business—one that will democratize electricity in a fundamentally new way. To judge this, however, it is important to see Musk's efforts in the context of the lessons of history.

HISTORY'S LESSONS

Does Musk's vision reflect history's lessons for truly revolutionary inventors? Generally, the answer is yes. Indeed, he may be channeling Edison, Westinghouse, and Tesla with his vision for technology, and Insull with his policy vision and, perhaps, his financing strategy. In addition, Musk shows, in his own way, a respect for science and scientists that was at the core of the breakthroughs by Franklin, Faraday, Maxwell, and Einstein. Although Musk's vision offers solutions to only a particular subset of problems, it is still fascinating to find that Musk's ambitions reflect the lessons learned in the most important events in the discovery and application of electricity.

Start at the End

In the Age of Edison, both Edison with his direct-current system and Westinghouse with his alternating-current system started their design at the end; that is, they first focused on the actual use or "end use" for electricity. Edison focused first on his lightbulb with painstaking research on the best material to use. He then backtracked to create all the elements of his system from the dynamo (his power plant) to the transmission lines that brought the power to light his bulb. Westinghouse focused first on the alternating-current electric motor that Nikola Tesla conceived and built. He worked back from there, especially on the alternating-current transmission system that would carry the power from his power plants, however distant, and deliver it to his customers, however widely dispersed.

The electricity business lost that focus on end use over the years because cost-plus rate making centered the profits on big power plants; that is where the money was made and that is where the attention was paid. Less, if any, attention was paid to innovative ways to serve particular end uses. In the Age of Harm, attention refocused on end use but only because the goal was to use less electricity given the harm electricity generation was thought to cause the environment. Attention even returned to lightbulbs: indeed, Edison's incandescent bulb was banned under new federal law because it was deemed to be less efficient than alternatives like compact fluorescent bulbs.[12]

Until Musk's ventures, however, few innovators in the electricity business had successfully focused on end use in a positive way. Now Musk has revived and developed technologies to create a new end use—even a new customer—in Tesla Motors' electric vehicles and the people who invest in them. He also brought the focus back to end use with the combination of solar rooftops from SolarCity, plus storage with Powerwall—thus creating a modern version of "personal" power generation. In this sense, Musk started at the end and found the greatest motivation for innovation there.

Cross-Pollinate and Collaborate (a Culture of Invention)

Edison and Westinghouse each filled their laboratories with people who contributed either general genius or specific experience. Edison created a new culture of invention at his Menlo Park lab, and, of course, made major contributions to the technology for a wide range of businesses: the telegraph, the telephone, and motion pictures, in addition to electricity. Westinghouse also created a culture of invention at his Garrison Alley lab and, in addition to his own genius, he cultivated the talents of, among others, Nikola Tesla, who felt cheated by Edison. Westinghouse similarly served a range of businesses, making especially important contributions to railroad safety. That experience gave him an understanding of long-distance power transmission, which is central to the alternating-current system. These early efforts established a culture that promoted cross-pollination and collaboration. They also brought us technologies that have become permanently woven into ordinary life.

Musk has since brought his own genius to the business and, at the same time, has cultivated the talents of others. It could be argued that his form of cultivation is a bit tougher than the mentorship of gentle genius Westinghouse, but it is cultivation nonetheless.

Dolly Singh, who headed SpaceX's recruitment for five years, makes the point with this metaphor: "Diamonds are created under pressure, and Elon Musk is a master diamond maker." Yes, she admits, Musk applies the pressure directly and forcefully on everyone, but he inspires many. Singh tells the story of what happened immediately after the SpaceX rocket failed on August 2, 2008. Musk walked right past the press to talk to his team. He reminded them that they all knew it was hard—literally, it was rocket science. He reassured them that SpaceX had the financial backing to try again, and again. And he inspired them by saying, "For my part, I will never give up and I mean never." In less than two months, on September 28, 2008, SpaceX became the first private company to orbit Earth with a liquid-fueled rocket. Leadership is not about being liked by everyone; it is about achieving success for everyone.[13]

Musk has also cross-pollinated industry-specific knowledge, as advances in lithium-ion batteries in laptop computers and other mobile devices clearly gave him a powerful foundation for both his electric vehicles and his electric storage devices. Now he intends to realize economies of scale in manufacturing at his Gigafactory. That strategy will serve both Tesla Motors and Powerwall, inevitably leading to technological overlap between both companies. The cross-pollination goes even further: Musk already took some of the major manufacturing innovations used by SpaceX to inspire and inform the manufacture of his electric vehicles. As we take a wide-angle lens on the history of invention and electricity, it seems that Elon Musk has the ability to inspire his team and to draw inspiration across his ventures just as Edison and Westinghouse did more than a century before him.[14]

Make Science (and Scientists) Matter

Man's understanding of electricity is based on some of the most stunning scientific discoveries of all time. That alone makes it different from other conventional energy forms; not that science did not and does not matter to oil, coal, and natural gas, but scientific endeavor is not the essence of these as it is with electricity. As Einstein showed, electricity was not just one stop along the path of scientific discovery in the nineteenth and twentieth centuries; the study of electricity was the path itself. This gives even greater hope for scientific discovery with electricity. The first burst of scientific discovery for electricity was in the Age of Franklin and encompassed Franklin, Faraday, and Maxwell. After a long hiatus, in the Age of Edison and the Age of Big, Tesla and Einstein led two further surges in scientific breakthroughs. Along the way, the electricity business continued to motivate technological innovation even as it became embattled on the regulatory front. In 2016 Elon Musk is perhaps the most fitting torchbearer for our latest revolution in both the science and business of electricity.

Today, out of the thousands of pages written on the electricity business, few even mention science. It is as if all that groundbreaking science has been tapped out, as if there can be no further bursts of discovery. Surely this cannot be true. As we face challenges around the United States' growing energy consumption and concerns over environmental sustainability, we cannot presume this deep well is dry. Science did and does matter to the electricity business, and serious respect for and interest in it must be revived.

Scientists matter, too. There is no more compelling example of this than Franklin. His stature as a scientist was, in large part, the reason he could negotiate with the French, and those negotiations led to an alliance that eventually ensured the American colonies' victory at Yorktown. Perhaps such a far-reaching public payoff cannot be expected ever again, but the manner in which scientists conduct themselves still matters. Signature examples lie in Maxwell's great respect for Faraday, and Einstein's respect for both Faraday and Maxwell. The manner and tone of scientists reflect whether society can maintain a civil discourse on matters of importance. With Franklin, Faraday, Maxwell, and Einstein, scientists were thought to bring wisdom to scientific inquiry; their inventions, meanwhile, deeply affected many other debates and conflicts of the day. Just think what might be achieved through a nuanced and civil discourse on issues such as global climate change and shale gas development. Add the breakthroughs in medical sciences and the development of artificial intelligence and it is clear that science and scientists may matter to society today more than ever before.

Like Edison, Musk is committed to using science to develop his products rather than developing science as an end in itself. Those products, in their very ambition, embed advanced scientific thought, from rocket science to personal power storage. Even Musk's motives are based in science, as far as we can tell. For him, SpaceX is a path to colonizing Mars, while Tesla Motors and SolarCity contribute to mitigating global climate change. His choice of solar photovoltaics is an additional bet on the advancement of science; the hoped-for dramatic reduction in cost in the future depends on significant advances in the science of manufacturing and materials in the present. Musk seems to have doubled down on science with SolarCity's business strategy of building, rather than buying, its solar panels. Furthermore, Musk is not shy about speaking out as a scientist on the major issues of the day—including, interestingly, his opposition to the development of artificial intelligence.[15]

Open a Door to Outsized Economic Impact

The economic impact of new electricity technologies is not demonstrated strictly by the amount of money spent on electricity. Technological change is not all about getting cheaper energy. Economic impact is also tied up with how that new technology transforms the economy. The move from water power to steam power transformed the economy as factories no longer had to be situated next to water; instead, the factory could be located much closer to both customers and to employees. The same indirect, outsized economic impact was realized with Morse's telegraph—the "Victorian Internet"; the telegraph allowed the scale and scope of many other industries to grow. The significant economic impact of Nikola Tesla's alternating-current motor was that it allowed the redesign of the factory floor. No longer were workers tethered to the steam engine in the basement; the work flow would be designed instead around what is best for product manufacture. On the national stage, the government-created Hoover Dam brought water to allow the California desert to bloom, while the private innovation of George Mitchell altered the geopolitics of energy and built a bridge to future mitigation for global climate change.

However, predicting the impact of technological change is more difficult in the Age of Harm than it was in the Age of Edison and the Age of Big. The driving force for change in the Age of Harm is to minimize the environmental harm caused by air pollutants. Given that, it is important to ask whether there will be any positive economic impact from these new technologies. In the Age of Uncompromising Belief—more an age of partisanship and outrage than an approach to science—it can be challenging to locate evenhanded assessments of the benefits versus the harms of new technologies.

Musk's approach does not immediately promise outsized economic impact, but it may open the door to it. The most enticing opportunity for such change might occur because the new technologies are designed with particular end uses in mind and built from scratch. For example, Musk is producing a *new* vehicle, not just a new *electric* vehicle. Vance notes that Tesla's "Model S was named *Motor Trend's* Car of the Year" in 2012, and concludes that "the Model S was not just the best electric car; it was the best car, period, and *the* car people desired." Moreover, Tesla is a strategic competitor to the current industry leaders because its cars are built on a completely different platform, and, therefore, can be manufactured in a fundamentally new way, including the manufacture of the "engine," the lithium-ion battery. The broader impact of this new platform isn't known yet, but it could lead to an interest in new materials and new manufacturing methods that could stimulate new industries. The ultimate, outsized economic impact, however, may come from the fact that the Tesla and other electric cars are the essential tool for a revolution in the transportation industry: the emergence of driverless or autonomous vehicles.[16]

Moreover, Musk (for now) manufactures *in the United States*. This means all three of his businesses are meant to gain competitive advantage for America in industries targeted by strategic competitors in other countries—China in particular. For rockets, Musk asserts, "Our primary long-term competition is in China." China also has targeted electric vehicles and solar panels. There are no guarantees, but there may be a significant economic impact from Musk reviving American manufacturing with a unique mix, as Vance puts it, of "software, electronics, advanced materials, and computing horsepower."[17]

Count on an Intrinsic Trust in Business

Edison and Westinghouse used private capital and private companies to achieve their technological breakthroughs. This gave good reason for the public to trust private investment in the electricity business during the Age of Edison. Insull circumscribed that trust when he asked for geographic monopolies to be carved out and invited regulation of the profit of those monopolies. The monopolization ended competition within the geographic area covered by each regional monopoly, but, still, these were privately owned, not government-owned, companies.

The Great Depression had the potential to tip the scales toward government ownership of the electricity business; after all, the Great Depression was seen as the result of a great failure of capitalism in the 1920s. President Franklin D. Roosevelt put the blame for the Depression on big business owners, including Insull, but he did not call for government ownership; indeed, he spoke against it. While President Roosevelt broke apart all the major electric holding companies of the day, he was restrained in his explicit choice not to seek government ownership. The TVA (Tennessee Valley Authority) project was one major exception: it was a rare, large-scale experiment with state capitalism, but it never was replicated, and it did not set a template for government control.

The 1970s and 1980s brought successful efforts to deregulate the electricity business, or, at least, to open the door to some competition through regulatory reform. Then confidence in that competitive reform was undermined by three high-profile events: the California crisis in 2000, the collapse of Enron in 2001, and the blackout of 2003. These three events may have comprised the biggest challenge to the intrinsic trust in private business—an even greater specific challenge than the Great Depression. Any one of these events could have derailed competitive reform and, yet, not even the three in combination did that.

All three of Musk's corporations embody direct, disruptive competition for established industries: SpaceX in the aerospace industry, Tesla Motors in the automobile industry, and SolarCity plus Powerwall in the electricity industry. Moreover, each of these companies pursues explicit, strategic competitive advantage; effective strategic competition is at the core of Musk's success. In this way Musk is counting on the American public's continued, intrinsic trust in private business.

Acknowledge That Environmental Policy Will Dominate

The tragic "episodes" of pollution in Donora, New York City, and London in the mid-twentieth century were the early impetus for environmental regulation; they gave dramatic, physical evidence of the harm industry could cause. Next came Rachel Carson's *Silent Spring* in 1962, which created a template for environmental activism. It, too, emphasized physical evidence and added an implied burden of proof on industry to show that its operations caused no harm. In this sense Rachel Carson is the mother of the modern environmental movement and the herald of the Age of Harm.

Recent efforts to combat global climate change prove the resilience of environmentalists in the Age of Harm. The most widely known effort came in the form of the Waxman-Markey bill calling for an elaborate cap-and-trade program to cut carbon dioxide emissions. Those opposed to action to combat global climate change thought the defeat of cap-and-trade was the end of the story, but environmentalists simply followed a plan B that went through the courts. This avenue, which included the US Supreme Court, gave the US EPA the right to regulate carbon dioxide emissions from all sources, cars as well as power plants, under existing law. President Obama's Clean Power Plan is one result of this plan B. Beyond resilience, the environmentalists' success shows that the tightening of environmental regulations for the electricity business may be inevitable—with or without a vote by the people or their elected representatives.

Musk's enterprises are based on this inevitability of tighter regulations for greenhouse gas emissions. Both Tesla Motors and SolarCity are businesses driven in large part by global climate concerns, as well as the money and mandates offered in the name of those concerns. That is, these two ventures are betting on the view that environmental policy will drive the electricity business. That bet, however, may be jeopardized by pending court challenges and by the policies of the Trump administration.

Draw a New Jurisdictional Boundary

State regulation of the electricity business was Insull's idea, and it soon was the norm across America. Over time, however, the courts would not allow the states to regulate transactions that crossed state lines—so-called interstate sales. The holding companies were the real concern in this regard; the three largest controlled 44.5 percent of US generation in 1932. They also controlled the bulk of the interstate sales of electricity, although those sales accounted for only about 14 percent of all electricity generation. In 1935 with the Public Utility Holding Company Act and the Federal Power Act, the federal government gained the authority to both dismantle the holding companies and regulate any remaining interstate sales. These laws then set the jurisdictional boundary between the federal and state governments: generally, the federal government had jurisdiction over interstate sales, and states had jurisdiction over all the rest.[18]

The placement of that boundary has been argued out in the courts over many years, and the jurisdictional boundary has moved accordingly. Then, in 2013, two federal courts ruled that two state commissions in Maryland and New Jersey did not have the right to order their utilities to conduct competitive solicitations to get new power plants built that the states thought were needed to assure reliable electricity service. The Supreme Court ruled on the Maryland case in April 2016 and upheld the lower courts.

With all the uncertainty looming over the future, and all the technologies that can compete to serve future needs, the United States' priority is to achieve a diversified portfolio of technologies for the electricity business. If the states retained their jurisdiction to order utilities to build or buy new power resources of the sort the state preferred, the United States would be assured of a diversified portfolio of technologies simply because the states are so different.[19] The Supreme Court ruling puts that diversity in doubt. Moreover, had the court applied the same principles it had applied just three months earlier, in another federal preemption case, it might have reversed the lower court and restored states' rights.[20]

The jurisdictional boundary matters for Elon Musk's companies because they participate in both state and federal programs that give substantial incentives to technologies including electric vehicles, solar photovoltaics, and electric storage. These preemption cases threaten to an unknown extent the state programs that give incentives for products from Tesla Motors, SolarCity, and Powerwall. Moreover, the cases tee up the issue of when courts should defer to elected officials on matters of policy, rather than rush to judge solely on matters of law. The right policy is that federal and state programs can coexist.

Discern a Purposeful Role for Government

A long-held theory of the business world is that the role of government, if such a role is needed, should be larger in the early stages of technological development and smaller in the later stages. Specifically, the theory holds that technologies evolve through research, development, demonstration, and commercialization—an RDD&C process—and the government's role should be largest in the research stage and taper off from there. It should be more involved in research because the uncertainties of a payoff make private investment in research less likely. Looking at the history of the electricity business, however, it seems clear that this theory has been implemented backward.

Government involvement, at least historically, was lightest in the early stages. In the Age of Franklin, the science poured out of Franklin, Faraday, and Maxwell at their own initiative. The same was true for the Age of Edison: both Edison and Westinghouse (with Tesla) either funded their own efforts or secured private investors. This may have simply reflected the tone of the times—it was before the role of government was expanded significantly in the Great Depression—but it is worth highlighting here because it undermines the widely accepted RDD&C process view.

The first major expansion of government involvement came with Insull's regulatory deal. With this arrangement, the country was carved into geographic monopolies that would then charge for electricity at cost-plus-profit prices. So the expanded government involvement not only came later in the RDD&C process—actually in the commercialization stage—but also came in a fundamentally different form. This was not government funding of any stage of the development process; it was *economic* regulation governing who got to sell electricity and at what price. Insull discerned this role for state government because he believed—correctly, as it turns out—that the resulting economies of scale and scope would help to achieve the overarching goal of the day: making electricity available and affordable to all.

That goal has changed over time and so have the tactics to pursue it. The most frequently stated overarching goal today is to mitigate global climate change. The most common tactic, meanwhile, is designation of "official technologies" along with the use of a wide range of direct and indirect subsidies. Today, renewable technologies such as wind and solar are the official technologies, and they are subsidized with both money from the federal government and mandates from the states.

Musk's companies have employed and enjoyed the full benefit of this money and these mandates, and they have been criticized heavily for doing so. The *Los Angeles Times* claimed the government subsidies to all three of Musk's companies added up over time to $4.9 billion. Musk disputes how the article implies that subsidies are the reason for his success, and he argues that the subsidies he enjoys are not out of line with subsidies he sees in other energy industries. Still, legitimate concern exists about the government "picking winners" rather than allowing competitors to win or lose on a level playing field. A new purposeful role for government has to be discerned with the overall goal well stated and with tactics that serve that goal fully specified.[21]

Use a Wide-Angle Lens

To understand the past and to anticipate the future, it is essential that decision makers consider events (and possible eventualities) through a wide-angle lens. Hopefully the previous eighteen chapters have illustrated the value of this proposition.

For Musk and his ventures, that means several, less immediately evident factors must be taken into account to understand their significance. Tesla's electric vehicles represent more than a head-to-head competition between conventional internal combustion engines and the lithium-ion battery "engine." The industry-altering possibilities for buying electric vehicles include the broad move to driverless vehicles; electricity may turn out to be the preferred fuel for these vehicles, and their success would then drive Tesla's sales.[22]

Law, policy, geopolitics, and finance matter, too. Tesla's direct sales to retail customers have been challenged in the courts by traditional auto distributors; the grounds are that long-standing state law requires manufacturers to sell their cars through franchised dealerships. In policy, meanwhile, ever tightening fuel efficiency standards for conventional cars create a competitive advantage for electric vehicles, and Musk supports them. Around the world, surging demand for lithium raises concern about the sources of supply. Some say 70 percent of long-term lithium supplies are located in three countries—Chile, Argentina, and Bolivia—while others point to new sources emerging, including some in America. There is also increasing attention on China's interest in lithium and, as a consequence, the possible negative effects on availability and price. Finally, given the history of personal financial involvement by Insull and Enron, some have raised questions about Musk's generous investment in his companies—despite the fact that much of his early profits were beneficially put at risk when they went toward his ambitious new projects.[23]

A PATH FORWARD

A Second Battle of the Systems (a Second Democratization)

Besides the Huxley quotation at the start of this chapter, there are two other famous thoughts on the lessons of history that are illustrative of the story here. The first is George Santayana's view that "those who cannot remember the past are condemned to repeat it." A counter of sorts is attributed to Mark Twain, in which he supposedly said, "History does not repeat itself but it rhymes." The two most intriguing rhymes in the context of the electricity industry would be a second Battle of the Currents and a second democratization of electricity.[24]

Decades ago the titans of technology in the electricity business—Edison and Westinghouse with Tesla—fought it out to see whose system would rule. Westinghouse and Tesla won, with their alternating-current system proving more valuable than Edison's direct-current system. An important subplot to the story is that the Westinghouse and Tesla system, implemented by Insull, also defeated self-generation by big customers such as the trolley cars and manufacturing plants. The alternating-current system won because it gave customers what they wanted: electricity available and affordable to all. The price of electricity was lowered dramatically by the system's economies of scale and scope, and the choice of cost-plus regulation accommodated the realization of those economies. The alternating-current system was a big success in this regard, and it was a big success even if one focuses just on the engineering: in 2003 the National Academy of Engineers chose electrification as the number one engineering achievement of the twentieth century.[25]

Assuming that today's emerging system technologies prove themselves, will customers still turn to the alternating-current system? Is that system, with its far-flung power plants and an intricate transmission grid, still the best? Customers will continue to turn to that system if it continues to provide what they want. Today customers still want electricity to be available and affordable, but they also want more. These new demands include more local—or "doorstep"—reliability, better environmental performance, and a clear choice between options.

Doorstep reliability is a concern exacerbated by extreme weather events. While the alternating-current system has been remarkably reliable overall, today's customers want to be assured that the lights will stay on at their homes, offices, and factories even in the middle of a storm. People point to the damage done in 2012 by Superstorm Sandy in New Jersey as an example, including the damage due to the loss of electric service. Despite the widespread awareness of this damage, when one local utility proposed an expensive plan to prevent loss of region-wide electric service in the future, consumer groups settled on giving only 31 percent of the funds requested. It might be that the consumer groups simply did not see a payoff for doorstep reliability from investment in the *region-wide* grid. More broadly, the media feeds these doubts about the region-wide grid with frequent stories about possible electric service outages due to volatile weather, cyber attacks, and even low-tech terrorist attacks.[26]

How do the competitors to the grid score on these three customer demands? One alternative is often referred to as the microgrid. It is still a grid—it matches a need for electricity with resources to supply that electricity—but it operates on a much smaller scale. The microgrid at Princeton University has gotten substantial attention in recent years. On the doorstep reliability front, it scores well; that microgrid kept the lights on during Superstorm Sandy. In terms of environmental performance, microgrids including Princeton's appear to score well because they are open to using a full range of resources to serve customers. The resources include renewables like solar, demand-side resources like shutting off power use when supplies are tight, and conventional resources like modern natural gas–fired power plants. Consumer choice, meanwhile, was the primary motivation for the Princeton microgrid: the university can choose whether to take power from the grid or the microgrid.[27]

Personal power, another competitor to the grid, also scores well; Musk's companies are prominent examples. As we have seen, the combination of SolarCity rooftop solar and Powerwall would function as a personal power source. In terms of local reliability, the intent of Powerwall is to be ready when power from the grid is curtailed, so personal power scores well on this basis. Personal power scores well in terms of environmental performance, too, if the bulk of the power is solar rooftop. Consumer choice is inherent to this system because the customer can choose to buy power from the grid or take it from the personal power system.

These emerging technologies create the opportunity for a second battle of the systems. Indeed, it is something of a rematch between the smaller, local power of Edison's system (now represented by modern microgrids and personal power) and Westinghouse's expansive grid. The first such battle marked a major, beneficial move to the alternating-current system. Clearing a path for a second battle of the systems may bring the same magnitude of beneficial gain because it improves doorstep reliability and environmental performance in the customer's eye and, by definition, offers customers choice. This is not to say microgrids and personal power are inexpensive or that all the technology and policy issues have been worked out—or that they are likely to win. However, history shows that regulation and law must make way for a fair fight.

This second battle of the systems may also be characterized as a second democratization of electricity. Insull drove the first by producing electricity for everyone. In the near future, microgrids and personal power might democratize electricity in the sense that they allow individual customers to "vote for" the level of local reliability, environmental performance, and choice that is tailored to their needs.

Carbon Tax, Tax Reform, and Policy Sweep

Cost-plus-profit prices eliminated the incentive to invest in the development of a full range of new technologies. Why would a utility make a risky investment in new technology if the utility could earn no more than a standard profit if the new technology worked, and possibly earn no profit at all or suffer a loss if the technology did not work? In response to this lack of incentive, governments designated "official technologies." Nuclear power was once such technology. The government subsidized the development of nuclear power consistent with the overarching goal expressed by President Eisenhower in his Atoms for Peace speech at the United Nations: nuclear power was to be an economic weapon in the Cold War.

The overarching goal has changed since then and, as already noted, is now best defined as mitigation of global climate change. There are new official technologies today, too, ranging from wind to solar to clean coal to demand reduction to microgrids and personal power. The tools used to implement this designation of official technologies also are wide ranging. In terms of money, they include tax incentives such as production tax credits or investment tax credits, loan guaranties, direct cash subsidies, and some research and development funds. Beyond the money, there are mandates. For example, twenty-nine states plus the District of Columbia have rules that require their local utilities to produce a specified share of their electricity from renewable fuels.[28]

Placing a price—a tax—on carbon would be an alternative to the subsidies and mandates from federal and state governments. Shifting from money and mandates to a carbon tax would address head-on the concern about government "picking winners" when so many of those "winners" actually fail. These oft-cited failures cut across the spectrum of technologies from electric vehicles to solar photovoltaics to clean coal. So, too, a carbon tax puts in place a uniform incentive to control carbon emissions—rather than one that varies widely across the official technologies. And, notably, it allows a choice between investing in preventive controls now or just paying the tax and waiting to adapt if and when it is necessary later.

For those greatly concerned about global climate change, a carbon tax would establish a broad and permanent incentive to cut carbon dioxide emissions. Equally important, those less concerned with global climate change might be brought into a compromise if the funds from the carbon tax were used to fund income tax reform; that is, a new tax placed on something undesirable—carbon emissions—could be substituted for an old tax on something that is wanted—smart investment and hard work. In the wake of this carbon-tax-and-tax-reform compromise, a third prong of the policy could emerge: a "policy sweep" by the federal and state governments. That is, the mandates and the money targeted to official technologies would be swept away.

Influential parties ranging from Bill Gates to ExxonMobil to George P. Shultz and James A. Baker III have suggested putting a price on carbon. The carbon tax, tax reform, and policy sweep concept are just a few of many policy visions for the future. It is important to move past the stymied ideological fight in the Age of Uncompromising Belief. The government must take on a positive role, based solely on what works to achieve the goals of the day, and not based on abstractions about the right size of government. Similarly, as reflected in Gates' proposals, the future must be based on a full range of new technologies—some unimagined as of yet—not limited to today's wind and solar technologies.[29]

History's Legacy (How It Helps)

History can help decision makers in business and government by leading them to take a broader view. This can function in two ways: history leads decision makers to look across time; surely what happened yesterday influences what happens today. History also leads decision makers to look across all the factors that drive change, including those documented in this book: science and technology; politics and geopolitics; regulatory policy; law and the courts; business strategy; economics; and the catchall factor called "culture."

Examples of the benefits of such broadness are plentiful in the electricity business. If the narrative of electricity was wrongly said to begin with Edison, it would have missed the heart and soul of the original story—the science of Franklin, Faraday, and Maxwell. Failing to consider that science would lead to a faulty understanding of the past as well as an inability to see the great hope of the future.

The full significance of Watt's steam engine cannot be gauged until Nikola Tesla's electric motor comes into view. Watt's steam engine made manufacturers dependent upon—some might say addicted to—using mechanical power. That addiction meant manufacturers were eager to try new and better sources of such power; Tesla's alternating-current electric motor was just that. In this sense, Watt's innovation set the table for Tesla's innovation. Untethered from the steam engine in the basement, the electric motor allowed a productive redesign of the factory floor, which, in turn, led to dramatic economic impact.

Similarly, one cannot support an intrinsic trust in business without considering a lengthy series of historical events, ranging from the first

Battle of the Currents to Enron's collapse. This broadening perspective reveals that it took four decades to gain traction for the deregulation of the electricity business, extending from the Chicago School's intellectual pressure in the 1960s to the Arab oil embargo in the 1970s to the surge of mergers in the early 1990s to the call for regional transmission organizations in the 2000s. Nor without a broadened view could the roots of today's shale gas revolution be seen in the deregulation of natural gas and electricity in the 1980s.

So, too, a look back can give access to unique evidence on a particular point. For instance, some say that science and spiritual faith cannot coexist and, further, that faith impedes scientific progress. Michael Faraday's life and work give ample counterevidence to the claim. Faraday's faith was hardly an impediment to his groundbreaking scientific advances. Indeed, his faith in *one God* led him to search with all his heart and skill to find *one force*. The link Faraday discovered between electricity and magnetism is still the foundation of both the electricity and telecommunication businesses. Diversity of views of every sort is needed to create fundamental change.

To many, too much of the money spent in science and technology is aimed at toys by the rich and for the rich. This criticism is directed at Musk's ventures, for instance. But a broader view shows that what might start as toys by the rich, for the rich, can end up as significant advances in science and technology for everybody. Franklin's wealth allowed him to retire and devote five years to the science of electricity—leading to breakthroughs that made his views the accepted standard. Edison placed some of his first power plants in the homes and even in a ship owned by rich investors. And, of course, in that era nearly all investment came from the wealthy. Musk has argued that the development of the expensive Tesla Model S will lead to the development of more affordable electric vehicles—and it has. It is not how and where the revolution begins that matters; it is whether it ends with true advances, judged in the cultural context of the times.[30]

Similarly, the ideological battle over the role of government might be defused if evidence from the history of electricity was considered. That role has varied greatly over time and, yet, across these variations, government was frequently successful in defining and serving the overarching goal of the day. The key question is what works, and the answers are far more nuanced than a purely partisan ideological debate allows.

The history told here shows how the electricity business arrived at the present through both successes and failures. It can serve as a guide to the future if it is assessed through civil discourse that reflects a wide-angle lens on history. Still, there is no way to accurately predict how the future will unfold, nor is it guaranteed that man will try to know and apply history's lessons. The only guarantee is that whatever happens will be, well . . . simply electrifying.

EPILOGUE

Just Imagine

We Need Energy Miracles
—Bill Gates (2014)[1]

Benjamin Franklin accepted the challenge to see the unseen—to imagine elec-tricity. Today's challenge is no different: to imagine the uses and sources of electricity in the future. To create a second Age of Franklin, a time of awe and discovery, the same boundless imagination is required, unconstrained by past ideologies of science, technology, politics, government policy, economics, business strategy, and culture. The one legitimate constraint is to pursue only what works to achieve clearly stated goals.

AN ABUNDANCE OF TECHNOLOGIES

Thankfully there is no shortage of imagination around the world. One reason is that visionary, hardworking scientists like Faraday and Maxwell forged a path. They showed how to make electricity by simply pushing and pulling a magnet inside a coil of wire. The mechanical power to do the pushing and pulling most often comes from steam produced by a range of methods, from burning coal to incinerating garbage to controlling a nuclear reaction. Compressing natural gas to drive a combustion turbine—essentially a jet engine—is well understood and widely applied today, too. And the push and pull can be driven just as easily by other forces; wind, or ocean waves, or river currents, for example.[2]

Some suggest an old technology can be taught new tricks. Specifically, the old technology is nuclear power and the new trick is to manufacture small reactors. A University of Chicago study in 2011 constructively and carefully made the case for small modular reactors (SMRs). The strongest case is that nuclear power has no carbon emissions and can be built on a large scale. The challenges to SMRs were laid out fairly by the authors: "safety, nonproliferation, waste management, and economic competitiveness.[3]

Readily understood and available, but a bit harder to explain, are technologies such as solar photovoltaics and fuel cells. Photovoltaics convert light into electricity at an atomic level. This technology echoes Einstein, who was awarded the Nobel Prize in Physics for his work on the photoelectric effect. The space program was the first serious use of the technology to provide power on spacecraft. Solar photovoltaics rely on the same types of semiconductor material used in computers; indeed, the link to the science of computers is part of the reason for optimism for solar photovoltaics to greatly improve performance and lower costs.[4]

A fuel cell converts hydrogen and oxygen into water and, by doing that, produces electricity: a "flow" of electrons. Since it is a chemical reaction, it produces few of the air pollution emissions that occur with fossil fuel combustion—and that is a primary appeal. However, the full impact on emissions, including greenhouse gases, depends importantly on the source of the hydrogen for the fuel cell. Also appealing is the fact that cells can be stacked to meet any scale of electric demand. What is less appealing is the relatively high cost.[5]

Photovoltaics and fuel cells could be important warriors in the second Battle of the Currents because both can be used at a scale suitable for personal power. They are important, too, in the sense that the potential for dramatic performance improvements and cost reductions lies in the manufacture of the devices themselves—economies of scale in manufacturing, not in the power plant.

New science is at the heart of new technology. Consider as one example the newly created material called graphene. The 2010 Nobel Prize in Physics was awarded to Andre K. Geim and Konstantin S. Novoselov for "producing, isolating, identifying and characterizing graphene." Graphene is a single atomic layer of carbon. The Royal Swedish Academy of Science calls carbon "the most fascinating element in the periodic table," with graphite being the most common form; graphite is used in the ordinary pencil, and the academy points out that anyone who has used such a pencil "has probably produced graphene-like structures without knowing it." The excitement is palpable, with the two Nobel winners themselves writing that graphene "has already revealed a cornucopia of new physics and potential applications." For the purposes of this narrative, the potential application for electricity technology is most important—and the excitement remains undiminished since the Nobel win. The University of Manchester asks us to "imagine fully charging a smartphone in seconds, or an electric car in minutes. That's the power of graphene." And further, that "graphene membranes could be used to sieve hydrogen gas from the atmosphere—a development that could pave the way for electric generation powered by air."[6]

Another example is the work of Jean-Pierre Sauvage, Sir J. Fraser Stoddart, and Bernard L. Feringa, who together won the 2016 Nobel Prize in Chemistry "for the design and synthesis of molecular machines." In awarding the prize, the Academy went on to liken the advance to the science behind the electric motor: "In terms of development, the molecular motor is at the same stage as the electric motor was in the 1830s, when scientists displayed various spinning cranks and wheels, unaware that they would lead to washing machines, fans and food processors. Molecular machines will most likely be used in the development of things such as new materials, sensors and energy storage systems."

The stories of graphene and molecular machines are meant only to illustrate the possible impact of new science; these are by no means the only paths forward, but they show what can be done if we step back and just imagine.[7]

OPEN THE DOOR TO ALL

This history is focused on the technologies used to produce and distribute electricity. And yet it would clearly be wrong to identify technology alone as the driving force behind the electricity industry. The choice of technology is itself driven by so many factors, including government laws, policies, and regulations. A wide-angle lens must be used to fully see the nature of the choices made in the past. The Edison versus Westinghouse/Tesla Battle of the Currents was perhaps the clearest instance of the more advantageous technology emerging victorious, while its competitor was abandoned. The alternating-current system won because it best served the goals of the day and, therefore, was encouraged and accommodated by policies such as establishing geographic monopolies.

After that fair fight, the choice of technology was driven substantially by Insull's cost-plus rate making. This structure promoted capital-intensive technologies like coal-fired power, although oil and natural gas–fired power plants remained a constructive, competitive threat.

The designation of official technologies then began. Nuclear power is the prime example. Although it is still used for up to 20 percent of America's electricity generation, in the end, nuclear did not help make electricity available and affordable to all. The designation of nuclear as an official technology was probably misguided in the long run, because it may have pushed nuclear to come on line too fast and at too big of a scale; after all, it was born small-scale and used to power submarines.[8]

Today, wind and solar have been designated official technologies. State governments mandate their use and the federal government provides financial subsidies. One significant, positive difference is that, unlike with nuclear power, the states typically require wind and solar producers to compete on price and performance to win the mandates; the competition has driven improvements in performance and price.

No one designated shale gas as an official technology, yet today it is the most important technological innovation for the electricity business. It is a true innovation—it made natural gas abundant and cheap and, thereby, has kept electricity prices lower than they would have been. For shale gas, government policy simply opened the door through the deregulation of natural gas production and electricity generation. Deregulation then opened the door to Mitchell's and Devon Energy's hard work and smart investment. Despite or perhaps because it was *not* an official technology, shale gas–fired power helped achieve America's long-standing goal of making electricity available and affordable to all. Moreover, it is positioned to help achieve the added goals of doorstep reliability, environmental protection, and choice—at least as a bridge to a fundamentally different future.

THE RICHEST MAN IN THE WORLD

Even with the abundance of technologies we have today, we cannot claim to have opened the door to all technologies. There are always forks in the road and choices to be made. What matters—a lot—is who chooses and how the choices are made.

What would the richest man in the world do? We can guess at that with some precision because Bill Gates has committed both his money and his genius to addressing global climate change. His plan for the future includes both the "pull" of a carbon tax and the "push" of accelerated research and development. In between his pull and push are a great number of bets on deployment of specific technologies by private investors.[9]

What are Gates' bets? He sets ambitious standards for the ultimate winning technologies: they must have zero carbon emissions, be reliable around the clock, and cost less than fossil fuels (presumably with the carbon tax in place). Gates says bluntly that, while progress has been made, current technologies, including wind and solar, don't do the job. And neither is shale gas a final solution. In this spirit of frankness, Gates even warns about exaggeration when ascribing today's weather events solely to global climate change; El Niño is a bigger factor right now.[10]

Gates has invested about $1 billion so far, with direct investments in fifteen ventures and thirty indirect investments through venture capital.

Four investments reveal the type and range of his choices. First, nuclear power, because it is the only technology with zero carbon emissions that is already built at a large scale. The technology Gates backs is the Traveling Wave Reactor by TerraPower. It uses "waste" nuclear fuel without enrichment—so proliferation concerns are addressed. TerraPower claims its automatic safety features make a Fukushima incident impossible and the technology is ready to build now.[11]

A second bet is solar-chemical power. Gates points to Nate Lewis, at the California Institute of Technology, who uses artificial photosynthesis to produce hydrogen fuel from water. A third bet is high-altitude wind, which taps into the energy of the jet stream. A fourth bet is on carbon engineering that pulls carbon dioxide out of the air as an alternative to carbon capture and sequestration.[12]

Gates and other billionaires have also generously committed to new technology through the creation of the Breakthrough Energy Coalition.[13]

No plan opens all doors to the future. Choosing mitigation of global climate change as the overarching goal cuts out some technologies: those that would improve rather than remove fossil fuels. Moreover, the broad policies that set the stage for new technologies must actually be put in place. Cost-plus rate making has to be eliminated; it has been a disincentive to invest in new technologies for a hundred years. The temptation and ability to designate official technologies has to be eliminated, too. A way to productively tap into both old and new basic science has to be developed. And, most broadly, the arena for a second battle of the currents and a second democratization must be built.

An essential part of that arena is environmental policy. Can a constructive compromise on climate change policy be struck in the Age of Uncompromising Belief? Those who say no may point to President Trump's cabinet appointments: Oklahoma attorney general Scott Pruitt to head the EPA; former Texas governor Rick Perry to lead the Department of Energy; and ExxonMobil CEO Rex Tillerson as secretary of state. The lessons of history, however, tell us now is the time for compromise. The 2010 derailment of the Waxman-Markey bill was such a time, too, but those opposed to climate change policy chose to savor the tactical win and let the strategic opportunity pass. The result was President Obama's Clean Power Plan. Recollection of this lost opportunity by those on both sides of the issue might lead to a true compromise built, perhaps, on innovative ideas such as the mix of a carbon tax, tax reform, and policy sweep. With the right policies in place, plus an open door to new technologies, we can look forward to a fundamentally and remarkably improved electricity business for the twenty-first century.

LIST OF ACRONYMS

AC—alternating current
AEC—US Atomic Energy Commission
AFL—American Federation of Labor
CBO—Congressional Budget Office
CFCs—chlorofluorocarbons
CIO—Congress of Industrial Organizations
CPP—Clean Power Plan
CS—coordinate systems
DC—direct current
DDT—dichlorodiphenyltrichloroethane
EIA—Energy Information Administration
EPA—Environmental Protection Agency
EPAct of 1992—Energy Policy Act of 1992
EPAct of 2005—Energy Policy Act of 2005
ERDA—Energy Research and Development Administration
ERO—Electric Reliability Organization
EWGs—exempt wholesale generators
FERC—Federal Energy Regulatory Commission
FPA—Federal Power Act
FPC—Federal Power Commission
FTC—Federal Trade Commission
GE—General Electric
GHGs—greenhouse gases
IAEA—International Atomic Energy Agency
IEA—International Energy Agency
IPCC—United Nations' Intergovernmental Panel on Climate Change
ISO—independent system operator
IUI—Insull Utility Investments
kW—kilowatt
kWh—kilowatt-hour
LILCO—Long Island Lighting Company
LIPA—Long Island Power Authority

LNG—liquefied natural gas
MIT—Massachusetts Institute of Technology
MMBtu—Million British thermal units
MW—megawatt
NAAQS—National Ambient Air Quality Standards
NASA—National Aeronautics and Space Administration
NERC—North American Electric Reliability Council
NOAA—National Oceanic and Atmospheric Administration
NRC—National Research Council
PG&E—Pacific Gas & Electric
$PM_{2.5}$—particulate materials no more than 2.5 microns in width
PUHCA—Public Utility Holding Company Act of 1935
PURPA—Public Utility Regulatory Policies Act of 1978
PX—California Power Exchange
QFs—qualifying facilities
RDD&C—research, development, demonstration, and commercialization
RTO—regional transmission organization
SCE—Southern California Edison
SEC—Securities and Exchange Commission
SMR—small modular reactor
SPE—special-purpose entity
STR—Submarine Thermal Reactor
TVA—Tennessee Valley Authority
UMW—United Mine Workers of America

BIBLIOGRAPHY

SELECT PRIMARY, INSTITUTIONAL, AND UNIQUE SOURCES

"An Account of the Death of Mr. George William Richman, Professor of Experimental Philosophy, a Member of the Imperial Academy of Sciences at Petersburg. Translated from the High-Dutch." *Philosophical Transactions* 1755–1756, no. 49 (1755): 61–69.

Addison, Joseph. *Spectator* 1, no. 10, 1711.

"American Electromagnetic Telegraph: A Classic Invention." *The Science News* 18 (1930): 54–55.

"Apology for Printers." *The Pennsylvania Gazette,* June 10, 1731.

Arrhenius, Svante. "On the Influence of Carbonic Acid in the Air upon the Temperature of the Ground." *Philosophical Magazine and Journal of Science* 41 (1896): 237–76.

Bell, Michelle L., and Devra Lee Davis. "Reassessment of the Lethal London Fog of 1952: Novel Indicators of Acute and Chronic Consequences of Acute Exposure to Air Pollution." *Environmental Health Perspectives* 109 (2001): 389–94.

Bill Number: AB 1890 (passed Sept. 23, 1996). *California Legislative Information.* Accessed October 1, 2016. <ftp://www.leginfo.ca.gov/pub/95-96/bill/asm/ab_1851-1900/ab _1890_bill_960924_chaptered.html>

Birol, Fatih. "Golden Rules for a Golden Age of Gas: World Energy Outlook Special Report on Unconventional Gas." *International Energy Agency,* May 29, 2012. Accessed October 1, 2016. <http://www.worldenergyoutlook.org/media /weowebsite/2012/goldenrules/WEO2012_GoldenRulesReport.pdf>

"The Bituminous Coal Conservation Act of 1935." *Yale Journal of Law* XLV (1935): 293–314.

Bush, George W. "Letter to Members of the Senate on the Kyoto Protocol on Climate Change." *The American Presidency Project,* March 13, 2001. <http://www .presidency.ucsb.edu/ws/?pid=45811>

Callendar, G. S. "The Artificial Production of Carbon Dioxide and Its Influence on Temperature." *Journal of the Royal Meteorological Society* 64 (1937): 223–40.

Cawley, Kim. "The Federal Government's Responsibilities and Liabilities Under the Nuclear Waste Policy Act." *Congressional Budget Office*, December 3, 2014. Accessed October 1, 2016. <https://www.cbo.gov/sites/default/files/114th-congress-2015-2016/reports/51035-NuclearWaste_Testimony.pdf>

Center for Climate and Energy Solutions. "Comparison Chart of Waxman-Markey and Kerry-Lieberman." *Center for Climate and Energy Solutions*. Accessed October 1, 2016. <http://www.c2es.org/federal/congress/111/comparison-waxman-markey-kerry-lieberman>

"Coal Strike Is Called Off." *Chicago Daily Tribune,* June 23, 1943.

The Conservation Foundation. *Implications of Rising Carbon Dioxide Content of the Atmosphere*. New York: The Conservation Foundation, 1963.

Cook, Donald C. "Some Comments on the Current Utility Scene, New York Society of Security Analysts, New York City, New York." Speech, March 29, 1950. *Securities and Exchange Commission*. <http://www.sec.gov/news/speech/speecharchive/1950speech.shtml>

Cornell University Law School. "Sixteen U.S. Code Chapter Twelve—Federal Regulation and Development of Power." *Cornell University Law School*. Accessed October 1, 2016. <https://www.law.cornell.edu/uscode/text/16/chapter-12?qt-us>

Critic. Vol. 29 (1896).

Davis, Devra L., Michelle L. Bell, and Tony Fletcher. "A Look Back at the London Smog of 1952 and the Half Century Since." *Environmental Health Perspectives* 110 (2002): A734–A735.

"Death of a Genius," *Time*, May 2, 1955.

"Death of a Titan." *Time*, October 26, 1931.

Dewar, James A. "The Information Age and the Printing Press: Looking Backward to See Ahead." *Rand Corporation*, 2005. Accessed October 1, 2016. <http://www.rand.org/pubs/papers/P8014/index2.html#fn0>

"Edison's U.S. Patents, 1880–1882." *Thomas Edison Papers, Rutgers University*. Accessed October 1, 2016. <http://edison.rutgers.edu/patente2.htm>

Eicher, Edward C. "Address, Annual Convention, Edison Electric Institute, Buffalo, New York." June 5, 1941. *Securities and Exchange Commission*. <https://www.sec.gov/news/speech/1941/060541eicher.pdf>

Einstein, Albert. "Does the Inertia of a Body Depend upon Its Energy Content?" *Annalen der Physik* 18 (1905): 639–41.

———. "On the Electrodynamics of Moving Bodies." *Annalen der Physik* 17 (1905): 891–921.

———. "On a Heuristic Point of View Concerning the Production and Transformation of Light." *Annalen der Physik* 17 (1905): 132–48.

———. "On the Movement of Small Particles Suspended in Stationary Liquids Required by the Molecular-Kinetic Theory of Heat." *Annalen der Physik* 17 (1905): 549–60.

Eisenhower, Dwight. "Atoms for Peace Speech." New York, NY, December 8, 1953. *Dwight D. Eisenhower Presidential Library, Museum and Boyhood Home.* <http://www.eisenhower.archives.gov/all_about_ike/speeches/atoms_for_peace.pdf>

Electric Energy Market Competition Task Force. "Report to Congress on Competition in Wholesale and Retail Markets for Electric Energy." *Electric Energy Market Competition Task Force.* Accessed October 1, 2016. <http://www.ferc.gov/legal/fed-sta/ene-pol-act/epact-final-rpt.pdf>

"Endangerment and Cause or Contribute Findings for Greenhouse Gases under Section 202(a) of the Clean Air Act." *Environmental Protection Agency*, 2009. Accessed October 1, 2016. <www.epa.gov/climatechange/endangerment/>

Energy Information Administration. "Annual Energy Review." Accessed October 1, 2016. <www.eia.gov/totalenergy/data/annual/>

Energy Information Administration. "International Energy Outlook 2016." Accessed October 1, 2016. <www.eia.gov/totalenergy/data/monthly/pdf/mer.pdf>

Energy Information Administration. "Monthly Energy Review." Accessed October 1, 2016. <www.eia.gov/totalenergy/data/monthly/pdf/mer.pdf>

"Enron Trial Exhibits and Releases." *Department of Justice.* Accessed June 31, 2006. <https://www.justice.gov/archive/index-enron.html>

Environmental Protection Agency. "Assessment of the Potential Impacts of Hydraulic Fracturing for Oil and Gas on Drinking Water Resources." *Environmental Protection Agency*, June 2015. Accessed October 1, 2016. <http://www2.epa.gov/sites/production/files/2015-07/documents/hf_es_erd_jun2015.pdf>

———. "Clean Power Plan for Existing Power Plants." *Environmental Protection Agency.* Accessed October 1, 2016. <http://www2.epa.gov/cleanpowerplan/clean-power-plan-existing-power-plants>

———. "Clean Power Plan State-Specific Fact Sheets." *Environmental Protection Agency*, September 16, 2016. Accessed October 1, 2016. <https://www.epa.gov/cleanpowerplantoolbox/clean-power-plan-state-specific-fact-sheets>

———. "Fact Sheet: Overview of the Clean Power Plan," *Environmental Protection Agency.* Accessed October 1, 2016. <http://www.epa.gov/airquality/cpp/fs-cpp-overview.pdf>

"Eyewitness to Murder: Recounting the Ludlow Massacre." *History Matters, George Mason University.* Accessed October 1, 2016. <http://historymatters.gmu.edu/d/5737/>

"Far Worse Than Hanging; Kemmler's Death Proves an Awful Spectacle. The Electric Current Had to Be Turned On Twice before the Deed Was Fully Accomplished." *New York Times*, August 7, 1890.

Federal Energy Regulatory Commission. Docket No. RM99-2-000; Order No. 2000. *Federal Energy Regulatory Commission*, December 20, 1999. Accessed October 1, 2016. <http://www.ferc.gov/legal/maj-ord-reg/land-docs/RM99-2A.pdf>

———. "Energy Policy Act of 2005." *Federal Energy Regulatory Commission, Fact Sheet*, August 8, 2006. Accessed October 1, 2016. <www.ferc.gov/legal/fed-sta/epact-fact-sheet.pdf>

————. "Final Report on Price Manipulation in Western Markets: Fact-Finding Investigation of Potential Manipulation of Electric and Natural Gas Prices." Docket No. PA02-2-000. *Federal Energy Regulatory Commission*, March 2003. Accessed October 1, 2016. <http://www.ferc.gov/legal/maj-ord-reg/land-docs /part-2-03-26-03.pdf>

————. Order No. 636: Restructuring of Pipeline Services (1992). *Federal Energy Regulatory Commission*. Accessed October 1, 2016. <http://www.ferc.gov/legal/maj -ord-reg/land-docs/restruct.asp>

————. Order No. 888: Final Rule (1996). *Federal Energy Regulatory Commission*. Accessed October 1, 2016. <https://www.ferc.gov/legal/maj-ord-reg/land-docs /rm95-8-00w.txt>

————. "Technical and Support Document for Endangerment and Cause or Contribute Findings for Greenhouse Gases." *Environmental Protection Agency*. Accessed October 1, 2016. <https://www3.epa.gov/climatechange/Downloads /endangerment/TSD_Endangerment.pdf>

————. "Twentieth Annual Report." No. 21 (1940).

————. "Twenty-Sixth Annual Report." No. 23 (1946).

Franklin, Benjamin. *Experiment and Observations on Electricity*. London, 1751. Smithsonian Institution. Accessed October 1, 2016. <http://library.si.edu/digital -library/book/experimentsobser00fran>

————. "A Letter of Benjamin Franklin, Esq; to Mr. Peter Collinson, F.R.S. concerning an electrical Kite." *Philosophical Transactions* 47 (1752): 565–67. Accessed October 1, 2016. <http://rstl.royalsocietypublishing.org/content/47/565>

————. "The Papers of Benjamin Franklin." Yale University. Accessed October 1, 2016.

Franklin Institute. Thomas P. Jones (ed). "Journal of the Franklin Institute Volumes 25–26." *Franklin Institute*, 1838.

Fukushima Nuclear Accident Independent Investigation Commission. Official Report. *National Diet of Japan*, 2012. Accessed October 1, 2016. <www.nirs.org /fukushima/naiic_report.pdf>

"A Full and True Relation of the Death and Slaughter of a Man and His Son at Plough together with Four Horses . . . Slain by the Thunder and Lightning." London (1680).

Gates, Bill. "We Need Energy Miracles." *Gatesnotes* (blog), June 25, 2014. Accessed October 1, 2016. <https://www.gatesnotes.com/Energy/Energy-Miracles>

Geim, A. K., and K. S. Novoselov. "The Rise of Graphene." *Nature Materials* 6 (2007): 183–91. Accessed October 1, 2016. doi: 10.1038/nmat1849.

Gelinas, Donald J. [Associate Director, FERC Office of Markets, Tariffs, and Rates]. Memo to Sam Behrends, LeBoeuf, Lamb, Greene & MacRae LLP [Enron Counsel]. May 6, 2002.

Gore, Al, Lawrence Bender, Davis Guggenheim, Laurie David, Scott Z. Burns, Jeff Skoll, and Laurie Chilcott. *An Inconvenient Truth: A Global Warning*. DVD. Hollywood, CA: Paramount Classics, 2006.

Greenburg, Leonard, Morris B. Jacobs, Bernadette M. Drolette, Franklyn Field, and M. M. Braverman. "Report of an Air Pollution Incident in New York City, November 1953." *Public Health Reports* 77 (1962): 7–16.

Healy, Robert E. "Section 11 of the Public Utility Holding Company Act." Speech, November 5, 1941. *Securities and Exchange Commission*. <www.sec.gov/news/speech /1940/091040healy.pdf>

Heap, David Porter. "Report on the International Exhibition of Electricity, Paris." *Engineer Department, U.S. Army*, 1884. Accessed October 1, 2016. <archive.org/ stream/reportoninterna01heapgoog#page/n194/mode/2up>

Holly, Israel. "Youth liable to sudden death; excited seriously to consider thereof, and speedily to prepare therefor. The substance of a discourse, delivered on the day of the funeral of three young men, who were killed by lightning, at Suffield," printed by Thomas Green (1766).

Holt, Mark, and Gene Whitney. "Greenhouse Gas Legislation: Summary and Analysis of H.R. 2454 as Passed by the House of Representatives." *Congressional Research Service*. Accessed October 1, 2016. <http://research.policyarchive.org /18878.pdf>

Holzworth, George. "Vertical Temperature Structure during the 1966 Thanksgiving Week Air Pollution Episode in New York City." *Monthly Weather Review* 100 (1972): 445–50.

"Hoover Dam: Fatalities at Hoover Dam." *Reclamation: Managing Water in the West*. Last updated March 12, 2015. <http://www.usbr.gov/lc/hooverdam/history /essays/fatal.html>

Hoover, Herbert. "Address at Madison Square Garden in New York City." *The American Presidency Project*, October 31, 1932. <http://www.presidency.ucsb.edu /ws/?pid=23317>

———. "Statement on a National Tribute to Thomas Alva Edison." *The American Presidency Project*, October 20, 1931. <http://www.presidency.ucsb.edu/ws/?pid =22861>

Hughes, J. David. "Drilling Deeper: A Reality Check on U.S. Government Forecasts for a Lasting Tight Oil & Shale Gas Boom." *Post Carbon Institute*, October 2014. Accessed October 1, 2016. <http://www.postcarbon.org/wp-content /uploads/2014/10/Drilling-Deeper_FULL.pdf>

Hunt, Andrew, Jerrold L. Abraham, Bret Judson, and Colin L. Berry. "Toxicologic and Epidemiologic Clues from the Characterization of the 1952 London Smog Fine Particulate Matter in Archival Autopsy Lung Tissues." *Environmental Health Perspectives* 111 (2003): 1209–14.

Insull, Samuel. *The Memoirs of Samuel Insull: An Autobiography*. Edited by Larry Plachno. Polo, IL: Transportation Trails, 1992.

InterAcademy Council. "Climate Change Assessments, Review of the Processes and Procedures of the IPCC." *InterAcademy Council*, October 2010. Accessed October 1, 2016. <http://reviewipcc.interacademycouncil.net/report.html>

Intergovernmental Panel on Climate Change. *Climate Change: The IPCC Scientific Assessment.* New York: Cambridge University Press, 1990. Accessed October 1, 2016. <https://www.ipcc.ch/ipccreports/far/wg_I/ipcc_far_wg_I_full_report.pdf>

———. *Climate Change 2014: Mitigation of Climate Change.* New York: Cambridge University Press, 2014. Accessed October 1, 2016. <http://www.ipcc.ch/report/ar5/wg3/>

———. *IPCC Second Assessment: Climate Change 1995.* Accessed October 1, 2016. <https://www.ipcc.ch/pdf/climate-changes-1995/ipcc-2nd-assessment/2nd-assessment-en.pdf>

———. "IPCC Statement on the Melting of Himalayan Glaciers." January 20, 2010. Accessed October 1, 2016. <https://www.ipcc.ch/pdf/presentations/himalaya-statement-20january2010.pdf>

International Atomic Energy Agency. "Additional Report of Japanese Government to IAEA—Accident at TEPCO's Fukushima Power Stations, 15 September 2011." Accessed October 1, 2016. <https://www.iaea.org/newscenter/focus/fukushima/additional-japan-report>

———. "Operational & Long-Term Shutdown Reactors." *International Atomic Energy Agency*, September 24, 2016. Accessed October 1, 2016. <https://www.iaea.org/PRIS/WorldStatistics/OperationalReactorsByCountry.aspx>

Jickling, Mark. "The Enron Collapse: An Overview of Financial Issues." *CRS Report for Congress*, March 28, 2002. Accessed October 1, 2016. <http://fpc.state.gov/documents/organization/9267.pdf>

Johnson, Lyndon B. "Remarks at the Presentation of the 1964 Presidential Medal of Freedom Awards." Speech, Washington, DC, September 14, 1964. *The American Presidency Project.* <http://www.presidency.ucsb.edu/ws/?pid=26496>

Kemeny, John G. "Report of the President's Commission on the Accident at Three Mile Island." 1979.

Komanoff, Charles. *Power Plant Cost Escalation: Nuclear and Capital Costs, Regulation, and Economics.* New York: Van Nostrand Reinholdt Publishing, 1981.

Larabee, Leonard W., Ralph L. Ketcham, Helen C. Boatfield, and Helene H. Fineman, eds. *The Autobiography of Benjamin Franklin.* New Haven: Yale University Press, 1964.

Lay, George C. "Benjamin Franklin." *Godey's Magazine* CXXXIII (1796).

Lilienthal, David E. "Business and Government in the Tennessee Valley." *Annals of the American Academy of Political and Social Science* 172 (1934): 45–49.

———. *TVA: Democracy on the March.* New York: Harper & Brothers, 1944.

Massachusetts et al. v. Environmental Protection Agency et al. 05 U.S. 1120 (2007).

Maxwell, James Clerk. "A Dynamical Theory of the Electromagnetic Field." *Philosophical Transactions of the Royal Society of London* 155 (1865): 459–512.

Mershon, Sherie, and Tim Palucka. *A Century of Innovation: From the U.S. Bureau of Mines to the National Energy Technology Laboratory.* National Energy Technology Laboratory. October 2010.

Mills, Clarence A. "The Donora Episode." *Science* 111 (1950): 67–68.

Mills, Mark P. "SHALE 2.0: Technology and the Coming Big-Data Revolution in America's Shale Oil Fields." *Manhattan Institute*, May 2015. Accessed October 1, 2016. <http://www.manhattan-institute.org/pdf/eper_16.pdf>

MIT. "The Future of Nuclear Power." *MIT*, 2003. Accessed October 1, 2016. <http://web.mit.edu/nuclearpower/>

Mitchell, George P., and Mark D. Zoback. "George Mitchell: The Duty to Fracture Responsibly." *Fuelfix*, February 20, 2012. Accessed October 1, 2016. <http://fuelfix.com/blog/2012/02/20/opinion-the-duty-to-fracture-responsibly/>

Moniz, Ernest J., Henry D. Jacoby, and Anthony J. M. Meggs. "The Future of Natural Gas: An Interdisciplinary MIT Study." *MIT Energy Initiative*, June 9, 2011. Accessed October 1, 2016. <http://www.mit.edu/~jparsons/publications /NaturalGas_Report_Final.pdf>

Morens, David M. "Death of a President," *New England Journal of Medicine* 341 (1999): 1845–50. Accessed October 1, 2016. doi: 10.1056/NEJM199912093412413.

Morse, Samuel F. B. *Imminent Dangers to the Free Institutions of the United States through Foreign Immigration*. 1835.

———. "Samuel F. B. Morse Papers at the Library of Congress, 1793 to 1919." Accessed October 1, 2016. <www.loc.gov/collections/samuel-morse-papers /articles-and-essays/timeline/1791-1839/>

Murray, W. S. *A Superpower System for the Region between Boston and Washington*. Washington: Government Printing Office, 1921.

Musk, Elon. "Master Plan, Part Deux." *Tesla* (blog), July 20, 2016. Accessed October 1, 2016. <https://www.tesla.com/blog/master-plan-part-deux>

Nathanson, Milton N. "Updating the Hoover Dam Documents." *U.S. Department of the Interior*, 1978. Accessed October 1, 2016. <http://www.onthecolorado.com /Resources/LawOfTheRiver/HooverDamDocumentsUpdated.pdf>

National Assembly of France. "To George Washington from the National Assembly of France, 20 June 1790." *National Archives, Founders Online, Franklin Papers*. Accessed October 1, 2016. <http://founders.archives.gov/documents /Washington/05-05-02-0348>

National Center for Public Policy Research. "Byrd-Hagel Resolution." *The National Center for Public Policy Research*. Accessed October 1, 2016. <http://www.national center.org/KyotoSenate.html>

National Park Service. "Shippingport Atomic Power Station." *Historic American Engineering Record*, HAER No. PA-81. Accessed October 1, 2016. <http://cdn.loc. gov/master/pnp/habshaer/pa/pa1600/pa1658/data/pa1658data.pdf>

National Power Policy Committee. "Report with Respect to the Treatment of Holding Companies." H.R. Rep. No. 137 (1935).

Navigant Consulting. *Assessment of the Nuclear Power Industry*. July 2013. Accessed October 1, 2016. <http://pubs.naruc.org/pub/536D6E09-2354-D714-5175-4F938E94ADB5>

New York City Landmarks Preservation Commission. "Madison Square North Historic District Designation Report." *New York City Landmarks Preservation Commission*, June 26, 2011. Accessed October 1, 2016. <http://www.nyc.gov/html /lpc/downloads/pdf/reports/MadisonSquareNorth.pdf>

Nuclear Energy Institute. "World Statistics," *Nuclear Energy Institute*, May 2016. Accessed October 1, 2016. <http://www.nei.org/Knowledge-Center/Nuclear-Statistics/World-Statistics>

Obama, Barack. "Full Transcript: Obama's 2012 State of the Union Address." Washington, DC, January 25, 2012. *USA Today*. <http://www.usatoday.com/news/washington/story/2012-01-24/state-of-the-union-transcript/52780694/1>

———. "Remarks by the President in Announcing the Clean Power Plan." August 3, 2015. *Whitehouse.org*. <https://www.whitehouse.gov/the-press-office/2015/08/03/remarks-president-announcing-clean-power-plan>

Office of Management and Budget, White House. "Table 1.2—Summary of Receipts, Outlays, and Surpluses or Deficits (-) as Percentages of GDP: 1930–2021." *Office of Management and Budget, White House.* Accessed October 1, 2016. <https://obamawhitehouse.archives.gov/omb/budget/Historicals>

Office of the Secretary of Energy and Environment, Oklahoma. "Earthquakes in Oklahoma." *Office of the Secretary of Energy and Environment, Oklahoma.* Accessed October 1, 2016. <https://earthquakes.ok.gov/>

O'Reilly et al. v. Morse et al. 56 U.S. 62 (1853). <https://supreme.justia.com/cases/federal/us/56/62/case.html>

Pew Center on Global Climate Change. "Fifteenth Session of the Conference of the Parties to the United Nations Framework Convention on Climate Change and Fifth Session of the Meeting of the Parties to the Kyoto Protocol." *Center for Climate and Energy Solutions.* Accessed October 1, 2016. <http://www.c2es.org/docUploads/copenhagen-cop15-summary.pdf>

Phillips Petroleum Co. v. Wisconsin 347 U.S. 672 (1954).

Plass, Gilbert. "The Carbon Dioxide Theory of Climatic Change." *Tellus* 8 (1955): 140–54.

Pope, C. Arden III, Richard T. Burnett, Michael J. Thun, Eugenia E. Calle, Daniel Krewski, Kazuhiko Ito, and George D. Thurston. "Lung Cancer, Cardiopulmonary Mortality, and Long-Term Exposure to Fine Particulate Air Pollution." *Journal of the American Medical Association* 287 (2002): 1132–41.

Pope Francis. *Laudato sì*. May 24, 2015. *Libreria Editrice Vaticana*. Accessed October 1, 2016. <http://w2.vatican.va/content/francesco/en/encyclicals/documents/papa-francesco_20150524_enciclica-laudato-si.html>

Powers, William C. *Report of Investigation by the Special Investigative Committee of the Board of Directors of Enron Corp.* February 1, 2002. Accessed October 1, 2016. <http://picker.uchicago.edu/Enron/PowersReport(2-2-02).pdf>

President's Science Advisory Committee (PSAC). Pesticides report, 15 May 1963. John F. Kennedy Presidential Library and Museum. Accessed October 1, 2016. <https://www.jfklibrary.org/Asset-Viewer/Archives/JFKPOF-087-003.aspx>

"Public Utilities Regulatory Policies Act of 1978." *United States Bureau of Reclamation*, November 9, 1978. Accessed October 1, 2016. <http://www.usbr.gov/power/legislation/purpa.pdf>

Public Utility Holding Company Act of 1935: Hearings before the Committee on Interstate Commerce. 1935.

"Restoring the Quality of Our Environment." *President's Science Advisory Committee*, 1965.

Revelle, Roger, and Hans E. Seuss. "Carbon Dioxide Exchange Between Atmosphere and Ocean and the Question of an Increase of Atmospheric CO_2 during the Past Decades." *Tellus* (1957): 18–27.

Roosevelt, Franklin D. "Address at the Dedication of Boulder Dam." *The American Presidency Project*, September 30, 1945. <http://www.presidency.ucsb.edu/ws/?pid=14952>

———. "Annual Message to Congress." *The American Presidency Project*, January 4, 1935. <http://www.presidency.ucsb.edu/ws/?pid=14890>

———. "Campaign Address in Portland, Oregon on Public Utilities and Development of Hydro-Electric Power." *The American Presidency Project*, September 21, 1932. <http://www.presidency.ucsb.edu/ws/?pid=88390>

———. "Campaign Address on Progressive Government at the Commonwealth Club in San Francisco, California." *The American Presidency Project*, September 23, 1932. <http://www.presidency.ucsb.edu/ws/?pid=88391>

———. "Fireside Chat 24: On the Coal Crisis." Speech, May 2, 1943. *Miller Center: University of Virginia*. <millercenter.org/president/fdroosevelt/speeches/speech-3330>

Rosner, Robert, and Stephen Goldberg. "Small Modular Reactors—Key to Future Nuclear Power Generation in the U.S." *Energy Policy Institute at Chicago*, November 2011. Accessed October 1, 2016. <www.eenews.net/assets/2013/03/13/document_gw_01.pdf>

The Royal Swedish Academy of Science. "Scientific Background on the Nobel Prize in Physics 2010: Graphene." *The Royal Swedish Academy of Science*, November 29, 2010. Accessed October 1, 2016. <https://www.nobelprize.org/nobel_prizes/physics/laureates/2010/advanced-physicsprize2010.pdf>

Rutgers University. "Electric Lamp." *The Thomas Edison Papers*. Accessed October 1, 2016. <http://edison.rutgers.edu/lamp.htm>

"Samuel Insull Will Make Fight for Complete Vindication." *Ottawa Citizen*, May 7, 1934.

Securities and Exchange Commission. "The Regulation of Public-Utility Companies." *Securities and Exchange Commission*, 1995. Accessed October 1, 2016. <http://www.sec.gov/news/studies/puhc.txt>

Securities and Exchange Commission. "Twenty-Fifth Annual Report of the Securities and Exchange Commission: Fiscal Year Ended June 30, 1959." *Securities and Exchange Commission*, January 6, 1960. <https://www.sec.gov/about/annual_report/1959.pdf>

SpaceX. "Completed Missions." *SpaceX*. Accessed October 1, 2016. <http://www.spacex.com/missions>

———. "SpaceX Successfully Launches Falcon 1 to Orbit." *SpaceX*, September 28, 2008. Accessed October 1, 2016. <http://www.spacex.com/press/2012/12/19/spacex-successfully-launches-falcon-1-orbit>.

State of West Virginia, State of Texas, et al. v. United States Environmental Protection Agency and Regina A. McCarthy. Application February 9, 2016. Accessed October 1, 2016. <http://www.scotusblog.com/wp-content/uploads/2016/01/15A773-application.pdf>

Steward, Dan B. "George P. Mitchell and the Barnett Shale." *Journal of Petroleum Technology* (2013). Accessed October 1, 2016. <http://www.mydigitalpublication.com/article/George_P._Mitchell_And_The_Barnett_Shale/1535436/179598/article.html>

Tesla, Nikola. "A New System of Alternate Current—Motors and Transformers." Speech, New York, NY, May 16, 1888. *PBS*. <http://www.pbs.org/tesla/res/res_art02.html>

———. "The Transmission of Electrical Energy without Wires." *Electrical World and Engineer* (1904).

Townsend, Jonathan. "God's terrible doings are to be observed. A sermon preach'd at Needham, and occasion'd by the sudden and awful death of Mr. Thomas Gardner, by Lightning." Printed for Kneeland and Green (1746).

Toynbee, Arnold. *Lectures on the Industrial Revolution of the Eighteenth Century in England: Popular Addresses, Notes and Other Fragments.* New York: Longmans, Green, and Company, 1908.

Tyndall, John. "The Bakerian Lecture: On the Absorption and Radiation of Heat by Gases and Vapours, and on the Physical Connexion of Radiation, Absorption, and Conduction." *Philosophical Transactions of the Royal Society of London* 151 (1861): 1–36.

United Nations Environment Programme. "Brief Primer on the Montreal Protocol." *United Nations Environment Programme.* Accessed October 1, 2016. <http://ozone.unep.org/Publications/MP_Brief_Primer_on_MP-E.pdf>

United States of America v. Timothy N. Belden. Plea agreement. *Findlaw.* Accessed October 1, 2016. <http://news.findlaw.com/hdocs/docs/enron/usbelden101702plea.pdf>

United States Congress. *Memorial of Samuel Finley Breese Morse, Including Appropriate Ceremonies of Respect at the National Capitol, and Elsewhere.* Published by Order of Congress. Washington: Government Printing Office, 1875.

United States Geological Survey. "Oklahoma Earthquakes Magnitude 3.0 and Greater." *United States Geological Survey,* February 17, 2016. Accessed October 1, 2016. <http://earthquake.usgs.gov/earthquakes/states/oklahoma/images/OklahomaEQsBarGraph.png>

United States Nuclear Regulatory Commission. "Information Digest, 2014–2015: Appendix A: U.S. Commercial Nuclear Power Reactors—Operating Reactors." *United States Nuclear Regulatory Commission,* August 2014. Accessed October 1, 2016. <http://www.nrc.gov/reading-rm/doc-collections/nuregs/staff/sr1350/>

United States Senate. "The Role of the Board of Directors in Enron's Collapse." *Permanent Subcommittee on Investigations of the Committee of Governmental Affairs United States Senate,* July 8, 2002. Accessed October 1, 2016. <https://www.gpo.gov/fdsys/pkg/CPRT-107SPRT80393/pdf/CPRT-107SPRT80393.pdf>

U.S.-Canada Power System Outage Task Force. "Final Report on the August 14, 2003, Blackout on the United States and Canada: Causes and Recommendations." *U.S.-Canada Power System Outage Task Force*, April 2004. Accessed October 1, 2016. <http://energy.gov/sites/prod/files/oeprod/DocumentsandMedia/Blackout Final-Web.pdf>

U.S. Department of Commerce. "Historical Statistics of the United States, Colonial Times to 1970." *U.S. Department of Commerce*, 1975. Accessed October 1, 2016. <http://www2.census.gov/library/publications/1975/compendia/hist_stats _colonial-1970/hist_stats_colonial-1970p2-chS.pdf>

U.S. Department of Energy and Natural Resources Canada. "The Relationship between Competitive Power Markets and Grid Reliability." *Office of Energy Efficiency and Renewable Energy*, July 2006. Accessed October 1, 2016. <https:// energy.gov/sites/prod/files/oeprod/DocumentsandMedia/Blackout_Rec_12 _Final_Report_and_Transmittal.pdf>

"Utility Corporations: Summary Report of the Federal Trade Commission." Document 92, part 72-A. 1935.

West Virginia Office of Miner's Health Safety and Training. "Upper Big Branch Mine Disaster Investigative Report Summary." Accessed October 1, 2016. <http://www.wvminesafety.org/PDFs/Performance/EXECUTIVE%20 SUMMARY.pdf>

Westinghouse, George, Jr. "A Reply to Mr. Edison." *The North American Review* 149 (1889): 653–64.

———. "Sir Wm. Thomson and Electric Lighting." *The North American Review* 150 (1890): 321–29.

White House Office of the Press Secretary. "Key Excerpts of the Joint Comprehensive Plan of Action (JCPOA)." July 14, 2015. Accessed October 1, 2016. <https://www .whitehouse.gov/the-press-office/2015/07/14/key-excerpts-joint-comprehensive -plan-action-jcpoa>

World Nuclear Association. "Three Mile Accident." *World Nuclear Association*, January 2012. Accessed October 1, 2016. <http://www.world-nuclear.org/info /Safety-and-Security/Safety-of-Plants/Three-Mile-Island-accident/>

Yoder, Christian [of Enron Power Marketing], and Steven Hall [of Stoel Rives, LLP]. Memoranda to Richard Saunders [of Enron] from Christian Yoder. December 6, 2000.

Zanobetti, Antonella, and Joel Schwartz. "The Effect of Fine and Coarse Particulate Air Pollution on Mortality: A National Analysis." *Environmental Health Perspectives* 117 (2009): 898–903.

SELECT JOURNAL ARTICLES

Bernanke, Ben S. "The Macroeconomics of the Great Depression: A Comparative Approach." *National Bureau of Economic Research* 4813 (1994): 1–28. doi: 10.3386 /w4814.

Bishop, Morris. "Franklin in France." *Daedalus* 86 (1957): 214–30.

Bragg, William H. "Michael Faraday." *The Scientific Monthly* 33 (1931): 481–99.

Buchanan, Norman S. "The Origin and Development of the Public Utility Holding Company." *The Journal of Political Economy* 44 (1936): 31–53.

Chandler, Alfred D., Jr. "Anthracite Coal and the Beginnings of the Industrial Revolution in the United States." *The Business History Review* 46 (1972): 141–81.

Cohen, I. Bernard. "Benjamin Franklin: An Experimental Newtonian Scientist." *Bulletin of the American Academy of Arts and Sciences* 5 (1952): 2–6.

———. "The Two Hundredth Anniversary of Benjamin Franklin's Two Lightning Experiments and the Introduction of the Lightning Rod." *Proceedings of the American Philosophical Society* 96 (1952): 331–66.

Cohn, Steve. "The Political Economy of Nuclear Power (1945–1990): The Rise and Fall of an Official Technology." *Journal of Economic Issues* 24 (1990): 781–811.

Cowan, Robin. "Nuclear Power Reactors: A Study in Technological Lock-In." *The Journal of Economic History* 50 (1990): 541–67.

Crafts, Nicholas. "Steam as a General Purpose Technology: A Growth Accounting Perspective." *The Economic Journal* 114 (2004): 338–51.

Crawford, John W. and Steven L. Krahn. "The Naval Nuclear Propulsion Program: A Brief Case Study in Institutional Constancy." *Public Administration Review* 58 (1998): 159–66.

Dean, Gordon. "Atoms for Peace: An American View." *International Journal* 9 (1954): 253–60.

Debnath, Lokenath. "A Short Biography of Joseph Fourier and Historical Development of Fourier Series and Fourier Transforms." *International Journal of Mathematical Education in Science and Technology* 43 (2011): 589–612. doi: 10.1080/0020739X.2011.633712.

Demsetz, Harold. "Why Regulate Utilities?" *The Journal of Law and Economics* 11 (April 1968): 55–65.

Dull, Jonathan R. "Franklin the Diplomat: The French Mission." *Transactions of the American Philosophical Society* 72 (1982): 1–76.

Eisenstein, Elizabeth L. "In the Wake of the Printing Press." *The Quarterly Journal of the Library of Congress* 35 (1978): 183–97.

Emmons III, William M. "Franklin D. Roosevelt. Electric Utilities and the Power of Competition." *The Journal of Economic History* 53 (1993): 880–907.

Frickel, Scott. "Engineering Heterogeneous Accounts: The Case of Submarine Thermal Reactor Mark-I." *Science, Technology, & Human Values* 21 (1996): 28–53.

Greenstone, Michael, and Adam Looney. "Paying Too Much for Energy? The True Costs of Our Energy Choices." *Daedalus* 141 (2012): 10–30.

Hare, E. H. "Michael Faraday's Loss of Memory." *Proceedings of the Royal Society of Medicine* 67 (1974): 617–18.

Heilbron, J. L. "Plus and Minus: Franklin's Zero-Sum Way of Thinking." *Proceedings of the American Philosophical Society* 150 (2006): 607–17.

Hughes, Thomas P. "Harold P. Brown and the Executioner's Current: An Incident in the AC-DC Controversy." *The Business History Review* 32 (1958): 143–65.

———. "Technology and Public Policy: The Failure of Giant Power." *Institute of Electrical and Electronics Engineers* 64 (1976): 1361–71. Accessed October 1, 2015. doi: 10.1109/PROC.1976.10327.

Hulme, Mike. "On the Origin of 'the Greenhouse Effect': John Tyndall's 1859 Interrogation of Nature." *Weather, Royal Meteorological Society* 64 (2009): 121–23. Accessed October 1, 2016. <http://onlinelibrary.wiley.com/doi/10.1002/wea.386/pdf>

Joskow, Paul L. "California's Electricity Crisis." *Oxford Review of Economic Policy* 17 (2001): 365–88. Accessed October 1, 2016. doi: 10.1093/oxrep/17.3.365.

Krige, John. "Atoms for Peace. Scientific Internationalism and Scientific Intelligence." *Osiris* 21 (2006): 161–81.

Lear, Linda. "Rachel Carson's *Silent Spring*." *Environmental History Review* 17 (1993): 23–48.

Martin, Roscoe C. "The Tennessee Valley Authority: A Study of Federal Control." *Law and Contemporary Problems* 22 (1957): 351–77.

McCormick, Blaine, and Burton W. Folsom Jr. "A Survey of Business Historians on America's Greatest Entrepreneurs." *The Business History Review* 77 (2003): 703–16.

McKinney, Robert. "Atomic Energy: Its Contribution to World Peace Through Economic Self-Help." *Proceedings of the Academy of Political Science* 26 (1957): 72–82.

Nonnenmacher, Tomas. "State Promotion and Regulation of the Telegraph Industry." *The Journal of Economic History* 61 (2001): 19–36.

Posner, Richard A. "Theories of Economic Regulation." *The Bell Journal of Economics and Management Science* 5 (Autumn 1974): 335–58.

Priest, George L. "The Origins of Utility Regulation and the 'Theories of Regulation' Debate." *The Journal of Law and Economics* 36 (1993): 289–323.

Rosenberg, Nathan, and Manuel Trajtenberg. "A General Purpose Technology at Work: The Corliss Steam Engine in the Late-Nineteenth-Century United States." *Journal of Economic History* 64 (2004): 61–99.

Ross, Jerry, and Barry M. Staw. "Organizational Escalation and Exit: Lessons from the Shoreham Nuclear Power Plant." *The Academy of Management Journal* 36 (1993): 701–32.

Schmalensee, Richard, and Robert N. Stavins. "The SO_2 Allowance Trading System: The Ironic History of a Grand Policy Experiment." *The Journal of Economic Perspectives* 27 (2013): 103–21.

Schrag, Daniel P. "Is Shale Gas Good for Climate Change?" *Daedalus* 141 (2012): 72–80.

Sharlin, Harold I. "The First Niagara Falls Power Project." *The Business History Review* 35 (1961): 59–74.

Shirk, Willis L. "'Atoms for Peace' in Pennsylvania." *Pennsylvania Heritage* 35 (2009).

Snyder, Lynne Page. "The Death-Dealing Smog over Donora, Pennsylvania: Industrial Air Pollution, Public Health Policy, and the Politics of Expertise, 1948–1949." *Environmental History Review* 18 (1994): 117–39.

Stigler, George J., and Claire Friedland. "What Can Regulators Regulate? The Case of Electricity." *The Journal of Law and Economics* 5 (1962): 1–16.

Taylor, Arthur R. "Losses to the Public in the Insull Collapse: 1932–1946." *Business History Review* 36 (1962): 188–204.

Thomas, John Meurig. "Faraday and Franklin." *Proceedings of the American Philosophical Society* 150 (2006): 523–41.

Wengert, Norman. "TVA—Symbol and Reality." *The Journal of Politics* 13 (1951): 369–92.

Yates, JoAnne. "The Telegraph's Effect on Nineteenth Century Markets and Firms." *Business and Economic History* 15 (1986): 149–63.

SELECT BOOKS

Alinsky, Saul D. *John L. Lewis: An Unauthorized Biography*. New York: Cornwall Press, 1949.

Alvarado, Rudolph. *Thomas Edison*. Indianapolis: Alpha Books, 2002.

Banner, Stuart. *The Death Penalty: An American History*. Cambridge, MA: Harvard University Press, 2009.

Bodanis, David. *Electric Universe: The Shocking True Story of Electricity*. New York: Random House, 2005.

———. *E=mc²: A Biography of the World's Most Famous Equation*. New York: Walker Publishing Company, 2000.

Brands, H. W. *The First American: The Life and Times of Benjamin Franklin*. New York: Doubleday, 2000.

Busch, Francis X. *Guilty or Not Guilty: An Account of the Trials of the Leo Frank Case, the D.C. Stephenson Case, the Samuel Insull Case, the Alger Hiss Case*. New York: The Bobbs-Merrill Company, Inc., 1952.

Carnegie, Andrew. *James Watt*. Honolulu: University Press of the Pacific, 1913.

Carson, Rachel. *Silent Spring*. New York: Houghton Mifflin Harcourt, 2002.

Chaplin, Joyce. *The First Scientific American: Benjamin Franklin and the Pursuit of Genius*. New York: Basic Books, 2006.

Cheney, Margaret. *Tesla: Man Out of Time*. New York: Simon & Schuster, 2001.

Cheney, Margaret, and Robert Uth. *Tesla: Master of Lightning*. New York: MetroBooks, 2001.

Cohen, I. Bernard. *Science and the Founding Fathers: Science in the Political Thought of Thomas Jefferson, Benjamin Franklin, John Adams, and James Madison*. New York: W. W. Norton and Company, 1995.

Doren, Carl Van. *Benjamin Franklin (1706–1790)*. New York: Viking Press, 1938.

Dubofsky, Melvyn, and Warren Van Tine. *John L. Lewis: A Biography*. Urbana: University of Illinois Press, 1986.

Dunar, Andrew J., and Dennis McBride, eds. *Building Hoover Dam: An Oral History of the Great Depression*. Reno: University of Nevada Press, 1993.

Einstein, Albert. *Ideas and Opinions*. New York: Three Rivers Press, 1954.

Einstein, Albert, and Leopold Infeld. *The Evolution of Physics: The Growth of Ideas from Early Concepts to Relativity and Quanta*. New York: Simon & Schuster, 1938.

Fisher, Howard J. *Faraday's Experimental Research in Electricity: Guide to First Reading*. Santa Fe: Green Lion Press, 2001.

Friedel, Robert, and Paul Israel. *Edison's Electric Light*. New Jersey: Rutgers University Press, 1987.

Hamilton, James. *A Life of Discovery: Michael Faraday, Giant of the Scientific Revolution*. New York: Random House, 2002.

Hargrove, Erwin C. *Prisoners of Myth: The Leadership of the Tennessee Valley Authority 1933–1990*. Knoxville: University of Tennessee Press, 2001.

Heilbron, J. L. *Electricity in the 17th and 18th Centuries: A Study in Early Modern Physics*. New York: Dover Publications, Inc., 1999.

Hiltzik, Michael. *Colossus: The Turbulent, Thrilling Saga of the Building of Hoover Dam*. New York: Free Press, 2010.

Hirsch, Robert F. *Power Loss: The Origins of Deregulation and Restructuring in the American Electric Utility System*. Cambridge, MA: MIT Press, 2001.

Howard, Don, and John Stachel, eds. *Einstein: The Formative Years 1879–1909*. Boston: Birkhaüser, 2000.

Hughes, Thomas P. *Networks of Power: Electrification in Western Society 1880–1930*. Baltimore: Johns Hopkins University Press, 1983.

Huxley, Aldous. *Collected Essays*. New York: Harper Brothers, 1959.

Isaacson, Walter. *Benjamin Franklin: An American Life*. New York: Simon and Schuster, 2004.

————. *Einstein: His Life and Universe*. New York: Simon & Schuster, 2007.

Israel, Paul. *Edison: A Life of Invention*. New York: John Wiley and Sons, 1998.

Komanoff, Charles. *Power Plant Cost Escalation: Nuclear and Capital Costs, Regulation, and Economics*. New York: Van Nostrand Reinholdt Publishing, 1981.

Layzer, Judith A. *The Environmental Case: Translating Values into Policy*, 2nd ed. Washington, DC: CQ Press, 2006.

Lear, Linda. *Rachel Carson: Witness for Nature*. New York: Houghton Mifflin Harcourt, 2009.

Lomborg, Bjorn. *Cool It: The Skeptical Environmentalist's Guide to Global Warming*. New York: Vintage Books, 2007.

Lynn, Michael R. *Popular Science and Public Opinion in 18th Century France*. Manchester, UK: Manchester University Press, 2006.

Mabee, Carleton. *The American Leonardo: A Life of Samuel F. B. Morse*. New York: Purple Mountain Press, 2000.

Mahon, Basil. *The Man Who Changed Everything: The Life of James Clerk Maxwell.* Chichester, UK: John Wiley and Sons, 2003.

McDonald, Forrest. *Insull.* Chicago: University of Chicago Press, 1962.

McLean, Bethany, and Peter Elkind. *The Smartest Guys in the Room: The Amazing Rise and Scandalous Fall of Enron.* New York: Portfolio/Penguin, 2003.

Muller, Richard A. *Physics for Future Presidents: The Science Behind the Headlines.* New York: W. W. Norton & Company, 2008.

Neuse, Steven M. *David E. Lilienthal: The Journey of an American Liberal.* Knoxville: University of Tennessee Press, 1996.

Olson, James Stuart. *Historical Dictionary of the Great Depression, 1929–1940.* Westport, CT: Greenwood Press, 2001.

Papanikolas, Zeese, and Wallace Stenger. Foreword to *Buried Unsung: Louis Tikas and the Ludlow Massacre.* Lincoln: University of Nebraska Press, 1991.

Prout, Henry Goslee. *A Life of George Westinghouse.* Ann Arbor: University of Michigan, 1921.

Reid, James D. *The Telegraph in America: Its Founders, Promoters, and Noted Men.* New York: Derby Brothers, 1879.

Robinson, Eric, and A. E. Musson. *James Watt and the Steam Revolution.* New York: Augustus M. Kelley, 1969.

Salzman, James, and Barton H. Thompson Jr. *Environmental Law and Policy.* New York: Hudson Press, 2007.

Schiffer, Michael Brian. *Draw the Lightning Down: Benjamin Franklin and Electrical Technology in the Age of Enlightenment.* Berkeley: University of California Press, 2003.

Schlozman, Daniel. *When Movements Anchor Parties: Electoral Alignments in American History.* Princeton, NJ: Princeton University Press, 2015.

Schurr, Sam H., and Bruce S. Netschert. *Energy in the American Economy.* Baltimore: Johns Hopkins University Press, 1960.

Silverman, Kenneth. *Lightning Man: The Accursed Life of Samuel F. B. Morse.* New York: Random House, 2003.

Simpson, Thomas K. *Figures of Thought.* Santa Fe: Green Lion Press, 2006.

Skrabec, Quentin, Jr. *George Westinghouse: Gentle Genius.* New York: Algora Publishing, 2007.

Spulber, Daniel F. *Regulation and Markets.* Boston: MIT Press, 1989.

Standage, Tom. *The Victorian Internet: The Remarkable Story of the Telegraph and the Nineteenth Century's On-Line Pioneers.* New York: Walker Publishing Company, 1999.

Sternglass, Ernest. *Secret Fallout: Low-Level Radiation from Hiroshima to Three-Mile Island.* New York: McGraw-Hill, 1981.

Stevens, Joseph E. *Hoover Dam: An American Adventure.* Norman, OK: University of Oklahoma Press, 1988.

Taleb, Nassim Nicholas. *The Black Swan: The Impact of the Highly Improbable.* New York: Random House, 2007.

Thompson, Robert Luther. *Wiring a Continent: The History of the Telegraph Industry in the United States 1832–1866.* Princeton, NJ: Princeton University Press, 1947.

Trefil, James, and Robert M. Hazen. *The Sciences: An Integrated Approach*. Hoboken, NJ: John Wiley and Sons, 2004.

Tucker, Tom. *The Bolt of Fate: Benjamin Franklin & His Electric Kite Hoax*. New York: Public Affairs, 2003.

Uglow, Jenny. *The Lunar Men: Five Friends Whose Curiosity Changed the World*. New York: Farrar, Straus and Giroux, 2002.

Vance, Ashlee. *Elon Musk: Tesla, SpaceX, and the Quest for a Fantastic Future*. New York: HarperCollins, 2015.

Walters, Kerry S. *Benjamin Franklin and His Gods*. Urbana: University of Illinois Press, 1999.

Weart, Spencer R. *The Discovery of Global Warming*. Cambridge, MA: Harvard University Press, 2003.

Williams, L. Pierce. *Michael Faraday: A Biography*. London: Basic Books, 1965.

ENDNOTES

Introduction

1. Carl Van Doren, *Benjamin Franklin (1706–1790)* (New York: Viking Press, 1938), 171.
2. Assuming a 100-watt incandescent bulb has 1600-lumen output, and one candlepower equals 12.57 lumens, the bulb is the equivalent of 127 candles.
3. Tom Tucker, *Bolt of Fate: Benjamin Franklin & His Electric Kite Hoax* (New York: Public Affairs, 2003).
4. David M. Morens, MD, "Death of a President," *New England Journal of Medicine*, 341 (1999), accessed October 1, 2016. doi: 10.1056/NEJM199912093412413.
5. Utility Corporations: Summary Report of the Federal Trade Commission, Document 92, Part 72-A, Introduction, 35–39; Franklin D. Roosevelt; "Campaign Address on Progressive Government at the Commonwealth Club in San Francisco, California," *The American Presidency Project*, accessed August 30, 2015, <http://www.presidency.ucsb.edu/ws/?pid=88391>.

Chapter 1

1. I. Bernard Cohen, *Science and the Founding Fathers: Science in the Political Thought of Thomas Jefferson, Benjamin Franklin, John Adams, and James Madison* (New York: W. W. Norton and Company, 1995), 184–85.
2. "A Full and True Relation of the Death and Slaughter of a Man and His Son at Plough together with Four Horses . . . Slain by the Thunder and Lightning." London (1680); Israel Holly, "Youth liable to sudden death; excited seriously to consider thereof, and speedily to prepare therefor. The substance of a discourse, delivered on the day of the funeral of three young men, who were killed by lightning, at Suffield," printed by Thomas Green (1766), accessed October 1, 2016, <http://quod.lib.umich.edu/e/evans/N08338.0001.001/1:1?rgn=div1;view=fulltext>; Jonathan Townsend, "God's terrible doings are to be observed. A sermon preach'd at Needham, and occasion'd by the sudden and awful death of Mr. Thomas Gardner, by Lightning," printed for Kneeland and Green (Boston, 1746): 10.

3. I. Bernard Cohen, "Benjamin Franklin: An Experimental Newtonian Scientist," *Bulletin of the American Academy of Arts and Sciences* 5, no. 4 (January 1952): 2; Joyce Chaplin, *The First Scientific American: Benjamin Franklin and the Pursuit of Genius* (New York: Basic Books, 2006), 5; Cohen, *Science and the Founding Fathers*, 174–95.

4. Michael R. Lynn, *Popular Science and Public Opinion in 18th Century France* (Manchester, UK: Manchester University Press, 2006), 43–44.

5. H. W. Brands, *The First American: The Life and Times of Benjamin Franklin* (New York: Doubleday Publishing Company, 2000), 22; James A. Dewar, "The Information Age and the Printing Press: Looking Backward to See Ahead." *Rand Corporation*, 2005, accessed October 1, 2016, <http://www.rand.org/pubs /papers/P8014/index2.html#fn0>; Elizabeth L. Eisenstein, "In the Wake of the Printing Press," *The Quarterly Journal of the Library of Congress* 35, no. 3 (July 1978): 184, 187–191; "Apology for Printers," *The Pennsylvania Gazette*, June 10, 1731, online at *National Humanities*, accessed October 1, 2016, <http://nation-alhumanitiescenter.org/pds/becomingamer/ideas/text5/franklinprinting .pdf>.

6. Brands, *The First American*, 33–34, 88, 166–67; Benjamin Franklin, "An Account of the New Invented Pennsylvanian Fire-Places," 1744, online at National Archives, Founders Online, Franklin Papers, accessed October 1, 2016, <http://founders.archives.gov/documents/Franklin/01-02-02-011 4#BNFN-01-02-02-0114-fn-0027> .

7. George C. Lay, "Benjamin Franklin," *Godey's Magazine*, vol. 133, no. 796 (October 1796): 339–43; Kerry S. Walters, *Benjamin Franklin and His Gods* (Urbana: University of Illinois Press, 1999), 44–45.

8. Cohen, *Science and the Founding Fathers*, 136; Brands, *The First American*, 187–206; Benjamin Franklin, *The Autobiography of Benjamin Franklin*, 2nd ed., ed. Leonard W. Larabee, Ralph L. Ketcham, Helen C. Boatfield, and Helene H. Fineman (New Haven: Yale University Press, 1964), 240–44.

9. Chaplin, *First Scientific American*, 6, 18; J. L. Heilbron, *Electricity in the 17th and 18th Centuries: A Study in Early Modern Physics* (New York: Dover Publications, Inc., 1999), 1; Joseph Addison, *Spectator* 1, no. 10 (1711): 44.

10. Chaplin, *First Scientific American*, 94; Franklin, *Autobiography*, 241–42.

11. Cohen, "Benjamin Franklin: An Experimental Newtonian Scientist," 4, 6.

12. Cohen, *Science and the Founding Fathers*, 58; Heilbron, *Electricity in the 17th and 18th Centuries*, 5–6.

13. Franklin, *Autobiography*, 241, note 4; Cohen, *Science and the Founding Fathers*, 141–42.

14. Cohen, *Science and the Founding Fathers*, 142; J. L. Heilbron, "Plus and Minus: Franklin's Zero-Sum Way of Thinking," *Proceedings of the American Philosophical Society* 150 (2006): 608; E. Philip Krider, "Benjamin Franklin and Lightning Rods," *Benjamin Franklin House*, accessed October 1, 2016, <http://www .benjaminfranklinhouse.org/site/sections/about_franklin/physicstoday vol59no1p42_48.pdf>.

15. Heilbron, *Electricity in the 17th and 18th Centuries,* 312–13; Heilbron, "Plus and

Minus: Franklin's Zero-Sum Way of Thinking," 608; "Journal of the Franklin Institute," vol. 161 (January to June 1906): 253–56.

16. Heilbron, *Electricity in the 17th and 18th Centuries*, 313.

17. Heilbron, *Electricity in the 17th and 18th Centuries*, 287.

18. Heilbron, *Electricity in the 17th and 18th Centuries*, 314.

19. Cohen, *Benjamin Franklin's Science*, 70; Heilbron, *Electricity in the 17th and 18th Centuries*, 340–41.

20. Heilbron, *Electricity in the 17th and 18th Centuries*, 346–48.

21. Heilbron, *Electricity in the 17th and 18th Centuries*, 349; Cohen, "Benjamin Franklin: An Experimental Newtonian Scientist," 6.

22. I. Bernard Cohen, "The Two Hundredth Anniversary of Benjamin Franklin's Two Lightning Experiments and the Introduction of the Lightning Rod," *Proceedings of the American Philosophical Society* 96, no. 3 (1952): 332–36; Cohen, *Benjamin Franklin's Science*, 157.

23. "A Letter of Benjamin Franklin, Esq; to Mr. Peter Collinson, F.R.S. concerning an Electrical Kite," October 1, 1752, online at *Philosophical Transactions*, accessed October 1, 2016, <http://rstl.royalsocietypublishing.org/content/47/565>.

24. I. Bernard Cohen, "The Two Hundredth Anniversary," 333–36.

25. "An Account of the Death of Mr. George William Richman, Professor of Experimental Philosophy, a Member of the Imperial Academy of Sciences at Petersburg. Translated from the High-Dutch," *Philosophical Transactions* 1755–1756, no. 49 (1755): 62–63, accessed October 1, 2016, <http://rstl.royal societypublishing.org/content/49/61.full.pdf+html>.

26. Cohen, "The Two Hundredth Anniversary," 331–36; Tucker, *Bolt of Fate*, 135–56.

27. Tucker, *Bolt of Fate*, 186–87.

28. Tucker, *Bolt of Fate*, 186–87.

29. Michael Brian Schiffer, *Draw the Lightning Down: Benjamin Franklin and Electrical Technology in the Age of Enlightenment* (Berkeley: University of California Press, 2003), 195–200; Heilbron, *Electricity in the 17th and 18th Centuries*, 380–82; Cohen, *Science and the Founding Fathers*, 169–70.

30. Chaplin, *First Scientific American*, 5.

31. The full text of *Experiment and Observations on Electricity* is available online from the Smithsonian Institution, <http://library.si.edu/digital-library/book/experiment sobser00fran>.

32. Chaplin, *First Scientific American*, 156–57, 177–78; Brands, *First American*, 302.

33. Brands, *First American*, 472–74; Chaplin, *First Scientific American*, 237.

34. Jonathan R. Dull, "Franklin the Diplomat: The French Mission," *Transactions of the American Philosophical Society* 72, no. 1 (1982): 26–27.

35. Morris Bishop, "Franklin in France," *Daedalus* 86, no. 3 (May 1957): 228–30.

36. "Benjamin Franklin to William Franklin," *Franklin Papers*, August 16, 1784, accessed October 1, 2016, <https://journals.psu.edu/phj/article/viewFile/25468/25237>.

37. Cohen, *Science and the Founding Fathers*, 184–85.

38. Thomas Fleming, "Greatness Speaks to Greatness, One Last Time," *Journal of the American Revolution*, August 29, 2013, accessed August 26, 2015, <http://allthingsliberty.com/2013/08/greatness-speaks-to-greatness-one-last-time/>.

39. Brands, *First American*, 715; "To George Washington from the National Assembly of France, 20 June 1790." National Archives, Founders Online, Franklin Papers, accessed October 1, 2016, <http://founders.archives.gov/documents/Washington/05-05-02-0348>.

Chapter 2

1. Andrew Carnegie, *James Watt* (Honolulu: University Press of the Pacific, 1913), 91–2.

2. Eric Robinson and A. E. Musson, *James Watt and the Steam Revolution* (New York: Augustus M. Kelley Publishers, 1969), 22, 40.

3. Jenny Uglow, *The Lunar Men: Five Friends Whose Curiosity Changed the World* (New York: Farrar, Straus and Giroux, 2002), 26–31; Carnegie, *James Watt*, 19–20, 35–36, 38.

4. Uglow, *The Lunar Men*, 27.

5. Uglow, *The Lunar Men*, 98–99; Carnegie, *James Watt*, 48–51.

6. Uglow, *The Lunar Men*, 101–102; Robinson, *James Watt and the Steam Revolution*, 24–25.

7. Uglow, *The Lunar Men*, 103.

8. Uglow, *The Lunar Men*, 68, 103–104, 132.

9. Robinson, *James Watt and the Steam Revolution*, 11–13.

10. "Time to Fix Patents," *Economist*, August 8, 2015, accessed October 1, 2016, <www.economist.com/news/leaders/21660522-ideas-fuel-economy-todays-patent-systems-are-rotten-way-rewarding-them-time-fix>.

11. Carnegie, *James Watt*, 97, 105; Uglow, *The Lunar Men*, 17, 19, 21; Arnold Toynbee, *Lectures on the Industrial Revolution of the Eighteenth Century in England: Popular Addresses, Notes and Other Fragments* (New York: Longmans, Green, and Company, 1908), 52.

12. Uglow, *The Lunar Men*, 252–53, 289–91; Samuel Smiles, *Lives of Boulton and Watt* (London: William Clowes and Sons, 1865), 205.

13. Robinson, *James Watt and the Steam Revolution*, 62.

14. Carnegie, *James Watt*, Preface.

15. Robinson, *James Watt and the Steam Revolution*, 19; Nathan Rosenberg and Manuel Trajtenberg, "A General Purpose Technology at Work: The Corliss Steam Engine in the Late-Nineteenth-Century United States," *Journal of Economic History* 64, no. 1 (2004): 61, 63–65, 74–75, 92–94; "The Transcontinental Railroad: It's All About Steam," *Linda Hall Library*, accessed October 1, 2016, <http://railroad.lindahall.org/essays/locomotives.html>; "A History of Steamboats," *U.S. Army Corps of Engineers*, accessed October 1, 2016, <http://www.sam.usace.army.mil/Portals/46/docs/recreation/OPCO/montgomery/pdfs/10th and11th/ahistoryofsteamboats.pdf>.

16. Carnegie, *James Watt*, 195–96. 3.6 billion equals 150 million horsepower times 24 men per horsepower.
17. Nicholas Crafts, "Steam as a General Purpose Technology: A Growth Accounting Perspective," *The Economic Journal* 114, no. 495 (April 2004): 342.
18. Rosenberg, "A General Purpose Technology at Work," 74–75.
19. Rosenberg, "A General Purpose Technology at Work," 62–63, 80–83.
20. Carnegie, *James Watt*, 219.
21. "Bank of England to Issue New £50 note in 2011," *Banknote News*, June 22, 2009, accessed October 1, 2016, <http://www.bankofengland.co.uk/publications /Pages/news/2014/072.aspx>.
22. Robinson, *James Watt and the Steam Revolution*, 6, 8–9.

Chapter 3

1. L. Pierce Williams, *Michael Faraday: A Biography* (London: Basic Books, 1965), 196.
2. John Meurig Thomas, "Faraday and Franklin," *Proceedings of the American Philosophical Society* 150, no. 4 (2006): 532–33; William H. Bragg, "Michael Faraday," *The Scientific Monthly* 33, no. 6 (1931): 487–89.
3. Williams, *Michael Faraday*, xi; James Hamilton, *A Life of Discovery: Michael Faraday, Giant of the Scientific Revolution* (New York: Random House, 2002), xi.
4. Williams, *Michael Faraday*, 8, 14.
5. Williams, *Michael Faraday*, 14–15.
6. Williams, *Michael Faraday*, 10, 16.
7. Williams, *Michael Faraday*, 16, 25–26.
8. Williams, *Michael Faraday*, 20–25.
9. Howard J. Fisher, *Faraday's Experimental Research in Electricity: Guide to First Reading* (New Mexico: Green Lion Press, 2001), 15–16.
10. Hamilton, *A Life of Discovery*, 37–38, 50; Williams, *Michael Faraday*, 19, 31, 40–41.
11. Williams, *Michael Faraday*, 31, 34–35, 42.
12. Williams, *Michael Faraday*, 138–44, 151–53.
13. Williams, *Michael Faraday*, 153–54, 156.
14. Williams, *Michael Faraday*, 153–54, 156; Hamilton, *A Life of Discovery*, 165–66.
15. Hamilton, *A Life of Discovery*, 187–88; Williams, *Michael Faraday*, 160.
16. Fisher, *Faraday's Experimental Research in Electricity*, 29–30.
17. Fisher, *Faraday's Experimental Research in Electricity*, 29–31.
18. Fisher, *Faraday's Experimental Research in Electricity*, 30–34.
19. Fisher, *Faraday's Experimental Research in Electricity*, 37–38.
20. Fisher, *Faraday's Experimental Research in Electricity*, 44.
21. Fisher, *Faraday's Experimental Research in Electricity*, 40–41, 44, 49.
22. Williams, *Michael Faraday*, 196.
23. Bragg, "Michael Faraday," 489–90; Williams, *Michael Faraday*, 211.
24. Williams, *Michael Faraday*, 242–44, 284.
25. Williams, *Michael Faraday*, 283–84, 286–87; Thomas, "Faraday and Franklin," 532–33.

26. Bragg, "Michael Faraday," 490; Thomas, "Faraday and Franklin," 534.

27. Williams, *Michael Faraday*, 2, 4; Hamilton, *A Life of Discovery*, 4.

28. Williams, *Michael Faraday*, 103–4, 341.

29. Williams, *Michael Faraday*, 105.

30. E. H. Hare, "Michael Faraday's Loss of Memory." *Proceedings of the Royal Society of Medicine* 67 (1974): 618, accessed October 1, 2016, <http://www.ncbi.nlm.nih.gov/pmc/articles/PMC1651719/>; Williams, *Michael Faraday*, 492–93, 497.

31. Williams, *Michael Faraday*, 499.

32. Williams, *Michael Faraday*, 500.

33. Basil Mahon, *The Man Who Changed Everything: The Life of James Clerk Maxwell* (UK: John Wiley and Sons Ltd., 2003), 1.

34. Mahon, *Man Who Changed Everything*, 15–16.

35. Mahon, *Man Who Changed Everything*, 19–20.

36. Mahon, *Man Who Changed Everything*, 23–24, 35–36, 49; Michael Atiyah, "Foreword to James Clerk Maxwell Commemorative Booklet," *James Clerk Maxwell Foundation*, accessed October 1, 2016, <http://www.clerkmaxwellfoundation.org/ForewordSirMA_Booklet.pdf>.

37. Albert Einstein, *Ideas and Opinions* (New York: Three Rivers Press, 1954), 266–67, 269.

38. Mahon, *Man Who Changed Everything*, xv–xvi, 2.

39. Mahon, *Man Who Changed Everything*, 57, 59–60, 65.

40. Thomas K. Simpson, *Figures of Thought* (Santa Fe: Green Lion Press, 2006), 5, 128–29.

41. Thomas K. Simpson, *Figures of Thought*, 51–52, 56–57.

42. James Trefil and Robert M. Hazen, *The Sciences: An Integrated Approach* (Hoboken, NJ: Wiley and Sons, 2004), 126.

43. Thomas K. Simpson, *Figures of Thought*, 64–65.

44. Thomas K. Simpson, *Figures of Thought*, 67–68.

45. Thomas K. Simpson, *Figures of Thought*, 68–71.

46. Thomas K. Simpson, *Figures of Thought*, 70–71.

47. Thomas K. Simpson, *Figures of Thought*, 70–71.

48. Thomas K. Simpson, *Figures of Thought*, 118, 120–21.

49. Mahon, *The Man Who Changed Everything*, 1–2, 126–28.

50. James Clerk Maxwell, "A Dynamical Theory of the Electromagnetic Field," *Philosophical Transactions of the Royal Society of London* 155 (1865): 466, accessed October 1, 2016, <http://rstl.royalsocietypublishing.org/content/155/459.full.pdf+html>.

Chapter 4

1. *Memorial of Samuel Finley Breese Morse, Including Appropriate Ceremonies of Respect at the National Capitol, and Elsewhere*, published by order of Congress (Washington: Government Printing Office, 1875), 258.

2. Samuel F. B. Morse, "American Electromagnetic Telegraph: A Classic Invention," *Science News* 18, no. 485 (1930): 54–55.

3. O'Reilly et al. v. Morse et al. 56 U.S. 62 (1853), paragraph 434, accessed October 1, 2016, <https://h2o.law.harvard.edu/collages/3986>.

4. Kenneth Silverman, *Lightning Man: The Accursed Life of Samuel F. B. Morse* (New York: Random House, 2003) 4–6.

5. Silverman, *Lightning Man*, 37, 40–41, 47–48, 52, 60–61, 72–73; Carleton Mabee, *The American Leonardo* (New York: Purple Mountain Press, 2000), 47–48, 52.

6. Silverman, *Lightning Man*, 122–25, 133–34; Samuel F. B. Morse, *Imminent Dangers to the Free Institutions of the United States through Foreign Immigration* (1835; New York: Arno Press, Inc., 1969), accessed October 1, 2016, <http://www.wwnorton.com /college/history/eamerica/media/ch12/resources/documents/morse.htm>; Mabee, *The American Leonardo*, 169–70.

7. James D. Reid, *The Telegraph in America: Its Founders, Promoters, and Noted Men* (New York: Derby Brothers, 1879), 87.

8. Mabee, *The American Leonardo*, 148–54.

9. Mabee, *The American Leonardo*, 156–57, 189; O'Reilly v. Morse—56 U.S. 62 (1853) *Justia.com*, paragraph 112, accessed October 1, 2016, <http://supreme .justia.com/cases/federal/us/56/62/case.html>; "Samuel F. B. Morse Papers at the Library of Congress, 1793 to 1919," Library of Congress, accessed October 1, 2016, <www.loc.gov/collections/samuel-morse-papers/articles-and-essays /timeline/1791-1839/>; Thomas P. Jones (ed.), "Journal of the Franklin Institute," vols. 25–26, published by the Franklin Institute (1838): 108.

10. Mabee, *The American Leonardo*, 189–90, 194–95; Silverman, *Lightning Man*, 169.

11. Mabee, *The American Leonardo*, 210–11.

12. Silverman, *Lightning Man*, 174–78.

13. Silverman, *Lightning Man*, 192, 220–21.

14. "Invention of the Telegraph," *Library of Congress*, accessed October 1, 2016, <www.loc.gov/collection/samuel-morse-papers/articles-and-essays/invention -of-the-telegraph/>; Silverman, *Lightning Man*, 236.

15. Mabee, *The American Leonardo*, 277–78.

16. "Samuel F. B. Morse Papers at the Library of Congress, 1793 to 1919," *Library of Congress*, accessed October 1, 2016, <https://www.loc.gov/resource/mmorse .018001/?sp=324/>.

17. Silverman, *Lightning Man*, 283–84, 293–95; "O'Reilly v. Morse—56 U.S. 62 (1853)" paragraph 422.

18. Mabee, *The American Leonardo*, 309, 311; Silverman, *Lightning Man*, 437.

19. Tom Standage, *The Victorian Internet: The Remarkable Story of the Telegraph and the Nineteenth Century's On-Line Pioneers* (New York: Walker Publishing Company, 1999), 72–78.

20. Standage, *The Victorian Internet*, 79–80, 83–84; "The First Transatlantic Cable 1858," *The Institution of Engineering and Technology*, accessed October 1, 2016, <http:// www.theiet.org/resources/library/archives/featured/trans-cable1858.cfm>.

21. Standage, *The Victorian Internet*, 84–86, 88.

22. *Critic* 29 (1896); Standage, *The Victorian Internet*, 89–90.

23. Standage, *The Victorian Internet*, vii, 101–102, 106–107, 132, 136.

24. Standage, *The Victorian Internet*, 150–63.
25. Standage, *The Victorian Internet*, 114–15.
26. Robert Luther Thompson, *Wiring a Continent: The History of the Telegraph Industry in the United States 1832–1866* (Princeton, NJ: Princeton University Press, 1947), 440.
27. JoAnne Yates, "The Telegraph's Effect on Nineteenth Century Markets and Firms," *Business and Economic History* 15 (1986): 151–52.
28. Yates, "The Telegraph's Effect," 149–50.
29. Yates, "The Telegraph's Effect," 155.
30. Tomas Nonnenmacher, "History of the U.S. Telegraph Industry," *Economic History Association, EH.net*, accessed October 1, 2016, <https://eh.net/encyclopedia/history-of-the-u-s-telegraph-industry>; Yates, "The Telegraph's Effect," 154–58; Thompson, *Wiring a Continent*, 442–45.
31. Tomas Nonnenmacher, "State Promotion and Regulation of the Telegraph Industry," *The Journal of Economic History*, 61, no. 1 (2001): 20, 24, 29.
32. Nonnenmacher, "State Promotion and Regulation of the Telegraph Industry," 28; Nonnenmacher, "History of the U.S. Telegraph Industry"; Thompson, *Wiring a Continent*, 440–42.
33. Silverman, *Lightning Man*, 397, 441.
34. Reid, *The Telegraph in America*, 665; "Apotheosis of Washington," *Architect of the Capitol*, January 30, 2014, accessed October 1, 2016, <http://www.aoc.gov/capitol-hill/other-paintings-and-murals/apotheosis-washington>.

Chapter 5

1. "Death of a Titan," *Time*, October 26, 1931.
2. Paul Israel, *Edison: A Life of Invention* (New York: John Wiley and Sons, 1998), vii.
3. Israel, *Edison: A Life of Invention*, 119–20; Robert Friedel and Paul Israel, *Edison's Electric Light* (New Jersey: Rutgers University Press, 1987), x.
4. Rudolph Alvarado, *Thomas Edison* (Indianapolis: Alpha Books, 2002), 4–7.
5. Alvarado, *Thomas Edison*, 8, 13–14; Israel, *Edison: A Life of Invention*, 11–12, 18.
6. Israel, *Edison: A Life of Invention*, 20, 26–27; Alvarado, *Thomas Edison*, 21.
7. Israel, *Edison: A Life of Invention*, 35, 37–39.
8. Israel, *Edison: A Life of Invention*, 45; Alvarado, *Thomas Edison*, 42–44.
9. Alvarado, *Thomas Edison*, 45–46, 49, 55–56.
10. Alvarado, *Thomas Edison*, 59; Israel, *Edison: A Life of Invention*, 118–20.
11. Alvarado, *Thomas Edison*, 66, 69; Israel, *Edison: A Life of Invention*, 130–32.
12. Alvarado, *Thomas Edison*, 77, 79–80, 83.
13. Friedel and Israel, *Edison's Electric Light*, 3–8.
14. Friedel and Israel, *Edison's Electric Light*, 7; "Madison Square North Historic District Designation Report," *New York City Landmarks Preservation Commission*, June 26, 2011, 10–11, accessed October 1, 2016, <www.nyc.gov/html/lpc/downloads/pdf/reports/MadisonSquareNorth.pdf >.
15. Friedel and Israel, *Edison's Electric Light*, 7.

16. Friedel and Israel, *Edison's Electric Light*, 10, 13, 16.
17. Friedel and Israel, *Edison's Electric Light*, 19, 22, 40.
18. Trefil, *The Sciences: An Integrated Approach*, 116–17.
19. Thomas P. Hughes, *Networks of Power: Electrification in Western Society 1880–1930* (Baltimore: Johns Hopkins University Press, 1983), 35–36.
20. "Electric Lamp," *The Thomas Edison Papers, Rutgers University*, accessed October 1, 2016, <http://edison.rutgers.edu/lamp.htm>; Friedel and Israel, *Edison's Electric Light*, 100, 106.
21. Friedel and Israel, *Edison's Electric Light*, 106.
22. Friedel and Israel, *Edison's Electric Light*, 93, 112.
23. "Edison's First Successful Electric Lamp," *Spark Museum of Electrical Invention*, accessed October 1, 2016, <http://www.sparkmuseum.org/collections/electricity-sparks-invention-(1800-1900)/edisons-first-successful-electric-lamp/>; Friedel and Israel, *Edison's Electric Light*, 114.
24. "Milestones: Pearl Street Station, 1882," *ETHW*, accessed October 1, 2016, <http://ethw.org/Milestones:Pearl_Street_Station,_1882>; Austin Smith, "New York Post from Sept. 4, 1882," *New York Post*, September 4, 2001, accessed October 1, 2016, <http://nypost.com/2001/09/04/new-york-post-from-sept-4-1882/>.
25. "Promoting Edison's Lamp," *Smithsonian Institution*, accessed October 1, 2016, <http://americanhistory.si.edu/lighting/19thcent/promo19.htm>; *Electricity Journal* 3, no. 2 (July 27, 1892): 15n21.
26. David Porter Heap, "Report on the International Exhibition of Electricity, Paris," *Engineer Department, U.S. Army*, 1884, 188–89, accessed October 1, 2016, <https://archive.org/stream/reportoninterna01heapgoog#page/n194/mode/2up>; Friedel and Israel, *Edison's Electric Light*, 217.
27. "Edison's U.S. Patents, 1880–1882," *Thomas Edison Papers, Rutgers University*, accessed October 1, 2016, <http://edison.rutgers.edu/patente2.htm>; Friedel and Israel, *Edison's Electric Light*, 157, 165.
28. "Edison Companies," *Thomas Edison Papers, Rutgers University*, accessed October 1, 2016, <http://edison.rutgers.edu/list.htm>.
29. Friedel and Israel, *Edison's Electric Light*, 211–14.
30. Friedel and Israel, *Edison's Electric Light*, 206–7, 219, 222.
31. Hughes, *Networks of Power*, 76–77.
32. Gilbert King, "Edison vs. Westinghouse: A Shocking Rivalry," *Smithsonian*, October 11, 2011, accessed October 1, 2016, <http://www.smithsonianmag.com/ist/?next=/history/edison-vs-westinghouse-a-shocking-rivalry-102146036/>; Israel, *Edison: A Life of Invention*, 336.
33. "Henry Ford Museum & Greenfield Village," Library of Congress, The American Folklife Center, accessed October 1, 2016, <http://memory.loc.gov/diglib/legacies/loc.afc.afc-legacies.200003161/>; Alvarado, *Thomas Edison*, 209.
34. Herbert Hoover, "Statement on a National Tribute to Thomas Alva Edison," *The American Presidency Project*, October 20, 1931, accessed October 1, 2016, <http://www.presidency.ucsb.edu/ws/?pid=22861>; "Death of a Titan."

Chapter 6

1. George Westinghouse Jr., "A Reply to Mr. Edison," *The North American Review* 149, no. 397 (1889): 656.

2. Rebecca Jacobson, "8 Things You Didn't Know About Nikola Tesla." *PBS Newshour,* July 10, 2013, accessed October 1, 2016, <www.pbs.org/newshour /rundown/5-things-you-didnt-know-about-nikola-tesla/>.

3. Margaret Cheney and Robert Uth, *Tesla: Master of Lightning* (New York: Barnes & Noble, 2001), 16–20, 53.

4. "Tesla: Life and Legacy, Coming to America," *PBS*, accessed October 1, 2016, <www.pbs.org/tesla/ll/ll_america.html>.

5. George Westinghouse Jr., "Sir Wm. Thomson and Electric Lighting," *The North American Review* 150, no. 400 (1890): 323.

6. David Bodanis, *Electric Universe: The Shocking True Story of Electricity* (New York: Random House, 2005), 47–49.

7. Margaret Cheney, *Tesla: Man Out of Time* (New York: Simon & Schuster, 2001), 40.

8. Westinghouse Jr., "Sir Wm. Thomson and Electric Lighting," 324.

9. "Tesla: Life and Legacy, War of the Currents," *PBS*, accessed October 1, 2016, <www.pbs.org/tesla/ll/ll_warcur.html>; A New System of Alternate Current—Motors and Transformers," *PBS*, accessed October 1, 2016. Originally read before the American Institute of Electrical Engineers, May 16, 1888, accessed October 1, 2016, <www.pbs.org/tesla/res/res_art02.html>.

10. Quentin Skrabec, *George Westinghouse: Gentle Genius* (New York: Algora Publishing, 2007), 28–29, 32.

11. Skrabec, *Gentle Genius*, 33–38.

12. Skrabec, *Gentle Genius*, 37–38.

13. Skrabec, *Gentle Genius*, 39–41.

14. Skrabec, *Gentle Genius*, 45.

15. Skrabec, *Gentle Genius*, 42, 45, 47.

16. Skrabec, *Gentle Genius*, 48, 68.

17. Skrabec, *Gentle Genius*, 77–79, 82–83.

18. Henry Goslee Prout, *A Life of George Westinghouse* (Ann Arbor: University of Michigan, 1921), 100–102; Skrabec, *Gentle Genius*, 89, 100.

19. Prout, *A Life of George Westinghouse*, 10–11.

20. Prout, *A Life of George Westinghouse*, 10, 11, 19, 113; Skrabec, *Gentle Genius*, 77, 103, 106.

21. Thomas P. Hughes, "Harold P. Brown and the Executioner's Current: An Incident in the AC-DC Controversy." *The Business History Review* 32, no. 2 (1958): 146–58, 151.

22. Hughes, "Executioner's Current," 156–61; Stuart Banner, *The Death Penalty: An American History* (Cambridge, MA: Harvard University Press, 2009), 186. "Far Worse Than Hanging; Kemmler's Death Proves an Awful Spectacle.

The Electric Current Had to Be Turned on Twice before the Deed Was Fully Accomplished," *New York Times*, August 7, 1890.

23. Westinghouse, "A Reply to Mr. Edison," 658–59.

24. Prout, *A Life of George Westinghouse*, 4, 134–37; Skrabec, *Gentle Genius*, 139–40.

25. Harold I. Sharlin, "The First Niagara Falls Power Project," *The Business History Review* 35, no. 1 (1961): 62–63.

26. Sharlin, "The First Niagara Falls Power Project," 65–69.

27. Sharlin, "The First Niagara Falls Power Project," 71–73.

28. Sharlin, "The First Niagara Falls Power Project," 74.

29. Cheney and Uth, *Tesla: Master of Lightning*, 53, 63.

30. "Tesla: Life and Legacy, Who Invented Radio?" *PBS*, accessed October 1, 2016, <http://www.pbs.org/tesla/ll/ll_whoradio.html>.

31. Cheney and Uth, *Tesla: Master of Lightning*, 85–86, 93 (photo).

32. Cheney and Uth, *Tesla: Master of Lightning*, 36–37, 86–87.

33. Nikola Tesla. "The Transmission of Electrical Energy without Wires," *Scientific American*, supplement 1483 (June 4, 1904), p. 23761; Cheney and Uth, *Tesla: Master of Lightning*, 89–93, 95.

34. Cheney and Uth, *Tesla: Master of Lightning*, 97–100; Nikola Tesla, "The Problem of Increasing Human Energy," *PBS*, accessed October 1, 2016, <http://www.pbs.org/tesla/res/res_art09.html>.

35. Cheney and Uth, *Tesla: Master of Lightning*, 104–105, 107.

36. Cheney and Uth, *Tesla: Master of Lightning*, 121–22.

37. "Letters to Nikola Tesla on His 75th Birthday in 1931 from Pioneers in Science and Engineering," *Tesla Memorial Society of New York*, accessed October 1, 2016, <http://www.teslasociety.com/letters.htm>; "Tesla's Portrait Shown," *New York Times*, March 2, 1916.

38. Prout, *A Life of George Westinghouse*, 280–81.

39. "Westinghouse Memorial," *Westinghouse Memorial*, accessed October 1, 2016, <www.westinghousememorial.org/story.htm>.

40. Blaine McCormick and Burton W. Folsom Jr., "A Survey of Business Historians on America's Greatest Entrepreneurs," *The Business History Review* 77, no. 4 (2003): 703–16.

Chapter 7

1. Roosevelt, "Campaign Address on Progressive Government"

2. Roosevelt, "Campaign Address on Progressive Government."

3. Samuel Insull, *The Memoirs of Samuel Insull: An Autobiography*, ed. Larry Plachno (Polo, IL: Transportation Trails, 1992), 10, 16–18; Forrest McDonald, *Insull* (Chicago: University of Chicago Press, 1962), 14–15.

4. McDonald, *Insull*, 20–21.

5. McDonald, *Insull*, 21–22.

6. McDonald, *Insull*, 26–27, 37–38; Insull, *The Memoirs of Samuel Insull*, 46.

7. McDonald, *Insull*, 57, 63.

8. Insull, *The Memoirs of Samuel Insull*, 73; McDonald, *Insull*, 96–98, 103–5.
9. McDonald, *Insull*, 67–68, 103–5.
10. Insull, *The Memoirs of Samuel Insull*, 76–77; McDonald, *Insull*, 58.
11. McDonald, *Insull*, 99–100; Insull, *The Memoirs of Samuel Insull*, 78–79.
12. McDonald, *Insull*, 89.
13. McDonald, *Insull*, 73, 113; Insull, *The Memoirs of Samuel Insull*, 89–90.
14. McDonald, *Insull*, 121.
15. Insull, *The Memoirs of Samuel Insull*, 83.
16. McDonald, *Insull*, 141.
17. McDonald, *Insull*, 139–42. Assuming capacity cut from 437,000 kW to 270,000 kW.
18. McDonald, *Insull*, 148, 150–53.
19. McDonald, *Insull*, 158–61.
20. Insull, *The Memoirs of Samuel Insull*, 119.
21. McDonald, *Insull*, 173.
22. McDonald, *Insull*, 173.
23. McDonald, *Insull*, 173–75.
24. McDonald, *Insull*, 177.
25. McDonald, *Insull*, 177.
26. McDonald, *Insull*, 179.
27. McDonald, *Insull*, 179.
28. McDonald, *Insull*, 182.
29. McDonald, *Insull*, 237.
30. McDonald, *Insull*, 278–83.
31. Insull, *The Memoirs of Samuel Insull*, 195–201; McDonald, *Insull*, 294.
32. McDonald, *Insull*, 299.
33. McDonald, *Insull*, 299–301; Insull, *The Memoirs of Samuel Insull*, 222.
34. Insull, *The Memoirs of Samuel Insull*, 151; McDonald, *Insull*, 263–65, 267.
35. McDonald, *Insull*, 308–12.
36. McDonald, *Insull*, 310–12.
37. Insull, *The Memoirs of Samuel Insull*, 233; McDonald, *Insull*, 308, 312–13, 315.
38. McDonald, *Insull*, 316–17; Insull, *The Memoirs of Samuel Insull*, 251, 264.
39. McDonald, *Insull*, 319; Insull, *The Memoirs of Samuel Insull*, 268; "Samuel Insull Will Make Fight for Complete Vindication," *Ottawa Citizen*, May 7, 1934.
40. Francis X. Busch, *Guilty or Not Guilty: An Account of the Trials of the Leo Frank Case, the D.C. Stephenson Case, the Samuel Insull Case, the Alger Hiss Case* (New York: Bobbs-Merrill, 1952), 136–38.
41. Busch, *Guilty or Not Guilty*, 143–44, 146, 150–52.
42. Busch, *Guilty or Not Guilty*, 161–62.
43. Busch, *Guilty or Not Guilty*, 173–74.
44. Busch, *Guilty or Not Guilty*, 178–79.
45. Busch, *Guilty or Not Guilty*, 186–88, 190–92.
46. Insull, *The Memoirs of Samuel Insull*, cclxxxvi.

47. Norman S. Buchanan, "The Origin and Development of the Public Utility Holding Company," *The Journal of Political Economy* 44, no. 1 (1936): 52; Arthur R. Taylor, "Losses to the Public in the Insull Collapse: 1932–1946," *Business History Review* 36, no. 2 (1962): 188–89.

48. Taylor, "Losses to the Public in the Insull Collapse," 188, 194.

49. *Historical Statistics of the United States, Colonial Times to 1970.* U.S. Department of Commerce (1975): 821, 827, accessed October 1, 2016, <http://www2 .census.gov/library/publications/1975/compendia/hist_stats_colonial-1970 /hist_stats_colonial-1970p2-chS.pdf>.

50. *Historical Statistics of the United States, Colonial Times to 1970,* 827.

Chapter 8

1. Franklin D. Roosevelt. "Campaign Address in Portland, Oregon on Public Utilities and Development of Hydro-electric Power," *The American Presidency Project*, September 21, 1932, accessed July 20, 2014, <http://www.presidency .ucsb.edu/ws/?pid=88390>.

2. Roosevelt, "Campaign Address in Portland, Oregon on Public Utilities and Development of Hydro-electric Power"; William M. Emmons III, "Franklin D. Roosevelt, Electric Utilities, and the Power of Competition," *The Journal of Economic History* 53, 4 (1993): 880.

3. Roosevelt, "Hydro-electric Power."

4. Roosevelt, "Hydro-electric Power."

5. "The Regulation of Public-Utility Companies," *Securities and Exchange Commission,* 1995, 1, 19–20, accessed October 1, 2016, <http://www.sec.gov/news/studies /puhc.txt>.

6. W. S. Murray, *A Superpower System for the Region between Boston and Washington* (Washington, DC: Government Printing Office, 1921), 9–10.

7. Murray, *A Superpower System,* 14.

8. Murray, *A Superpower System,* 13–14.

9. Murray, *A Superpower System,* tables 71 and 63.

10. Gifford Pinchot, "Introduction," *Annals of the American Academy of Political and Social Science* 118 (1925): vii, ix.

11. Pinchot, "Introduction," ix–x.

12. Morris Llewellyn Cooke, "Giant Power and Coal," *Annals of the American Academy of Political and Social Science* 111 (1924): 213, 216–17; Thomas Park Hughes, "Technology and Public Policy: The Failure of Giant Power," *Institute of Electrical and Electronics Engineers* 64, no. 9 (1976): 1363–64.

13. Cooke, "Giant Power and Coal," 213; Hughes, "Technology and Public Policy," 1365.

14. Pinchot, "Introduction," ix–x.

15. Roosevelt, "Hydro-electric Power."

16. Roosevelt, "Hydro-electric Power."

17. Roosevelt, "Hydro-electric Power."

18. Roosevelt, "Hydro-electric Power."

19. Roosevelt, "Hydro-electric Power."

20. Roosevelt, "Hydro-electric Power."

21. Roosevelt, "Hydro-electric Power."

22. Roosevelt, "Hydro-electric Power."

23. Roosevelt, "Hydro-electric Power."

24. Roosevelt, "Hydro-electric Power."

25. Loretto Mersh, "The Presidential Campaign of 1932" (master's thesis, Loyola University Chicago, 1937), 79–80, 84–85, 87.

26. Franklin D. Roosevelt, "Annual Message to Congress," *The American Presidency Project*, January 4, 1935.

27. Ben S. Bernanke, "The Macroeconomics of the Great Depression: A Comparative Approach," *National Bureau of Economic Research*, working paper no. 4813, August 1994, accessed October 1, 2016, <http://www.nber.org/papers/w4814>.

28. "A Report of the National Power Policy Committee with Respect to the Treatment of Holding Companies," Document 137, (March 12, 1935): 1–3.

29. "A Report of the National Power Policy Committee with Respect to the Treatment of Holding Companies," 4.

30. "Utility Corporations: Summary Report of the Federal Trade Commission," letter of transmittal, chapter two, Growth and Importance of Electric and Gas Industries.

31. "Utility Corporations: Summary Report of the Federal Trade Commission," 38, table 12.

32. "Utility Corporations: Summary Report of the Federal Trade Commission," 40.

33. "Utility Corporations: Summary Report of the Federal Trade Commission," 43, table 13, 44, 56.

34. "Utility Corporations: Summary Report of the Federal Trade Commission," 43.

35. Public Utility Holding Company Act of 1935: Hearings before the Committee on Interstate Commerce, 9.

36. "Utility Corporations: Summary Report of the Federal Trade Commission," 838, 840–41, 844.

37. "Utility Corporations: Summary Report of the Federal Trade Commission," 849.

38. Robert E. Healy, "Section 11 of the Public Utility Holding Company Act," *Securities and Exchange Commission*, November 5, 1941, 3–5.

39. Edward C. Eicher, "Address, Annual Convention, Edison Electric Institute, Buffalo, New York," *Securities and Exchange Commission*, June 5, 1941, 1–2, accessed October 1, 2016, <https://www.sec.gov/news/speech/1941/060541e-icher.pdf>.

40. Eicher, "Address, Annual Convention" 2, 4, 6.

41. Eicher, "Address, Annual Convention" 7.

42. Eicher, "Address, Annual Convention" 6–8.

43. Donald C. Cook, "Some Comments on the Current Utility Scene, New York Society of Security Analysts, New York City, New York," *Securities and Exchange Commission*, March 29, 1950, 1–2, 4, accessed October 1, 2016, <http://www.sec.gov/news/speech/speecharchive/1950speech.shtml>.

44. Cook, "Some Comments on the Current Utility Scene," 10.

45. "Twenty-Fifth Annual Report of the Securities and Exchange Commission Fiscal Year Ended June 30, 1959," *Securities and Exchange Commission*, January 6, 1960, page 123, accessed October 1, 2016, <https://www.sec.gov/about/annual_report/1959.pdf>.

46. "Fifteenth Annual Report of the Federal Power Commission," no. 332 (1935), 1.

47. "Fifteenth Annual Report of the Federal Power Commission," 1 on the Attleboro Gap, see, for example, Frank R. Lindh and Thomas W. Bone Jr., "State Jurisdiction over Distributed Generators," *Energy Law Journal* 34, no. 2 (2013): 499–539.

48. "Fifteenth Annual Report of the Federal Power Commission," 2; for original law and language, see "16 U.S. Code Chapter 12—Federal Regulation and Development of Power," *Cornell University Law School*, (hereafter FPA), accessed October 1, 2016, <https://www.law.cornell.edu/uscode/text/16/824>. For updated language and amendments, see FPA at page 37, accessed October 1, 2016, <http://legcounsel.house.gov/Comps/Federal%20Power%20Act.pdf>.

49. "FPA," 42, 45.

50. "Twentieth Annual Report of the Federal Power Commission," no. 21 (1940), 1–2.

51. "Twentieth Annual Report of the Federal Power Commission," 9–10.

52. "Twentieth Annual Report of the Federal Power Commission," 13; Daniel F. Spulber, *Regulation and Markets* (Boston: MIT Press, 1989), 3–5.

53. "Twentieth Annual Report of the Federal Power Commission," 147–48; Twenty-Sixth Annual Report of the Federal Power Commission, no. 23 (1946): 1–2.

Chapter 9

1. Franklin D. Roosevelt, "Address at the Dedication of Boulder Dam," *The American Presidency Project*, September 30, 1945, accessed October 1, 2016, <http://www.presidency.ucsb.edu/ws/?pid=14952>.

2. Joseph E. Stevens, *Hoover Dam: An American Adventure* (University of Oklahoma Press, 1988), vii–viii.

3. Franklin D. Roosevelt, "Address at the Dedication of Boulder Dam."

4. Michael Hiltzik, *Colossus: The Turbulent, Thrilling Saga of the Building of Hoover Dam* (New York: Free Press, 2010), ix.

5. Hiltzik, *Colossus*, 8–12, 18.

6. Hiltzik, *Colossus*, 19, 30, 33; Stevens, *Hoover Dam: An American Adventure*, 11–12.

7. Hiltzik, *Colossus*, 33–34, 36–39.

8. Hiltzik, *Colossus*, 39–42; Stevens, *Hoover Dam: An American Adventure*, 14.

9. Hiltzik, *Colossus*, 61–62; Stevens, *Hoover Dam: An American Adventure*, 17.

10. Hiltzik, *Colossus*, 73–75.

11. Hiltzik, *Colossus*, 81–84, 88; Stevens, *Hoover Dam: An American Adventure*, 26.

12. Hiltzik, *Colossus*, 109–11.

13. Scott Harrison, "St. Francis Dam Collapse Left a Trail of Death and Destruction," *Los Angeles Times*, March 19, 2016, accessed October 1, 2016, <http://www .latimes.com/local/california/la-me-stfrancis-dam-retrospective-20160319 -story.html>; Matt Blitz, "On Occasions like This, I Envy the Dead: The St. Francis Dam Disaster," *Smithsonian*, March 12, 2015, accessed October 1, 2016, <http://www.smithsonianmag.com/history/occasions-i-envy-dead-st-francis -dam-disaster-180954543/>; Hiltzik, *Colossus*, 113–15, 119.

14. Milton N. Nathanson, "Updating the Hoover Dam Documents," *U.S. Department of the Interior*, 1978, 6, accessed October 1, 2016, <http://www.onthecolorado.com /Resources/LawOfTheRiver/HooverDamDocumentsUpdated.pdf>.

15. Hiltzik, *Colossus*, 147; "Table 1.2—Summary of Receipts, Outlays, and Surpluses or Deficits (-) as Percentages of GDP: 1930–2021," *Office of Management and Budget White House*, <https://obamawhitehouse.archives.gov/omb/budget/ Historicals>

16. Stevens, *Hoover Dam: An American Adventure*, 34; Hiltzik, *Colossus*, 161–62.

17. Hiltzik, *Colossus*, 141; Stevens, *Hoover Dam: An American Adventure*, 38.

18. Stevens, *Hoover Dam: An American Adventure*, 39–42.

19. Stevens, *Hoover Dam: An American Adventure*, 46.

20. Hiltzik, *Colossus*, 199.

21. Hiltzik, *Colossus*, 208–11. See also J. L. Savage and Ivan E. Houk, "Checking Arch Dam Designs with Models," *Civil Engineering*, May 1931, accessed October 1, 2016, <https://www.usbr.gov/lc/hooverdam/museum/clippings/clipart84.pdf>.

22. Hiltzik, *Colossus*, 275–76.

23. Stevens, *Hoover Dam: An American Adventure*, viii.

24. *Boulder City: The Documentary*, PBS, accessed October 1, 2016, <http://www.pbs. org/bouldercity/script.htm>; Andrew J. Dunar and Dennis McBride, *Building Hoover Dam: An Oral History of the Great Depression* (Reno: University of Nevada Press, 1993), 65; *Building Boulder City*, PBS American Experience, accessed October 1, 2016, <http://www.pbs.org/wgbh/americanexperience/features/general-article /hoover-building-boulder-city/>; "Boulder City," *UNLV University Libraries*, accessed October 1, 2016, <http://digital.library.unlv.edu/collections/hoover -dam/boulder-city>.

25. *Boulder City*; *Building Boulder City*; "Hoover Dam: Wages," *Reclamation: Managing Water in the West*, March 12, 2015, accessed October 1, 2016, <http://www.usbr .gov/lc/hooverdam/history/essays/wages.html>.

26. Ed Koch, "'Eighth Wonder of the World' Workers Suffered Hardships, Death to Build Historic Hoover Dam," *Las Vegas Sun*, May 15, 2008, accessed October 1, 2016, <http://lasvegassun.com/news/2008/may/15/what-dam-project/>; Dunar and McBride, *Building Hoover Dam: An Oral History of the Great Depression*, 42–44.

27. "Hoover Dam: Fatalities at Hoover Dam." *Reclamation: Managing Water in the West,* last updated March 12, 2015, accessed October 1, 2016, <http://www.usbr.gov /lc/hooverdam/history/essays/fatal.html>; "Lawsuit against Six Companies," *UNLV University Libraries,* accessed October 1, 2016, <http://digital.library.unlv. edu/collections/hoover-dam/lawsuit-against-six-companies>; "Hoover Dam: High Scalers." *Reclamation: Managing Water in the West,* last updated March 12, 2015, accessed October 1, 2016, <http://www.usbr.gov/lc/hooverdam/history /essays/hscaler.html>.

28. Hiltzik, *Colossus,* 127, 295–97.

29. Hiltzik, *Colossus,* 296–98.

30. Hiltzik, *Colossus,* 296–99.

31. Hiltzik, *Colossus,* 299, 303, 309–12, 315.

32. Hiltzik, *Colossus,* 340.

33. Roosevelt, "Address at the Dedication of Boulder Dam."

34. Roosevelt, "Address at the Dedication of Boulder Dam."

35. Herbert Hoover, "Address at Madison Square Garden in New York City," *The American Presidency Project,* October 31, 1932, accessed October 1, 2016, <http:// www.presidency.ucsb.edu/ws/?pid=23317>.

36. Hiltzik, *Colossus,* 382–83.

37. Hiltzik, *Colossus,* 382–83; Hiltzik, *Colossus,* 217.

Chapter 10

1. Lorena Hickok. "Dear Mr. Hopkins," *The New Deal Network,* June 6, 1934, accessed October 1, 2016, <http://newdeal.feri.org/hopkins/hop15.htm>.

2. "The Enduring Legacy," *Tennessee Valley Authority,* accessed October 1, 2016, <http://152.87.4.98/heritage/fdr/index.htm/>; "The 1930s," *Tennessee Valley Authority,* accessed October 1, 2016, <https://www.tva.com/About-TVA/Our -History/The-1930s>; Lorena Hickok, "Dear Mr. Hopkins."

3. "TVA Goes to War," *Tennessee Valley Authority,* accessed October 1, 2016, <http://152.87.4.98/heritage/war/index.htm>; "The 1950s," *Tennessee Valley Authority,* accessed October 1, 2016, <https://www.tva.com/About-TVA/Our -History/The-1950s>; "The 1960s," *Tennessee Valley Authority,* accessed October 1, 2016, <https://www.tva.com/About-TVA/Our-History/The-1960s>.

4. Steven M. Neuse, *David E. Lilienthal: The Journey of an American Liberal* (Knoxville: University of Tennessee Press, 1996).

5. Roscoe C. Martin, "The Tennessee Valley Authority: A Study of Federal Control," *Law and Contemporary Problems* (1957): 353–54, accessed October 1, 2016, <http://scholarship.law.duke.edu/cgi/viewcontent.cgi?article=2721&context =lcp>; Erwin C. Hargrove, *Prisoners of Myth: The Leadership of the Tennessee Valley Authority 1933–1990* (Knoxville: University of Tennessee Press, 2001), 19; Jack Neely, "Clash of the Titans," *Tennessee Valley Authority,* accessed October 1, 2016, <http://152.87.4.98/heritage/titans/index.htm>.

6. Neely, "Clash of the Titans"; "The Enduring Legacy."

7. "Tennessee Valley Authority Act," *U.S. National Archives Records Administration*, Transcript of Tennessee Valley Authority Act (1933), section 23.

8. "Tennessee Valley Authority Act," sections 2(a), 2(f), 4(i), 5(e), 10, 15.

9. "David E. Lilienthal Is Dead at 81; Led U.S. Effort in Atomic Power," *New York Times*, January 16, 1981, accessed October 1, 2016, <http://www.nytimes.com/1981/01/16/obituaries/david-e-lilienthal-is-dead-aty-81-led-us-effort-in-atmoic-power.html?pagewanted=all>; Neuse, *Journey of an American Liberal*, 6, 12.

10. Neuse, *Journey of an American Liberal*, 20, 23–25, 29; "Donald R. Richberg," *University of Minnesota*, accessed October 1, 2016, <http://darrow.law.umn.edu/photo.php?pid=1312>.

11. Neuse, *Journey of an American Liberal*, 27, 29, 32.

12. Neuse, *Journey of an American Liberal*, 37, 43.

13. Neuse, *Journey of an American Liberal*, 42, 44–47.

14. Neuse, *Journey of an American Liberal*, 51.

15. Neuse, *Journey of an American Liberal*, 55, 59–63.

16. David E. Lilienthal, "Business and Government in the Tennessee Valley," *Annals of the American Academy of Political and Social Science* 172 (March 1934): 46.

17. Lilienthal, "Business and Government in the Tennessee Valley," 48.

18. Neuse, *Journey of an American Liberal*, 68, 70.

19. Neuse, *Journey of an American Liberal*, 70–71.

20. Neuse, *Journey of an American Liberal*, 74–76.

21. Neuse, *Journey of an American Liberal*, 82, 84–85.

22. Neuse, *Journey of an American Liberal*, 88–89.

23. Neuse, *Journey of an American Liberal*, 89, 110–11.

24. Neuse, *Journey of an American Liberal*, 89, 110–11.

25. Neuse, *Journey of an American Liberal*, 99–100.

26. Neuse, *Journey of an American Liberal*, 100–101, 103–5, 108.

27. David Lilienthal, *TVA: Democracy on the March* (New York: Harper & Brothers, 1944) 11, 18, 45.

28. Lilienthal, *TVA: Democracy on the March*, 12.

29. "TVA Goes to War."

30. Hargrove, *Prisoners of Myth*, 60, 63; Neuse, *Journey of an American Liberal*, 104, 151, 160.

31. Lorena Hickok, "Dear Mr. Hopkins."

32. Lilienthal, *TVA: Democracy on the March*, dedication, ix, 3–4, 8.

33. Lilienthal, *TVA: Democracy on the March*, 6, 21.

34. Lilienthal, *TVA: Democracy on the March*, 144–46, 155, 185.

35. Neuse, *Journey of an American Liberal*, 139–46.

36. Norman Wengert. "TVA—Symbol and Reality," *The Journal of Politics* 13, no. 3 (1951): 369–70.

37. Wengert, "TVA—Symbol and Reality," 378–80, 384–89, 391.

38. Martin, "The Tennessee Valley Authority," 353–56, 375.

Chapter 11

1. Saul D. Alinsky, *John L. Lewis: An Unauthorized Biography* (New York: Cornwall Press, 1949), 3.
2. Sinclair's most famous novel was and still is *The Jungle* (1906), which brought the dangerous practices of the meatpacking industry to light.
3. Sam H. Schurr and Bruce S. Netschert, *Energy in the American Economy* (Baltimore: Johns Hopkins University Press, 1960), 500.
4. Alinsky, *John L. Lewis*, 3.
5. Schurr, *Energy in the American Economy*, 76, 500; "Monthly Energy Review," *Monthly Energy Review, Energy Information Administration*, September 2016, 98, 109, accessed October 1, 2016, <https://www.eia.gov/totalenergy/data/monthly/pdf/mer.pdf>.
6. Alfred D. Chandler Jr., "Anthracite Coal and the Beginnings of the Industrial Revolution in the United States," *The Business History Review* 46, no. 2 (1972): 141–42, 145–46.
7. Chandler, "Anthracite Coal and the Beginnings of the Industrial Revolution in the United States," 147–49.
8. Schurr, *Energy in the American Economy*, 491–92, 500.
9. Schurr, *Energy in the American Economy*, 65–68, 77.
10. Schurr, *Energy in the American Economy*, 500.
11. Schurr, *Energy in the American Economy*, 492–93.
12. Schurr, *Energy in the American Economy*, 76–77, 500.
13. Schurr, *Energy in the American Economy*, 71, 81–82.
14. Schurr, *Energy in the American Economy*, 74.
15. See also Saul Alinsky, "John L: Something of a Man," *Nation*, June 30, 1969, accessed October 1, 2016, <http://www.thenation.com/article/154512/john-l-something-man>.
16. Alinsky, *John L. Lewis*, 4–5.
17. Alinsky, *John L. Lewis*, 5–7; "Coal Fatalities for 1900 Through 2013," *United States Department of Labor, Mine Safety and Health Administration*, accessed October 1, 2016, <http://www.msha.gov/stats/centurystats/coalstats.asp>; Alinsky, *John L. Lewis*, 7.
18. Wallace Stegner, foreword to *Buried Unsung: Louis Tikas and the Ludlow Massacre*, by Zeese Papanikolas (Lincoln: University of Nebraska Press, 1991), xvii; Ben Mauk, "The Ludlow Massacre Still Matters," *New Yorker*, April 18, 2014, accessed October 1, 2016, <http://www.newyorker.com/business/currency/the-ludlow-massacre-still-matters>.
19. "Eyewitness to Murder: Recounting the Ludlow Massacre," *History Matters, George Mason University*, accessed October 1, 2016, <http://historymatters.gmu.edu/d/5737/>; Mauk, "The Ludlow Massacre Still Matters."
20. Alinksy, *John L. Lewis*, 11–13.
21. Alinksy, *John L. Lewis*, 14–16.

22. "The Bituminous Coal Conservation Act of 1935," *Yale Journal of Law* 45, no. 2 (December 1935): 293–94.

23. "The Bituminous Coal Conservation Act of 1935," 295–96.

24. "Basic Labor Laws (United States of America): The Norris-LaGuardia Act (1932)," *Industrial Workers of the World*, accessed October 1, 2016, <http://www.iww.org/organize/laborlaw/Lynd/Lynd3.shtml>; Barbara Alexander, "The National Recovery Administration," *Economic History Association, EH.net*, accessed October 1, 2016, <http://eh.net/encyclopedia/the-national-recovery-administration-2/>; Daniel Schlozman, *When Movements Anchor Parties: Electoral Alignments in American History* (Princeton, NJ: Princeton University Press, 2015), 56; James Stuart Olson, *Historical Dictionary of the Great Depression, 1929–1940* (Westport, CT: Greenwood Press, 2001), 174.

25. Alinsky, *John L. Lewis*, 42, 73, 81–84, 98.

26. Alinsky, *John L. Lewis*, 111, 143–46.

27. Alinsky, *John L. Lewis*, 164, 177, 188–90, 219–20.

28. Alinsky, *John L. Lewis*, 224, 238–39, 241, 247.

29. Alinsky, *John L. Lewis*, 281, 287, 299–302.

30. Franklin D. Roosevelt, "Fireside Chat 24: On the Coal Crisis," *Miller Center: University of Virginia*, May 2, 1943, accessed October 1, 2016, <http://millercenter.org/president/fdroosevelt/speeches/speech-3330>; "Coal Strike Is Called Off," *Chicago Daily Tribune*, June 23, 1943, accessed October 1, 2016, <http://archives.chicagotribune.com/1943/06/23/page/1/article/coal-strike-is-called-off>.

31. Alinsky, *John L. Lewis*, 327, 329.

32. "Annual Energy Review," *Energy Information Administration*, 2011, accessed October 1, 2016, <http://www.eia.gov/totalenergy/data/annual/>; "Which States Produce the Most Coal?" *Energy Information Administration*, updated April 4, 2016, accessed October 1, 2016, <https://www.eia.gov/tools/faqs/faq.cfm?id=69&t=2>.

33. "Annual Energy Review," table 7.7, table 7.2; "Coal Fatalities for 1900 Through 2015," *United States Department of Labor*, accessed October 1, 2016, <http://arlweb.msha.gov/stats/centurystats/coalstats.asp>.

34. "Annual Energy Review," table 7.2, table 7.9, table 7.3.

35. "Which States Produce the Most Coal?"; Timothy Cama, "Sierra Club Targets Half of US Coal-Fired plants," *Hill*, April 18, 2015, accessed October 1, 2016, <http://thehill.com/policy/energy-environment/238191-sierra-club-targets-half-of-us-coal-plants>.

36. "Upper Big Branch Mine Disaster Investigative Report Summary," *West Virginia Office of Miners' Health, Safety and Training*, accessed October 1, 2016, <http://www.wvminesafety.org/PDFs/Performance/EXECUTIVE%20SUMMARY.pdf>; Alan Blinder, "Mixed Verdict for Donald Blankenship, Ex-Chief of Massey Energy, After Coal Mine Blast," *New York Times*, December 3, 2015, accessed October 1, 2016, <http://www.nytimes.com/2015/12/04/us/donald-blankenship-massey-energy-upper-big-branch-mine.html>; Alan Blinder, "Donald Blankenship Sentenced to a Year in Prison in Mine Safety

Case," *New York Times*, April 6, 2016, accessed October 1, 2016, <http://www.nytimes.com/2016/04/07/us/donald-blankenship-sentenced-to-a-year-in-prison-in-mine-safety-case.html?_r=0>.

37. Blinder, "Donald Blankenship Sentenced to a Year in Prison in Mine Safety Case."

38. Schurr, *Energy in the American Economy*, 76, 500.

39. "Monthly Energy Review: Table 6.2 Coal Consumption by Sector," *Monthly Energy Review, Energy Information Administration*, September 2016, 98, accessed October 1, 2016, <https://www.eia.gov/totalenergy/data/monthly/pdf/mer.pdf>.

40. Lyndon B. Johnson, "Remarks at the Presentation of the 1964 Presidential Medal of Freedom Awards," *The American Presidency Project*, September 14, 1964, accessed October 1, 2016, <http://www.presidency.ucsb.edu/ws/?pid=26496>; Alinksy, *John L. Lewis*, 372.

41. Melvyn Dubofsky and Warren Van Tine, *John L. Lewis: A Biography* (Urbana: University of Illinois Press, 1986), 377.

Chapter 12

1. Albert Einstein and Leopold Infeld, *The Evolution of Physics: The Growth of Ideas from Early Concepts to Relativity and Quanta* (New York: Simon & Schuster, 1938), 197–98. (With all due respect to Professor Infeld, and for convenience only, the remaining cites only Einstein in the text.)

2. Walter Isaacson, *Einstein: His Life and Universe* (New York: Simon & Schuster, 2007), 438; Kristen Rogheh Ghodsee, "Einstein's Pacifism: A Conversation with Wolfram Wette," *Institute for Advanced Study*, 2015, accessed October 1, 2016, <https://www.ias.edu/ideas/2015/ghodsee-einstein-pacifism>.

3. Matthew Chalmers, "Five Papers That Shook the World," *Physicsworld.com*, January 5, 2005, accessed October 1, 2016, <http://physicsworld.com/cws/article/print/2005/jan/05/five-papers-that-shook-the-world>

4. Chalmers, "Five Papers That Shook the World." The speed of light is about 186,000 miles per second.

5. Isaacson, *Einstein: His Life and Universe*, 10–11, 18–19.

6. Don Howard and John Stachel, eds., *Einstein: The Formative Years, 1879–1909* (Boston: Birkhaüser, 2000), 91–92.

7. Isaacson, *Einstein: His Life and Universe*, 25, 32–34, 36, 48.

8. Isaacson, *Einstein: His Life and Universe*, 56–57, 71.

9. Isaacson, *Einstein: His Life and Universe*, 73, 77–79.

10. Chalmers, "Five Papers That Shook the World."

11. Albert Einstein, "On a Heuristic Point of View Concerning the Production and Transformation of Light," *Annalen der Physik* (1905), online at *Princeton University Press Einstein Papers*, accessed October 1, 2016, <http://einsteinpapers.press.princeton.edu/vol2-trans/100>; Chalmers, "Five Papers That Shook the World."

12. Chalmers, "Five Papers That Shook the World"; Albert Einstein, "On the Movement of Small Particles Suspended in Stationary Liquids Required by the Molecular-Kinetic Theory of Heat," *Annalen der Physik* (1905), online at *Princeton University Press Einstein Papers*, accessed October 1, 2016, <http://einsteinpapers.press.princeton.edu/vol2-trans/137>.

13. Albert Einstein, "On the Electrodynamics of Moving Bodies," *Annalen der Physik* (1905), online at *Princeton University Press Einstein Papers*, accessed October 1, 2016, <http://einsteinpapers.press.princeton.edu/vol2-trans/154>.

14. Chalmers, "Five Papers That Shook the World."

15. Chalmers, "Five Papers That Shook the World"; Albert Einstein, "Does the Inertia of a Body Depend upon Its Energy Content?," *Annalen der Physik* (1905), online at *Princeton University Press Einstein Papers*, accessed October 1, 2016, <http://einsteinpapers.press.princeton.edu/vol2-trans/186?ajax>.

16. Isaacson, *Einstein: His Life and Universe*, 140–41, 143–45, 149.

17. Stephen Hawking, "A Brief History of Relativity," *CNN*, December 27, 1999, accessed October 1, 2016, <http://www.cnn.com/ALLPOLITICS/time/1999/12/27/relativity.html>; "May 29, 1919: Eddington Observes Solar Eclipse to Test General Relativity," *American Physical Society*, accessed October 1, 2016, <https://www.aps.org/publications/apsnews/201605/physicshistory.cfm>.

18. Isaacson, *Einstein: His Life and Universe*, 289, 292.

19. Einstein, *Evolution of Physics*, xvi.

20. Einstein, *Evolution of Physics*, 295–96.

21. Einstein, *Evolution of Physics*, 5–6, 8.

22. Einstein, *Evolution of Physics*, 9–11.

23. Einstein, *Evolution of Physics*, 76, 121–22.

24. Einstein, *Evolution of Physics*, 125.

25. Einstein, *Evolution of Physics*, 134.

26. Einstein, *Evolution of Physics*, 148–49.

27. Einstein, *Evolution of Physics*, 156, 161–62.

28. Einstein, *Evolution of Physics*, 169, 177.

29. Einstein, *Evolution of Physics*, 186.

30. Einstein, *Evolution of Physics*, 193–94.

31. Einstein, *Evolution of Physics*, 196–97. Emphasis added.

32. Einstein, *Evolution of Physics*, 197–98.

33. Einstein, *Evolution of Physics*, 243.

34. Einstein, *Evolution of Physics*, 253, 256.

35. David Bodanis, *E=mc²: A Biography of the World's Most Famous Equation* (New York: Walker Publishing Company, 2000), 96–99.

36. Bodanis, *E=mc²*, 100–101, 103–106.

37. Bodanis, *E=mc²*, 105–10.

38. Bodanis, *E=mc²*, 112–13.

39. Isaacson, *Einstein: His Life and Universe*, 471–73, 475–76; "Primary Resources: Letter from Albert Einstein to FDR, 8/2/39," *PBS American Experience*, accessed

October 1, 2016, <http://www.pbs.org/wgbh/americanexperience/features /primary-resources/truman-ein39/>.

40. Bodanis, *E=mc²*, 119–20, 124, 134–40.

41. Isaacson, *Einstein: His Life and Universe*, 477–78, 485, 489–90, 494; "July 1, 1946, Cover," *Time*, accessed October 1, 2016, <http://content.time.com/time /covers/0,16641,19460701,00.html>.

42. Einstein, *Evolution of Physics*, 92.

43. "Dr. Albert Einstein Dies in Sleep at 76; World Mourns Loss of Great Scientist," *New York Times*, April 19, 1955, accessed October 1, 2016, <http://www.ny times.com/learning/general/onthisday/bday/0314.html>; Einstein, *Ideas and Opinions*, 41–42, 45.

44. "Death of a Genius," *Time*, May, 2, 1955.

45. "Death of a Genius."

Chapter 13

1. Dwight Eisenhower, "Atoms for Peace Speech," *Dwight D. Eisenhower Presidential Library, Museum and Boyhood Home*, December 8, 1953, accessed October 1, 2016, <http://www.eisenhower.archives.gov/all_about_ike/speeches/atoms_for _peace.pdf>.

2. "Nuclear Reaction: Why Do Americans Fear Nuclear Power?" *PBS Frontline*, accessed October 1, 2016, <http://www.pbs.org/wgbh/pages/frontline/shows /reaction/etc/faqs.html>.

3. John W. Finney, "Rickover, Father of Nuclear Navy, Dies at 86," *New York Times*, July 9, 1986, accessed October 1, 2016, <http://www.nytimes.com/1986/07/09 /obituaries/rickover-father-of-nuclear-navy-dies-at-86.html?pagewanted=all>.

4. "What Is U.S. Electricity Generation by Energy Source?" *Energy Information Administration*, updated April 1, 2016, accessed October 1, 2016, <https://www .eia.gov/tools/faqs/faq.cfm?id=427&t=3>; Steve Hargreaves, "First New Nuclear Reactors OK'd in over 30 Years," *CNN Money*, February 9, 2012, accessed October 1, 2016, <http://money.cnn.com/2012/02/09/news/economy/nuclear_reactors/>; "The Nuclear Renaissance," *World Nuclear Association*, updated September 2015, accessed October 1, 2016, <http://www.world-nuclear.org/information-library /current-and-future-generation/the-nuclear-renaissance.aspx>; John D. Sutter, "Bill Gates and the 'Nuclear Renaissance,'" *CNN*, February 17, 2010, accessed October 1, 2016, <http://www.cnn.com/2010/TECH/02/17/bill.gates.nuclear/>.

5. Richard A. Muller, *Physics for Future Presidents: The Science Behind the Headlines* (New York: W. W. Norton & Company, 2008), 125–26; "The Future of Nuclear Power," *MIT*, 2003, 102, accessed October 1, 2016, <http://web.mit.edu /nuclearpower/>.

6. Muller, *Physics for Future Presidents*, 155–58; "Uranium Enrichment," *United States Nuclear Regulatory Commission*, July 19, 2016, accessed October 1, 2016, <http://www .nrc.gov/materials/fuel-cycle-fac/ur-enrichment.html>; "The Future of Nuclear Power," 102; "Uranium: Its Uses and Hazards," *Institute for Energy and Environmental*

　　　　Research, updated May 2012, accessed October 1, 2016, <http://ieer.org/resource
　　　　/factsheets/uranium-its-uses-and-hazards/>.

7. Muller, *Physics for Future Presidents*, 156–59; "Uranium Enrichment," *United States
　　　Nuclear Regulatory Commission.

8. "Key Excerpts of the Joint Comprehensive Plan of Action (JCPOA)," *White
　　　House*, July 14, 2015, accessed October 1, 2016, <https://www.whitehouse.gov
　　　/the-press-office/2015/07/14/key-excerpts-joint-comprehensive-plan-action
　　　-jcpoa>.

9. "Atoms for Peace Speech."

10. "Atoms for Peace Speech."

11. "Atoms for Peace Speech."

12. Gordon Dean, "Atoms for Peace: An American View," *International Journal* 9, no.
　　　4 (1954): 254–56.

13. Robert McKinney, "Atomic Energy: Its Contribution to World Peace through
　　　Economic Self-Help," *Proceedings of the Academy of Political Science* 26, no. 3 (1957):
　　　266, 269.

14. John Krige, "Atoms for Peace, Scientific Internationalism, and Scientific
　　　Intelligence," *Osiris*, 2nd ser., vol. 21 (2006): 180–81.

15. "Assessment of the Nuclear Power Industry," *Navigant Consulting*, July
　　　2013, 9–11, 12–13, accessed October 1, 2016, <http://pubs.naruc.org/pub
　　　/536D6E09-2354-D714-5175-4F938E94ADB5>; Willis L. Shirk, "'Atoms for
　　　Peace' in Pennsylvania," *Pennsylvania Heritage* 35, no. 2 (2009), accessed October 1,
　　　2016, <http://www.phmc.state.pa.us/portal/communities/pa-heritage/atoms-for
　　　-peace-pennsylvania.html>; "Shippingport Atomic Power Station," *Historic
　　　American Engineering Record, National Park Service*, HAER No. PA-81, 8, accessed
　　　October 1, 2016, <http://cdn.loc.gov/master/pnp/habshaer/pa/pa1600/pa1658
　　　/data/pa1658data.pdf>.

16. Shirk, "'Atoms for Peace' in Pennsylvania"; William Beaver, "Duquesne Light
　　　and Shippingport: Nuclear Power Is Born in Western Pennsylvania," *The Western
　　　Pennsylvania Historical Magazine* 70, no. 4 (October 1987): 339; Ernest Sternglass,
　　　Secret Fallout: Low-Level Radiation from Hiroshima to Three-Mile Island (New York:
　　　McGraw-Hill, 1981), 104.

17. "Assessment of the Nuclear Power Industry," 13–17, 19.

18. "Assessment of the Nuclear Power Industry," 23.

19. "Assessment of the Nuclear Power Industry," 23–24. ($693/$1,381); Ibid.
　　　($525/$1,218).

20. E. E. Kintner, "Admiral Rickover's Gamble." *Atlantic*, January 1, 1959, accessed Octo-
　　　ber 1, 2016, <http://www.theatlantic.com/magazine/archive/1959/01/admiral
　　　-rickovers-gamble/308436/>; Jason Reagle, "The First Icex: A Historical
　　　Journey by USS *Nautilus* (SSN-571)," *Undersea Warfare* 40, Summer 2009, accessed
　　　October 1, 2016, <http://www.public.navy.mil/subfor/underseawarfare
　　　magazine/Issues/Archives/issue_40/nautilus.html>; Shirk, "'Atoms for
　　　Peace' in Pennsylvania."

21. Kintner, "Admiral Rickover's Gamble"; Scott Frickel, "Engineering Heterogeneous Accounts: The Case of Submarine Thermal Reactor Mark-I," *Science, Technology & Human Values* 21, no. 1 (1996): 36–39.

22. Kintner, "Admiral Rickover's Gamble."

23. "Assessment of the Nuclear Power Industry," 12.

24. Frickel, "Engineering Heterogeneous Accounts," 39–40; Eric. P. Loewen, "The USS *Seawolf* Sodium-Cooled Reactor Submarine," *American Nuclear Society*, May 17, 2012.

25. Frickel, "Engineering Heterogeneous Accounts," 42, 48.

26. John W. Crawford and Steven L. Krahn, "The Naval Nuclear Propulsion Program: A Brief Case Study in Institutional Constancy," *Public Administration Review* 58, no. 2 (1998): 159; Robin Cowan, "Nuclear Power Reactors: A Study in Technological Lock-In," *The Journal of Economic History* 50, no. 3 (1990): 541.

27. Steve Cohn, "The Political Economy of Nuclear Power (1945–1990): The Rise and Fall of an Official Technology," *Journal of Economic Issues* 24, no. 3 (1990): 800; Eyder Peralta, "U.S. Regulators Approve First Nuclear Power Plant in a Generation," *NPR*, February 9, 2012, accessed October 1, 2016, <http://www.npr.org/sections/thetwo-way/2012/02/09/146646228/u-s-regulators-approve-first-nuclear-power-plant-in-a-generation>; "Information Digest, 2014–2015: Appendix A: U.S. Commercial Nuclear Power Reactors—Operating Reactors," *Nuclear Regulatory Commission*, August 2014, accessed October 1, 2016, <http://www.nrc.gov/reading-rm/doc-collections/nuregs/staff/sr1350/>.

28. "Emergency Preparedness at Nuclear Power Plants," *United States Nuclear Regulatory Commission*, April 2014, accessed October 1, 2016, <http://www.nrc.gov/reading-rm/doc-collections/fact-sheets/emerg-plan-prep-nuc-power.pdf>; "Power Reactors," *United States Nuclear Regulatory Commission*, updated June 28, 2016, accessed October 1, 2016, <http://www.nrc.gov/reactors/power.html>.

29. Charles Komanoff, *Power Plant Cost Escalation: Nuclear and Capital Costs, Regulation, and Economics* (New York: Van Nostrand Reinholdt Publishing, 1981), 16–17, 20; "Operational & Long-Term Shutdown Reactors," *International Atomic Energy Agency*, updated September 24, 2016, accessed October 1, 2016, <https://www.iaea.org/PRIS/WorldStatistics/OperationalReactorsByCountry.aspx>; "World Statistics," *Nuclear Energy Institute*, May 2016, accessed October 1, 2016, <http://www.nei.org/Knowledge-Center/Nuclear-Statistics/World-Statistics>; "International Energy Outlook 2016," chap. 5, "Electricity," *Energy Information Administration*, released May 11, 2016, figures 5–3, accessed October 1, 2016, <http://www.eia.gov/forecasts/ieo/electricity.cfm>.

30. Komanoff, *Power Plant Cost Escalation*, 24–26.

31. "Assessment of the Nuclear Power Industry," 34, 58.

32. Cohn, "The Political Economy of Nuclear Power," 781–83.

33. "Three Mile Accident," *World Nuclear Association*, January 2012, accessed October 1, 2016, <http://www.world-nuclear.org/info/Safety-and-Security/Safety-of-Plants/Three-Mile-Island-accident/>.

34. "Three Mile Accident."

35. John G. Kemeny, "Report of the President's Commission on the Accident at Three Mile Island," 1979, 1, 8.

36. Kemeny, "Accident at Three Mile Island," 7. Emphasis omitted.

37. Kemeny, "Accident at Three Mile Island," 18–19; "Backgrounder on the Three Mile Island Accident," *United States Nuclear Regulatory Commission*, December 12, 2014, accessed October 1, 2016, <http://www.nrc.gov/reading-rm/doc -collections/fact-sheets/3mile-isle.html>; "Three Mile Island: The Inside Story," *Smithsonian Natural Museum of American History*, accessed October 1, 2016, <http://americanhistory.si.edu/tmi/tmi12.htm>.

38. "Emergency Preparedness at Nuclear Power Plants"; "Backgrounder on the Three Mile Island Accident."

39. Clifford D. May, "Shoreham, Despite Plan to Scrap It, Gains Full License," *New York Times*, April 21, 1989, accessed October 1, 2016, <http://www. nytimes.com/1989/04/21/nyregion/shoreham-despite-plan-to-scrap-it-gains -full-license.html>; John Rather, "Planning the Fate of a Nuclear Plant's Land," *New York Times*, January 1, 2009, accessed October 1, 2016, <http://www .nytimes.com/2009/01/04/nyregion/long-island/04shorehamli.html?_r=0>.

40. "Introduction & History of the Shoreham Nuclear Power Plant," Long Island Power Authority, accessed September 4, 2013, <http://www.lipower.org/shoreham /history.html>.

41. Jerry Ross and Barry M. Staw, "Organizational Escalation and Exit: Lessons from the Shoreham Nuclear Power Plant," *The Academy of Management Journal* 36, no. 4 (August 1993): 708, accessed October 1, 2016, <http://citeseerx.ist.psu .edu/viewdoc/download?doi=10.1.1.657.7195&rep=rep1&type=pdf>.

42. Dennis Hevesi, "Nora Bredes, Who Fought Long Island Nuclear Plant, Dies at 60," *New York Times*, August 22, 2011, accessed October 1, 2016, <http://www. nytimes.com/2011/08/23/nyregion/nora-bredes-60-dies-fought-shoreham -nuclear-plant.html>.

43. "Assessment of the Nuclear Power Industry," 137.

44. "Processing of Used Nuclear Fuel," *World Nuclear Association*, September 2014, accessed October 1, 2016, <http://www.world-nuclear.org/info/Nuclear -Fuel-Cycle/Fuel-Recycling/Processing-of-Used-Nuclear-Fuel/>; William F. Shughart II, "Why Doesn't U.S. Recycle Nuclear Fuel?," *Forbes*, October 1, 2014, accessed October 1, 2016, <http://www.forbes.com/sites/realspin/2014/10/01 /why-doesnt-u-s-recycle-nuclear-fuel/#78330bfb7db4>.

45. "Assessment of the Nuclear Power Industry," 136–37; Ross and Staw, "Organizational Escalation and Exit," 708.

46. Kim Cawley, "The Federal Government's Responsibilities and Liabilities Under the Nuclear Waste Policy Act," *Congressional Budget Office*, December 3, 2014, 1, accessed October 1, 2016, <https://www.cbo.gov/sites/default /files/114th-congress-2015-2016/reports/51035-NuclearWaste_Testimony .pdf>; "Assessment of the Nuclear Power Industry," 138.

47. "The Official Report of the Fukushima Nuclear Accident Independent Investigation Commission," *National Diet of Japan*, 2012, 12, accessed October

1, 2016, <https://www.nirs.org/fukushima/naiic_report.pdf>; "Additional Report of Japanese Government to IAEA—Accident at TEPCO's Fukushima Power Stations, 15 September 2011," *International Atomic Energy Agency*, 5, accessed October 1, 2016, <https://www.iaea.org/newscenter/focus/fukushima/additional-japan-report>.

48. "Additional Report of Japanese Government to IAEA," 25–26, 31.
49. "The Official Report of the Fukushima Nuclear Accident Independent Investigation Commission," 7, 9, 20.
50. "The Official Report of the Fukushima Nuclear Accident Independent Investigation Commission," 7, 9, 20.
51. Nassim Nicholas Taleb, *The Black Swan: The Impact of the Highly Improbable* (New York: Random House, 2007).

Chapter 14

1. Rachel Carson, *Silent Spring*, rev. ed. with a new introduction by Linda Lear and an afterword by Edward O. Wilson (1962; New York: Houghton Mifflin Harcourt Publishing Company, repr. 2002), 1–3.
2. Eliza Griswold, "How *Silent Spring* Ignited the Environmental Movement," *New York Times*, September 21, 2012, accessed October 1, 2016, <http://www.nytimes.com/2012/09/23/magazine/how-silent-spring-ignited-the-environmental-movement.html?_r=0>.
3. Carson, *Silent Spring*, 1–3.
4. Carson, *Silent Spring*, 5, 6, 20.
5. Carson, *Silent Spring*, 41, 106–7, 133–35, 157–59.
6. Carson, *Silent Spring*, 127.
7. Carson, *Silent Spring*, 24, 26–27.
8. Carson, *Silent Spring*, 174.
9. Carson, *Silent Spring*, 201–4.
10. Griswold, "How *Silent Spring* Ignited the Environmental Movement."
11. Linda Lear, "Rachel Carson's *Silent Spring*," *Environmental History Review* 17, no. 2 (1993): 23–25.
12. Lear, "Rachel Carson's *Silent Spring*," 25, 27.
13. Lear, "Rachel Carson's *Silent Spring*," 29.
14. Lear, "Rachel Carson's *Silent Spring*," 37–38; Griswold, "How *Silent Spring* Ignited the Environmental Movement."
15. "President's Science Advisory Committee (PSAC): Pesticides Report, 15 May 1963," *John F. Kennedy Presidential Library and Museum*, accessed October 1, 2016, <https://www.jfklibrary.org/Asset-Viewer/Archives/JFKPOF-087-003.aspx>; Lear, "Rachel Carson's *Silent Spring*," 39.
16. Griswold, "How *Silent Spring* Ignited the Environmental Movement"; Linda Lear, *Rachel Carson: Witness for Nature* (New York: Houghton Mifflin Harcourt, 2009), 3.
17. Leonard Greenburg, MD, et al., "Report of an Air Pollution Incident in New York City, November 1953," *Public Health Reports* 77, no. 1 (1962): 7, accessed October

1, 2016, <http://www.ncbi.nlm.nih.gov/pmc/articles/PMC1914642/>; Devra L. Davis, Michelle L. Bell, and Tony Fletcher, "A Look Back at the London Smog of 1952 and the Half Century Since," *Environmental Health Perspectives* 110, no. 12 (2002): A734, accessed October 1, 2016, <http://www.ncbi.nlm.nih.gov/pmc/articles/PMC1241116/>.

18. Don Hopey and David Templeton, "In 1948, Smog Left Deadly Legacy in Donora," *Pittsburgh Post Gazette*,December 12, 2010, accessed October 1, 2016, <http://www.post-gazette.com/news/health/2010/12/12/In-1948-smog-left-deadly-legacy-in-Donora/stories/201012120248>; Lynne Page Snyder, "The Death-Dealing Smog over Donora, Pennsylvania: Industrial Air Pollution, Public Health Policy, and the Politics of Expertise, 1948–1949," *Environmental History Review* 18, no. 1 (1994): 117, 119–120.

19. Snyder, "The Death-Dealing Smog over Donora, Pennsylvania,"121–22.

20. Snyder, "The Death-Dealing Smog over Donora, Pennsylvania,"124–25, 129.

21. Snyder, "The Death-Dealing Smog over Donora, Pennsylvania," 123, 131–32; Clarence A. Mills, "The Donora Episode," *Science* 111 (1950): 67.

22. Percy Bysshe Shelley, quoted in Geoffrey Lean, "The Great Smog of London: The Air Was Thick with Apathy," *The Telegraph*, December 6, 2012, accessed October 1, 2016, <http://www.telegraph.co.uk/news/earth/countryside/9727128/The-Great-Smog-of-London-the-air-was-thick-with-apathy.html>.

23. Michelle L. Bell and Devra Lee Davis, "Reassessment of the Lethal London Fog of 1952: Novel Indicators of Acute and Chronic Consequences of Acute Exposure to Air Pollution," *Environmental Health Perspectives* 109, supplement 3 (June 2001): 389, accessed October 1, 2016, <http://www.ncbi.nlm.nih.gov/pmc/articles/PMC1240556/>.

24. Andrew Hunt, Jerrold L. Abraham, et al., "Toxicologic and Epidemiologic Clues from the Characterization of the 1952 London Smog Fine Particulate Matter in Archival Autopsy Lung Tissues," *Environmental Health Perspectives* 111, no. 9 (July 2003): 1209–10, 1213, accessed October 1, 2016, <http://www.ncbi.nlm.nih.gov/pmc/articles/PMC1241576/>.

25. George Holzworth, "Vertical Temperature Structure during the 1966 Thanksgiving Week Air Pollution Episode in New York City," *Monthly Weather Review* 100, no. 6 (June 1972): 445, 448–49, accessed October 1, 2016, <http://docs.lib.noaa.gov/rescue/mwr/100/mwr-100-06-0445.pdf>; Steve Tracton, "The Killer London Smog Event of December, 1952: A Reminder of Deadly Smog Events in U.S.," *Washington Post*, December 20, 2012, accessed October 1, 2016, <https://www.washingtonpost.com/blogs/capital-weather-gang/post/the-killer-london-smog-event-of-december-1952-a-reminder-of-deadly-smog-events-in-us/2012/12/19/452c66bc-498e-11e2-b6f0-e851e741d196_blog.html>.

26. Judith A. Layzer, *The Environmental Case: Translating Values into Policy, Second Edition* (Washington: CQ Press, 2006), 28.

27. Layzer, *Environmental Case*, 29, 34–37.

28. Layzer, *Environmental Case*, 26, 30–31.

29. Layzer, *Environmental Case*, 35–37; David Gerard and Lester B. Lave, "Implementing Technology-Forcing Policies: The 1970 Clean Air Act Amendments and the Introduction of Advanced Automotive Emissions Controls in the United States," *Technological Forecasting and Social Change* 72 (2005): 766, accessed October 1, 2016, <http://faculty.lawrence.edu/gerardd/wp-content/uploads/sites/9/2014/02/18-TFSC-Gerard-Lave.pdf>; James Salzman and Barton H. Thompson Jr., *Environmental Law and Policy* (New York: Hudson Press, 2007), 87–88.

30. Salzman, *Environmental Law and Policy*, 91, 94–97.

31. Layzer, *The Environmental Case*, 45–46.

32. Richard Schmalensee and Robert N. Stavins, "The SO_2 Allowance Trading System: The Ironic History of a Grand Policy Experiment," *Journal of Economic Perspectives* 27, no. 1 (2013): 103.

33. Schmalensee, "The SO_2 Allowance Trading System," 105–6.

34. Schmalensee, "The SO_2 Allowance Trading System," 106–7.

35. Schmalensee, "The SO_2 Allowance Trading System," 109–10.

36. Schmalensee, "The SO_2 Allowance Trading System," 111.

37. Schmalensee, "The SO_2 Allowance Trading System," 116–17.

38. C. Arden Pope III, PhD, et al., "Lung Cancer, Cardiopulmonary Mortality, and Long-Term Exposure to Fine Particulate Air Pollution," *Journal of the American Medical Association* 287, no. 9 (March 2002), accessed October 1, 2016, <http://www.ncbi.nlm.nih.gov/pmc/articles/PMC4037163/>; for further public health issues related to fine particulate matter and air pollution, see Antonella Zanobetti and Joel Schwartz, "The Effect of Fine and Coarse Particulate Air Pollution on Mortality: A National Analysis," *Environmental Health Perspectives* 117, no. 6 (June 2009), accessed October 1, 2016, <http://ehp.niehs.nih.gov/0800108>; J. G. Ayres, "The Mortality Effects of Long-Term Exposure to Particulate Air Pollution in the United Kingdom," *Crown* (2009), accessed October 1, 2016, <https://www.gov.uk/government/uploads/system/uploads/attachment_data/file/304641/COMEAP_mortality_effects_of_long_term_exposure.pdf>.

39. "Progress Cleaning the Air and Improving People's Health," *Environmental Protection Agency*, accessed October 1, 2016, <https://www.epa.gov/clean-air-act-overview/progress-cleaning-air-and-improving-peoples-health>.

Chapter 15

1. Final Report on Price Manipulation in Western Markets: Fact-Finding Investigation of Potential Manipulation of Electric and Natural Gas Prices, Docket No. PA02-2-000, *Federal Energy Regulatory Commission*, March 2003, accessed October 1, 2016, <http://www.ferc.gov/legal/maj-ord-reg/land-docs/part-2-03-26-03.pdf. VI-35>.

2. Dan Morain, "Deregulation Bill Signed by Wilson," *Los Angeles Times*, September 24, 1996, accessed October 1, 2016, <http://articles.latimes.com/1996-09-24/news/mn-47043_1_wilson-signed-legislation>.

3. Paul L. Joskow, "California's Electricity Crisis," *Oxford Review of Economic Policy* 17, no. 3 (2001): 366–75; Pete Wilson and David Freeman, "Energy Crisis, Electricity: Who's at Fault?" *Cal-Tax Digest*, July 2001, accessed October 10, 2014, <http://www.caltax.org/member/digest/july2001/july01-04.htm>.

4. "Fortune 500: 2001 Archive Full List 1–100," *CNN*, accessed October 1, 2016, <http://archive.fortune.com/magazines/fortune/fortune500_archive/full /2001/>; "Fortune 500: 2000 Archive Full List 1–100," *CNN*, accessed October 1, 2016, <http://archive.fortune.com/magazines/fortune/fortune500 _archive/full/2000/>; Bethany McLean and Peter Elkind, *The Smartest Guys in the Room: The Amazing Rise and Scandalous Fall of Enron* (New York: Portfolio /Penguin, 2003).

5. Bill Number: AB 1890 (passed September 23, 1996), *California Legislative Information*, accessed October 1, 2016, <ftp://www.leginfo.ca.gov/pub/95-96/bill/asm /ab_1851-1900/ab_1890_bill_960924_chaptered.html>; Electricity: Provisions of AB 1890, *Energy Information Administration*, accessed October 1, 2016, <http:// www.eia.gov/electricity/policies/legislation/california/assemblybill.html>.

6. Joskow, "California's Electricity Crisis," 368–69, 371; "Pacific Gas and Electric Company's Divesture of Electric Assets: Environmental Review," chap. 2, *State of California Public Utilities Commission*, updated February 26, 1999, accessed October 1, 2016, <http://www.cpuc.ca.gov/Environment/info/esa /divest-pge-two/eir/chapters/02-projd.htm>; Order 888 (issued April 24, 1996), *Federal Energy Regulatory Commission*, accessed October 1, 2016, <http:// www.ferc.gov/legal/maj-ord-reg/land-docs/order888.asp>.

7. Joskow, "California's Electricity Crisis," 375–77.

8. Joskow, "California's Electricity Crisis," 375, 377.

9. Joskow, "California's Electricity Crisis," 377–78.

10. Joskow, "California's Electricity Crisis," 378–79.

11. Joskow, "California's Electricity Crisis," 378–79.

12. Joskow, "California's Electricity Crisis," 379–80.

13. Joskow, "California's Electricity Crisis," 380–381.

14. Joskow, "California's Electricity Crisis," 381.

15. Joskow, "California's Electricity Crisis," 366, 383, 385–86.

16. "Final Report on Price Manipulation in Western Markets." *Federal Energy Regulatory Commission*, March 2003, vi, ES-1, accessed October 1, 2016, <http:// www.ferc.gov/legal/maj-ord-reg/land-docs/PART-I-3-26-03.pdf>.

17. "Final Report on Price Manipulation in Western Markets," ES-1, ES-6.

18. "Final Report on Price Manipulation in Western Markets," ES-4, ES-5, II-1, II-30, II-31, ES-5.

19. "Final Report on Price Manipulation in Western Markets," ES-5.

20. "Final Report on Price Manipulation in Western Markets," II-59, II-60.

21. "Final Report on Price Manipulation in Western Markets," ES-6, III-15.

22. "Final Report on Price Manipulation in Western Markets," I-1, I-2.

23. John R. Wilke and Robert Gavin, "Brazen Trade Marks New Path of Enron Probe for Regulators," *Wall Street Journal*, October 21, 2002, accessed October 1, 2016, <http://www.wsj.com/articles/SB1035159011191352671>.

24. Wilke, "Brazen Trade Marks New Path of Enron Probe for Regulators"; United States of America v. Timothy N. Belden, *Findlaw*, accessed October 1, 2016, <http://news.findlaw.com/hdocs/docs/enron/usbelden101702plea.pdf>.

25. Memo from Donald J. Gelinas [associate director, FERC Office of Markets, Tariffs, and Rates] to Sam Behrends, LeBoeuf, Lamb, Greene & MacRae LLP [Enron Counsel], May 6, 2002; Memoranda to Richard Saunders [of Enron] from Christian Yoder [of Enron Power Marketing, Inc.] and Stephen Hall [of Stoel Rives, LLP], December 6, 2000. Hereafter "Enron Smoking Gun Memo."

26. "Enron Smoking Gun Memo," p. 1 at A.1, page 2 at A.1, page 6 at 6.a.

27. "Enron Smoking Gun Memo," p. 2.

28. "Enron Smoking Gun Memo," p. 3, p. 5 at 1.a.

29. "Enron Smoking Gun Memo," p. 5 at 2.e, p. 6 at 4.e, p. 7 at 7.b.

30. "Final Report on Price Manipulation in Western Markets," VI-3, VI-35, VI-36.

31. "Final Report on Price Manipulation in Western Markets," I-13, I-14.

32. Wilson, "Energy Crisis, Electricity: Who's at Fault?"

33. Wilson, "Energy Crisis, Electricity: Who's at Fault?"

34. Wilson, "Energy Crisis, Electricity: Who's at Fault?"

35. Mark Jickling, "The Enron Collapse: An Overview of Financial Issues," *CRS Report for Congress*, updated March 28, 2002, CRS-2, accessed October 1, 2016, <http://fpc.state.gov/documents/organization/9267.pdf>; Alexei Barrionuevo and Vikas Bajaj, "What's at Heart of Enron Case, Accounting or 'Lies and Choices'?," *New York Times*, January 31, 2006, accessed October 1, 2016, <http://www.nytimes.com/2006/01/31/business/worldbusiness/31iht-enron.html>.

36. William C. Powers, "Report of Investigation by the Special Investigative Committee of the Board of Directors of Enron Corp," February 1, 2002, 2, 30–31, accessed October 1, 2016, <http://picker.uchicago.edu/Enron/Powers Report(2-2-02).pdf>.

37. Powers, "Report of Investigation of Enron Corp," 3–4, 18.

38. Powers, "Report of Investigation of Enron Corp," 5.

39. "The Role of the Board of Directors in Enron's Collapse," *Permanent Subcommittee on Investigations of the Committee of Governmental Affairs United States Senate*, July 8, 2002, 3, 18–19, accessed October 1, 2016, <https://www.gpo.gov/fdsys/pkg/CPRT -107SPRT80393/pdf/CPRT-107SPRT80393.pdf>.

40. "The Role of the Board of Directors in Enron's Collapse," 43–45.

41. Ken Brown and Ianthe Jeanne Dugan, "Arthur Andersen's Fall from Grace Is a Sad Tale of Greed and Miscues," *Wall Street Journal*, June 7, 2002, accessed October 1, 2016, <http://www.wsj.com/articles/SB1023409436545200>; "Ties to Enron Blinded Andersen," *Chicago Tribune*, September 3, 2002, accessed October 1, 2016, <http://www.chicagotribune.com/news/chi-0209030210sep03-story.html>.

42. "Ties to Enron Blinded Andersen."

43. "The Fall of Andersen," *Chicago Tribune*, September 1, 2002, accessed October 1, 2016, <http://www.chicagotribune.com/news/chi-0209010315sep01-story .html>; Linda Greenhouse, "Justices Unanimously Overturn Conviction of Arthur Andersen," *New York Times*, May 31, 2005, accessed October 1, 2016, <http://www.nytimes.com/2005/05/31/business/justices-unanimously -overturn-conviction-of-arthur-andersen.html?mtrref=www.google.com&gwh =3DF84BC2B8AC837475FB2146444DCE95&gwt=pay>.

44. Barrionuevo and Bajaj, "What's at Heart of Enron Case, Accounting or 'Lies and Choices'?"

45. "Enron Trial Exhibits and Releases," *Department of Justice*, accessed June 31, 2016, <https://www.justice.gov/archive/index-enron.html>.

46. "Enron Trial," Opening Statement of John Hueston, for the federal prosecutor, 375–77.

47. "Enron Trial," Opening Statement of John Hueston, for the federal prosecutor, 395.

48. "Enron Trial," Opening Statement of Mike Ramsey defending Ken Lay, 476–77.

49. "Enron Trial," Opening Statement of Mike Ramsey defending Ken Lay, 478.

50. "Enron Trial," Opening Statement of Mike Ramsey defending Ken Lay, 488–89.

51. Alexei Barrionuevo, "Enron's Skilling Is Sentenced to 24 Years," *New York Times*, October 24, 2006; "Enron: Where Are They Now?" *Houston Chronicle*, May 21, 2012.

52. David Shipley, "Editorial Notebook: Remembering Hoover," *New York Times*, August 10, 1992, accessed October 1, 2016, <http://www.nytimes.com /1992/08/10/opinion/editorial-notebook-remembering-herbert-hoover.html>.

Chapter 16

1. George J. Stigler and Claire Friedland, "What Can Regulators Regulate? The Case of Electricity," *The Journal of Law and Economics* 5 (1962): 1.

2. J. R. Minkel, "The 2003 Northeast Blackout—Five Years Later," *Scientific American*, August 13, 2008, accessed October 1, 2016, <http://www.scientific american.com/article/2003-blackout-five-years-later/>.

3. Order 888 (issued April 24,1996), *Federal Energy Regulatory Commission*, 13–14, accessed October 1, 2016, <http://www.ferc.gov/legal/maj-ord-reg/land-docs /order888.asp>.

4. Order 888, 14–16.

5. Order 888, 17–18.

6. Public Utilities Regulatory Policies Act of 1978, *United States Bureau of Reclamation*, November 9, 1978, accessed October 1, 2016, <http://www.usbr.gov/power /legislation/purpa.pdf>.

7. Russell Ray, "A Report on Combined Cycle Projects in North America," *Power Engineering*, February 3, 2014, accessed October 1, 2016, <http://www.power-eng .com/articles/2014/02/a-report-on-combined-cycle-projects-in-north-america

.html>; "Natural Gas 1998: Issues and Trends," *Energy Information Administration*, 53–54, accessed October 1, 2016, <https://news.duke-energy.com/releases /progress-energy-carolinas-to-retire-two-coal-fired-power-plants-oct-1>.

8. Report to Congress on Competition in Wholesale and Retail Markets for Electric Energy, *The Electric Energy Market Competition Task Force*, 23–24, accessed October 1, 2016, <http://www.ferc.gov/legal/fed-sta/ene-pol-act/epact-final-rpt.pdf>.

9. Order No. 888: Final Rule, *Federal Energy Regulatory Commission*, 75 FERC 61,080, 2, accessed October 1, 2016, <https://www.ferc.gov/legal/maj-ord-reg /land-docs/rm95-8-00w.txt>.

10. Order No. 888: Final Rule, 1, 5–6.

11. Order No. 888: Final Rule, 1, 5–6.

12. Order No. 888: Final Rule, 4, 8.

13. Docket No. RM99-2-000; Order No. 2000, *Federal Regulatory Commission*, December 20, 1999, 426–27, 706, accessed October 1, 2016, <http://www.ferc .gov/legal/maj-ord-reg/land-docs/RM99-2A.pdf>.

14. Phillips Petroleum Co. v. Wisconsin 347 U.S. 672 (1954), *Justia.com*, <https:// supreme.justia.com/cases/federal/us/347/672/>; "The History of Regulation," *Naturalgas.org*, September 20, 2013, accessed October 1, 2016, <http://natural gas.org/regulation/history/>.

15. Paul W. McAvoy, "The Natural Gas Policy Act of 1978," *Natural Gas Resources Journal* 19 (1979): 812, accessed October 1, 2016, <http://lawschool.unm.edu /nrj/volumes/19/4/03_macavoy_natural.pdf>; Natural Gas Policy Act of 1978, *Energy Information Administration*, accessed October 1, 2016, <http://www.eia.gov /oil_gas/natural_gas/analysis_publications/ngmajorleg/ngact1978.html>.

16. Order No. 636—Restructuring of Pipeline Services, 17–20, accessed October 1, 2016, <http://www.ferc.gov/legal/maj-ord-reg/land-docs/restruct.asp>.

17. Order No. 636—Restructuring of Pipeline Services, 9.

18. Robert F. Hirsch, *Power Loss: The Origins of Deregulation and Restructuring in the American Electric Utility System* (1999; Cambridge: MIT Press, repr. 2001), Appendix, tables A.2 and A.3.

19. Craig R. Roach, Stuart Rein, and Vincent Musco, *A Review of the Southwest Power Pool's Integrated Marketplace Proposal*. Washington, DC: *Boston Pacific*, December 30, 2010, 1–2.

20. Stigler and Friedland, "What Can Regulators Regulate?" 1, 11.

21. Stigler and Friedland, "What Can Regulators Regulate?" 8.

22. Harold Demsetz, "Why Regulate Utilities?" *The Journal of Law and Economics* 11, no. 1 (April 1968): 55–57, 65.

23. George J. Stigler, "The Theory of Economic Regulation," *The Bell Journal of Economics and Management Science* 2, no. 1 (1971): 3.

24. Richard A. Posner, "Theories of Economic Regulation," *The Bell Journal of Economics and Management Science* 5 (Autumn 1974): 335–36, 343.

25. George L. Priest, "The Origins of Utility Regulation and the 'Theories of Regulation' Debate," *The Journal of Law and Economics* 36 (April 1993): 289–90.

26. "Final Report on the August 14, 2003 Blackout on the United States and Canada: Causes and Recommendations," *U.S.-Canada Power System Outage Task Force*, April 2004, ii, accessed October 1, 2016, <http://energy.gov/sites/prod/files/oeprod /DocumentsandMedia/BlackoutFinal-Web.pdf>.

27. "The Relationship between Competitive Power Markets and Grid Reliability," *U.S. Department of Energy and Natural Resources Canada*, July 2006, iii, xiii–xvii, accessed October 1, 2016, <https://energy.gov/sites/prod/files /oeprod/DocumentsandMedia/Blackout_Rec_12_Final_Report_and _Transmittal.pdf>.

28. "The Relationship between Competitive Power Markets and Grid Reliability," 17, 19.

29. "The Relationship between Competitive Power Markets and Grid Reliability," 123–24.

30. Minkel, "The 2003 Northeast Blackout—Five Years Later," *Scientific American*, August 13, 2008; "Final Report on the August 14, 2003 Blackout on the United States and Canada," 4, 10, 14.

31. Energy Policy Act of 2005, *Federal Energy Regulatory Commission*, Fact Sheet, August 8, 2006, accessed October 1, 2016, <https://www.ferc.gov/legal/fed -sta/epact-fact-sheet.pdf>.

32. Energy Policy Act of 2005; "Commission Finalizes Electric Reliability Rulemaking Pursuant to the Energy Policy Act 2005," *Federal Energy Regulatory Commission*, February 2, 2006, accessed October 1, 2016, <https://www.ferc.gov /media/news-releases/2006/2006-1/02-02-06-E-1.asp>; "History of NERC," *North American Electric Reliability Corporation*, August 2013, 1–2, accessed October 1, 2016, <http://www.nerc.com/AboutNERC/Documents/History%20AUG 13.pdf>.

33. "History of NERC," 4.

34. "Commission Finalizes Rule Barring Market Manipulation"; "Final Rule Revises PURPA Mandatory Purchase Obligation for Electric Utilities, as Mandated by Energy Policy Act," *Federal Energy Regulatory Commission*, October 19, 2006, accessed October 1, 2016, <https://www.ferc.gov/media/news-releases /2006/2006-4/10-19-06-E-2.asp>; Energy Policy Act of 2005.

35. "About 60% of the U.S. Electric Power Supply Is Managed by RTOs," *Energy Information Administration*, April 4, 2011, accessed October 1, 2016, <http://www .eia.gov/todayinenergy/detail.cfm?id=790>.

Chapter 17

1. "Remarks by the President in Announcing the Clean Power Plan," *Whitehouse.gov*, August 3, 2015, accessed October 1, 2016, <https://www .whitehouse.gov/the-press-office/2015/08/03/remarks-president-announcing -clean-power-plan>.

2. "The Nobel Peace Prize 2007," *Nobelprize.org*, accessed October 1, 2016, <http://www. nobelprize.org/nobel_prizes/peace/laureates/2007/>; "Al Gore, Climate Panel Share Nobel Peace Prize," *National Geographic*, October 12, 2007, accessed October

1, 2016, <http://news.nationalgeographic.com/news/2007/10/071012-peace
-nobel.html>.

3. "Sources of Greenhouse Gas Emissions," *Environmental Protection Agency*, accessed
   October 1, 2016, <https://www.epa.gov/ghgemissions/sources-greenhouse-gas
   -emissions>.

4. John M. Broder, "Obama to Go to Copenhagen with Emissions Target,"
   New York Times, November 25, 2009, accessed October 1, 2016, <http://www
   .nytimes.com/2009/11/26/us/politics/26climate.html?_r=0>; "The American
   Clean Energy and Security Act (Waxman-Markey Bill)," *Center for Climate and
   Energy Solutions*, accessed October 1, 2016, <http://www.c2es.org/federal
   /congress/111/acesa>.

5. Lokenath Debnath, "A Short Biography of Joseph Fourier and Historical
   Development of Fourier Series and Fourier Transforms," *International Journal
   of Mathematical Education in Science and Technology* 43, no. 5 (2012): 595; John
   Tyndall, "The Bakerian Lecture: On the Absorption and Radiation of
   Heat by Gases and Vapours, and on the Physical Connexion of Radiation,
   Absorption, and Conduction," *Philosophical Transactions of the Royal Society of
   London* 151 (1861), accessed October 1, 2016, <http://web.gps.caltech.edu
   /~vijay/Papers/Spectroscopy/tyndall-1861.pdf>; Mike Hulme, "On the
   Origin of 'The Greenhouse Effect': John Tyndall's 1859 Interrogation of
   Nature," *Royal Meteorological Society* 64 (2009): 123, accessed October 1, 2016,
   doi: 10.1002/wea.386; Svante Arrhenius, "On the Influence of Carbonic Acid
   in the Air upon the Temperature of the Ground," *Philosophical Magazine and
   Journal of Science*, ser. 5, vol. 41 (April 1896): 267–69, accessed October 1, 2016,
   <http://www.rsc.org/images/Arrhenius1896_tcm18-173546.pdf>; Geoffrey
   Parker, "Lessons from the Little Ice Age," *New York Times*, March 22, 2014,
   accessed October 1, 2016, <http://www.nytimes.com/2014/03/23/opinion
   /sunday/lessons-from-the-little-ice-age.html?_r=0>.

6. Spencer R. Weart, *The Discovery of Global Warming* (Cambridge, MA: Harvard
   University Press, 2003), 8.

7. Andrew C. Revkin, "Earth Is Us," *New York Times*, January 28, 2008,
   accessed October 1, 2016, <http://dotearth.blogs.nytimes.com/2008/01/28
   /earth-is-us/?_r=0>; G. S. Callendar, "The Artificial Production of Carbon
   Dioxide and Its Influence on Temperature," *Journal of the Royal Meteorological
   Society*, May 19, 1937, 223, 236, accessed October 1, 2016, <http://online
   library.wiley.com/doi/10.1002/qj.49706427503/epdf>; Leo Hickman, "How
   the Burning of Fossil Fuels Was Linked to a Warming World in 1938," *The
   Guardian*, April 22, 2013, accessed October 1, 2016, <https://www.theguardian
   .com/environment/blog/2013/apr/22/guy-callendar-climate-fossil-fuels>;
   Gilbert Plass, "The Carbon Dioxide Theory of Climatic Change," *Tellus* 8,
   no. 2 (1955): 153, accessed October 1, 2016, <http://onlinelibrary.wiley.com
   /doi/10.1111/j.2153-3490.1956.tb01206.x/pdf>.

8. Roger Revelle and Hans E. Seuss, "Carbon Dioxide Exchange Between
   Atmosphere and Ocean and the Question of an Increase of Atmospheric CO2

during the Past Decades," *Tellus* 9 (1957): table 1, accessed October 1, 2016, <http://uscentrist.org/platform/positions/environment/context-environment /docs/Revelle-Suess1957.pdf>; "Restoring the Quality of Our Environment," *President's Science Advisory Committee* (1965): v, 126, accessed October 1, 2016, <http://dge.stanford.edu/labs/caldeiralab/Caldeira%20downloads/ PSAC,%201965,%20Restoring%20the%20Quality%20of%20Our%20 Environment.pdf>.

9. "Charles David Keeling Biography," *Scripps Institution of Oceanography*, accessed October 1, 2016, <http://scrippsco2.ucsd.edu/charles_david_keeling_ biography>; "Keeling Curve Lessons," *Scripps Institution of Oceanography*, <http://scrippsco2.ucsd.edu/history_legacy/keeling_curve_lessons>; Andrea Thompson, "2015 Begins with CO$_2$ above 400 PPM Mark," *Scientific American*, January 12, 2015, accessed October 1, 2016, <http://www.scientificamerican .com/article/2015-begins-with-co2-above-400-ppm-mark/>.

10. *The Conservation Foundation, Implications of Rising Carbon Dioxide Content of the Atmosphere* (New York: The Conservation Foundation, 1963), i; Howard E. Newell, "A Recommended National Program in Weather Modification," *National Aeronautics and Space Administration* (1966): I-3, accessed October 1, 2016, <http:// www.geoengineeringwatch.org/documents/19680002906_1968002906.pdf>.

11. Thomas H. Maugh II, "Willi Dansgaard Dies at 88; Scientist Who Recognized Climate Record in Ice Cap," *Los Angeles Times*, February 7, 2011, accessed October 1, 2016, <http://articles.latimes.com/2011/feb/07/local /la-me-willi-dansgaard-20110207>; William Ferguson, "Ice Core Data Help Solve a Global Warming Mystery," *Scientific American*, March 1, 2013, accessed October 1, 2016, <http://www.scientificamerican.com/article /ice-core-data-help-solve/>; Weart, *Discovery of Global Warming*, 130–31.

12. Weart, *Discovery of Global Warming*, 71–72, 96–97; Spencer R. Weart, "Government: The View from Washington, DC," *American Institute of Physics*, February 2016, accessed October 1, 2016, <https://www.aip.org/history/climate/Govt.htm>; "Bill Summary & Status: 95th Congress (1977–1978) H.R.6669," *Library of Congress*, September 17, 1978, accessed October 1, 2016, <http://thomas.loc .gov/cgi bin/bdquery/z?d095:HR06669:@@@D&summ2=m&>.

13. "Colder Winters Held Dawn of New Ice Age," *Washington Post*, January 11, 1970; "Another Ice Age?" *Time*, June 24, 1974; Robert C. Cowen, "New Ice Age Almost Upon Us?" *Christian Science Monitor*, November 14, 1979.

14. Weart, *Discovery of Global Warming*, 114–15, 120–21.

15. "Brief Primer on the Montreal Protocol," *United Nations Environment Programme*, accessed October 1, 2016, <http://ozone.unep.org/Publications/MP_Brief_Primer _on_MP-E.pdf>.

16. "History," *Intergovernmental Panel on Climate Change*, accessed October 1, 2016, <https://www.ipcc.ch/organization/organization_history.shtml>; J. T. Houghton, G. J. Jenkins, and J. J. Ephraums, eds., *Climate Change: The IPCC Scientific Assessment* (New York: Cambridge University Press, 1990) xi–xii, accessed October 1, 2016, <https://www.ipcc.ch/ipccreports/far/wg_I/ipcc_far_wg_I_full_report.pdf>.

17. "General Information." *United Nations Earth Summit +5,* accessed October 1, 2016, <http://www.un.org/esa/earthsummit/>; "IPCC Second Assessment: Climate Change 1995" *Intergovernmental Panel on Climate Change* (1995): 22, accessed October 1, 2016, <https://www.ipcc.ch/pdf/climate-changes-1995 /ipcc-2nd-assessment/2nd-assessment-en.pdf>.

18. Weart, *Discovery of Global Warming,* 173–74; Richard N. Cooper, "The Kyoto Protocol: A Flawed Concept," *Harvard University Center for the Environment,* 7–8, 11–12, accessed October 1, 2016, <http://www.environment.harvard.edu/docs /faculty_pubs/cooper_kyoto.pdf>; "Byrd-Hagel Resolution," *National Center for Public Policy Research,* accessed October 1, 2016, <http://www .nationalcenter.org/KyotoSenate.html>; George W. Bush, "Letter to Members of the Senate on the Kyoto Protocol on Climate Change," *The American Presidency Project,* March 13, 2001, accessed October 1, 2016, <http://www .presidency.ucsb.edu/ws/?pid=45811>.

19. "Endangerment and Cause or Contribute Findings for Greenhouse Gases under Section 202(a) of the Clean Air Act," *Environmental Protection Agency,* 2009, 66497, accessed October 1, 2016, <http://www.epa.gov/climatechange /endangerment/>; "Climate Change 2014: Mitigation of Climate Change," *Intergovernmental Panel on Climate Change,* 2014, accessed October 1, 2016, <http:// www.ipcc.ch/report/ar5/wg3/>.

20. "Climate Change 2014" 1, SPM 1, SPM 1.1, 4, SPM 1.2, accessed October 1, 2016, <http://www.ipcc.ch/report/ar5/wg3/>.

21. "Climate Change 2014" 21.

22. J. P. Stockholm, "The IPCC Climate-Change Report: It's Still Our Fault," *Economist,* September 27, 2013, accessed October 1, 2016, <http://www.economist .com/blogs/babbage/2013/09/ipcc-climate-change-report>.

23. "IPCC Statement on the Melting of Himalayan Glaciers," *Intergovernmental Panel on Climate Change,* January 20, 2010, accessed October 1, 2016, <https://www .ipcc.ch/pdf/presentations/himalaya-statement-20january2010.pdf>; Matthew Knight, "U.N. Climate Chiefs Apologize for Glacier Error," *CNN,* January 20, 2010, accessed October 1, 2016, <http://www.cnn.com/2010 /WORLD/asiapcf/01/20/glacier.himalayas.ipcc.error/>.

24. "Climate Change Assessments, Review of the Processes and Procedures of the IPCC," *InterAcademy Council,* October 2010, xii–xiv, 59, accessed October 1, 2016, <http://reviewipcc.interacademycouncil.net/report.html>.

25. Bjorn Lomborg, *Cool It: The Skeptical Environmentalist's Guide to Global Warming* (New York: Vintage Books, 2007); Bjorn Lomborg, "Cost-Effective Ways to Address Climate Change," *Washington Post,* November 17, 2010, accessed October 1, 2016, <http://www.washingtonpost.com/wp-dyn/content/article/2010/11/16 /AR2010111604973.html>.

26. Mark Holt and Gene Whitney, "Greenhouse Gas Legislation: Summary and Analysis of H.R. 2454 as Passed by the House of Representatives," *Congressional Research Service,* summary, accessed October 1, 2016, <http://research.policy archive.org/18878.pdf>.

27. "Comparison Chart of Waxman-Markey and Kerry-Lieberman," *Center for Climate and Energy Solutions,* accessed October 1, 2016, <http://www.c2es.org/federal/congress/111/comparison-waxman-markey-kerry-lieberman>.

28. Bryan Walsh, "Why the Climate Bill Died," *Time,* July 26, 2010, accessed October 1, 2016, <http://science.time.com/2010/07/26/why-the-climate-bill-died/>.

29. "EPA Denies Petition to Regulate Greenhouse Gas Emissions from Motor Vehicles," *Environmental Protection Agency,* August 28, 2003, accessed October 1, 2016, <https://yosemite.epa.gov/opa/admpress.nsf/b1ab9f485b098972852562 e7004dc686/694c8f3b7c16ff6085256d900065fdad!OpenDocument>; Massachusetts et al. v. Environmental Protection Agency et al., 05 U.S. 1120 (2007), accessed October 1, 2016, <https://www.law.cornell.edu/supct/html/05 -1120.ZS.html>.

30. "Endangerment and Cause or Contribute Findings for Greenhouse Gases," 40 CFR, chap. 1 at 66496 to 66498, 66517.

31. "Clean Air Act Permitting for Greenhouse Gas Emissions—Final Rules Fact Sheet," *Environmental Protection Agency,* April 17, 2009, 5, accessed October 1, 2016, <https://www.epa.gov/sites/production/files/2015-12/documents/20101223 factsheet.pdf>.

32. "Technical and Support Document for Endangerment and Cause or Contribute Findings for Greenhouse Gases," *Environmental Protection Agency,* ES-2, ES-3, accessed October 1, 2016, <https://www3.epa.gov/climatechange/Downloads /endangerment/TSD_Endangerment.pdf>.

33. "Fifteenth Session of the Conference of the Parties to the United Nations Framework Convention on Climate Change and Fifth Session of the Meeting of the Parties to the Kyoto Protocol," *Pew Center on Global Climate Change,* accessed October 1, 2016, <http://www.c2es.org/docUploads /copenhagen-cop15-summary.pdf>; "How to Live with Climate Change," *Economist,* November 25, 2010, <http://www.economist.com/node/17575027>.

34. George F. Will, "The Paris Agreement Is Another False 'Turning Point' on the Climate," *Washington Post,* December 16, 2015, accessed October 1, 2016, <https:// www.washingtonpost.com/opinions/another-false-urning-point-on-the -climate/2015/12/16/e16dbc36-a35b-11e5-9c4e-be37f66848bb_story.html>; Thomas L. Friedman, "Paris Climate Accord Is a Big, Big Deal," *New York Times,* December 16, 2015, accessed October 1, 2016, <http://www.nytimes.com /2015/12/16/opinion/paris-climate-accord-is-a-big-big-deal.html>.

35. Will, "The Paris Agreement Is Another False 'Turning Point' on the Climate"; Friedman, "Paris Climate Accord Is a Big, Big Deal."

36. Will, "The Paris Agreement Is Another False 'Turning Point' on the Climate."

37. Friedman, "Paris Climate Accord Is a Big, Big Deal."

38. "Paris Climate of Conformity," *Wall Street Journal,* December 13, 2015, accessed October 1, 2016, <http://www.wsj.com/articles/paris-climate-of-conformity -1450048095>.

39. "Remarks by the President in Announcing the Clean Power Plan."

40. Pope Francis, *"Laudato sì," Libreria Editrice Vaticana*, May 24, 2015, paragraphs 32–34, 165, 169, 175, accessed October 1, 2016, <http://w2.vatican.va/content/francesco/en/encyclicals/documents/papa-francesco_20150524_enciclica-laudato-si.html>.

41. "Clean Power Plan for Existing Power Plants," *Environmental Protection Agency*, October 23, 2015, accessed October 1, 2016, <http://www2.epa.gov/cleanpowerplan/clean-power-plan-existing-power-plants>; "Fact Sheet: Overview of the Clean Power Plan," *Environmental Protection Agency*, accessed October 1, 2016, <http://www.epa.gov/airquality/cpp/fs-cpp-overview.pdf>; Van Ness Feldman LLP, "EPA Issues Regulations to Control Carbon Dioxide Emissions from the Power Sector," August 5, 2015, accessed October 1, 2016, <http://www.vnf.com/epa-regulations-to-control-carbon-dioxide-emissions-from>.

42. "Clean Power Plan for Existing Power Plants," 64665; "Clean Power Plan State-Specific Fact Sheets," *Environmental Protection Agency*, updated September 16, 2016, accessed October 1, 2016, <https://www.epa.gov/cleanpowerplantoolbox/clean-power-plan-state-specific-fact-sheets>.

43. "Clean Power Plan for Existing Power Plants," 64667.

44. "Clean Power Plan for Existing Power Plants," 64685.

45. Craig R. Roach, PhD, and Vincent Musco, *Southwest Power Pool Annual Looking Forward Report: Strategic Issues Facing the Electricity Business*, Washington, DC: *Boston Pacific*, April 15, 2016, 29–34; Jonathan H. Adler, "Supreme Court Puts the Brakes on the EPA's Clean Power Plan," *Washington Post*, February 9, 2016, accessed October 1, 2016, <https://www.washingtonpost.com/news/volokh-conspiracy/wp/2016/02/09/supreme-court-puts-the-brakes-on-the-epas-clean-power-plan/>; State of West Virginia, State of Texas et al. v. United States Environmental Protection Agency and Regina A. McCarthy, application, February 9, 2016, accessed October 1, 2016, <http://www.scotusblog.com/wp-content/uploads/2016/01/15A773-application.pdf> and <http://www.supremecourt.gov/orders/court orders/020916zr3_hf5m.pdf>; Amy Harder and Brent Kendall, "Carbon-Rule Stay Puts Obama Environmental Legacy on the Line," *Wall Street Journal*, February 10, 2016, accessed October 1, 2016, <http://www.wsj.com/articles/carbon-rule-stay-puts-obama-environmental-legacy-on-the-line-1455150933>; Editorial Board, "The Court Blocks Efforts to Slow Climate Change," *New York Times*, February 11, 2016, accessed October 1, 2016, <http://www.nytimes.com/2016/02/11/opinion/the-court-blocks-efforts-to-slow-climate-change.html>.

46. Robert Barnes and Steven Mufson, "Supreme Court Freezes Obama Plan to Limit Carbon Emissions," *Washington Post*, February 9, 2016, accessed October 1, 2016, <https://www.washingtonpost.com/politics/courts_law/supreme-court-freezes-obama-plan-to-limit-carbon-emissions/2016/02/09/ac9dfad8-cf85-11e5-abc9-ea152f0b9561_story.html>.

Chapter 18

1. "The Father of Fracking," *The Economist*, August 3, 2015, accessed October 1, 2016, <http://www.economist.com/news/business/21582482-few-businesspeople-have-done-much-change-world-george-mitchell-father>.

2. Daniel Yergin, "'Energy Independence,'" *Wall Street Journal*, January 23, 2007, accessed October 1, 2016, <http://www.wsj.com/articles/SB11695195473928 4514>; "How Much Natural Gas Does the United States Have, and How Long Will It Last?," *Energy Information Administration*, November 18, 2015, accessed October 1, 2016, <http://www.eia.gov/tools/faqs/faq.cfm?id=58&t=8>; Alan Neuhauser, "The New U.S. Energy Era Will Be a Gas," *U.S. News*, May, 16, 2016, accessed October 1, 2016, <http://www.usnews.com/news/articles/2016-05-15/us-gas-exports-poised-to-surpass-imports-for-first-time-since-1957>.

3. "Natural Gas: Henry Hub Natural Gas Spot Price," *Energy Information Administration*, accessed October 1, 2016, <https://www.eia.gov/dnav/ng/hist/rngwhhdm.htm>; Chris Mooney, "How Super Low Natural Gas Prices Are Reshaping How We Get Our Power," *Washington Post*, October 28, 2015, accessed October 1, 2016, <https://www.washingtonpost.com/news/energy-environment/wp/2015/10/28/how-super-low-natural-gas-prices-are-reshaping-how-we-get-our-power/?utm_term=.9b397144a37a>.

4. Tom Fowler, "Exec Mitchell Laid Groundwork for Shale Gas Surge," *Houston Chronicle*, November 15, 2009, accessed October 1, 2016, <http://www.chron.com/business/energy/article/Exec-Mitchell-laid-groundwork-for-shale-gas-surge-1742206.php>.

5. Fatih Birol, "Golden Rules for a Golden Age of Gas: World Energy Outlook Special Report on Unconventional Gas," *International Energy Agency*, May 29, 2012, 25–26, 33–34, accessed October 1, 2016, <http://www.worldenergyoutlook.org/media/weowebsite/2012/goldenrules/WEO2012_GoldenRulesReport.pdf>; Ernest J. Moniz, Henry D. Jacoby, and Anthony J. M. Meggs, "The Future of Natural Gas: An Interdisciplinary MIT Study," *MIT Energy Initiative*, June 9, 2011, 42, accessed October 1, 2016, <http://www.mit.edu/~jparsons/publications/NaturalGas_Report_Final.pdf>.

6. Fowler, "Exec Mitchell Laid Groundwork for Shale Gas Surge"; Dan B. Steward, "George P. Mitchell and the Barnett Shale," *Journal of Petroleum Technology*, November 2013, accessed October 1, 2016, <http://www.mydigital publication.com/article/George_P._Mitchell_And_The_Barnett_Shale/1535436/179598/article.html>.

7. Yergin, "'Energy Independence.'"

8. Moniz, Jacoby, and Meggs, "The Future of Natural Gas," 5.

9. Daniel Yergin and Robert Ineson, "America's Natural Gas Revolution," *Wall Street Journal*, November 2, 2009, accessed October 1, 2016, <http://www.wsj.com/articles/SB10001424052748703399204574507440795971268>.

10. Moniz, Jacoby, and Meggs, "Future of Natural Gas," 29; "How Much Shale Gas Is Produced in the United States?," *Energy Information Administration*, June

14, 2016, accessed October 1, 2016, <http://www.eia.gov/tools/faqs/faq .cfm?id=907&t=8>; "U.S. Crude Oil and Natural Gas Proved Reserves, Year-End 2015," *Energy Information Administration*, released November 23, 2015, figure 5, accessed October 1, 2016, <https://www.eia.gov/naturalgas/crude oilreserves/>; "U.S. Crude Oil and Natural Gas Proved Reserves," Table 9: U.S. Proved Reserves of Total Natural Gas, Wet after Lease Separation, 2001–14, *Energy Information Administration*, November 23, 2015, accessed October 1, 2016, <http:// www.eia.gov/naturalgas/crudeoilreserves/pdf/table_9.pdf>.

11. "Natural Gas: Henry Hub Natural Gas Spot Price."

12. An MMBtu is a standard measure of energy content with a Btu defined as the amount of heat required to increase the temperature of a pound of water by one degree Fahrenheit; "Natural Gas: Henry Hub Natural Gas Spot Price"; Clifford Krauss and Eric Lipton, "After the Boom in Natural Gas," *New York Times*, October 20, 2012, accessed October 1, 2016, <http://www.nytimes .com/2012/10/21/business/energy-environment/in-a-natural-gas-glut-big -winners-and-losers.html>.

13. Moniz, Jacoby, and Meggs, "The Future of Natural Gas," 33.

14. Moniz, Jacoby, and Meggs, "The Future of Natural Gas," 31, 54.

15. Michael Greenstone and Adam Looney, "Paying Too Much for Energy? The True Costs of Our Energy Choices," *Daedalus* 141, no. 2 (Spring 2012) 19, 24, accessed October 1, 2016, <http://ceepr.mit.edu/files/papers/Reprint_243_WC.pdf>.

16. "Annual Energy Outlook 2014 with Projection to 2040," *Energy Information Administration*, April 2014, MT-21, MT-23, accessed October 1, 2016, <http:// www.eia.gov/forecasts/aeo/pdf/0383(2014).pdf>.

17. J. David Hughes, "Drilling Deeper: A Reality Check on U.S. Government Forecasts for a Lasting Tight Oil & Shale Gas Boom," *Post Carbon Institute*, October 2014, 6, 11, accessed October 1, 2016, <http://www.postcarbon.org /wp-content/uploads/2014/10/Drilling-Deeper_FULL.pdf>.

18. Mark P. Mills, "SHALE 2.0: Technology and the Coming Big-Data Revolution in America's Shale Oil Fields," *Manhattan Institute*, May 2015, Executive Summary, 6–9, 12–13, accessed October 1, 2016, <http://www.manhattan-institute.org/pdf /eper_16.pdf>.

19. Birol, "Golden Rules," 9–10.

20. Birol, "Golden Rules," 13–14, 17.

21. *Gasland*, dir. Josh Fox (HBO Documentary Films, 2011), DVD; "Gasland: About the Film," *Gasland*, accessed October 1, 2016, <http://one.gaslandthemovie .com/about-the-film>.

22. Chris Tucker, "EID Statement on *GasLand* Academy Award Nod: 'This nomination is fitting, as the Oscars are aimed at praising pure entertainment," *Energy in Depth*, January 25, 2011, accessed October 1, 2016, <https:// energyindepth.org/national/eid-statement-on-gasland-academy-award -nod-this-nomination-is-fitting-as-the-oscars-are-aimed-at-praising-pure -entertainment/>.

23. Ian Urbina, "Regulation Lax as Gas Wells' Tainted Water Hits Rivers," *New York Times*, February 26, 2011, accessed October 1, 2016, <http://www .nytimes.com/2011/02/27/us/27gas.html?_r=1&sq=fracking&st=cse&scp =2&pagewanted=print>.

24. "Assessment of the Potential Impacts of Hydraulic Fracturing for Oil and Gas on Drinking Water Resources," *Environmental Protection Agency*, June 2015, ES-1, ES-6, ES-10, ES-23, accessed October 1, 2016, <http://www2.epa.gov/sites /production/files/2015-07/documents/hf_es_erd_jun2015.pdf>.

25. "Potential Impacts of Hydraulic Fracturing," ES-6, ES-23.

26. "Potential Impacts of Hydraulic Fracturing," ES-6.

27. "Hydraulic Fracturing for Oil and Gas: Impacts from the Hydraulic Fracturing Water Cycle on Drinking Water Resources in the United States (Final Report)," *Environmental Protection Agency*, 2016: 40–41.

28. Daniel P. Schrag, "Is Shale Gas Good for Climate Change?," *Daedalus* 141, no. 2 (Spring 2012): 72.

29. Schrag, "Is Shale Gas Good for Climate Change?," 72–73, 77–78.

30. Schrag, "Is Shale Gas Good for Climate Change?," 79.

31. "EPA Proposes New Commonsense Measures to Cut Methane Emissions from the Oil and Gas Sector/Proposal Cuts GHG Emissions, Reduces Smog-Forming Air Pollution and Provides Certainty for Industry," *EPA Newsroom*, August, 18, 2015, accessed October 1, 2016, <https://www.epa.gov/newsreleases /epa-proposes-new-commonsense-measures-cut-methane-emissions-oil-and-gas -sectorproposal>.

32. "Earthquakes in Oklahoma," *Office of the Secretary of Energy and Environment, Oklahoma*, accessed October 1, 2016, <https://earthquakes.ok.gov/>; Roach and Musco, *Southwest Power Pool Annual Looking Forward Report*, 20–22; "Oklahoma Earthquakes Magnitude 3.0 and Greater," *United States Geological Survey*, February 17, 2016, accessed October 1, 2016, <http://earthquake.usgs .gov/earthquakes/states/oklahoma/images/OklahomaEQsBarGraph.png>.

33. "Full Transcript: Obama's 2012 State of the Union Address," *USA Today*, January 25, 2012, accessed October 1, 2016, <http://www.usatoday.com/news /washington/story/2012-01-24/state-of-the-union-transcript/52780694/1>.

34. Sherie Mershon and Tim Palucka, *A Century of Innovation: From the U.S Bureau Mines to the National Energy Technology Laboratory* (National Energy Technology Laboratory, 2010), 247–48, 253, 256–57.

35. Mershon and Palucka, *A Century of Innovation*, 260–61.

36. Mershon and Palucka, *A Century of Innovation*, 260–61, 295–96.

37. George P. Mitchell and Mark D. Zoback, "George Mitchell: The Duty to Fracture Responsibly," *Fuelfix*, February 20, 2012, accessed October 1, 2016, <http:// fuelfix.com/blog/2012/02/20/opinion-the-duty-to-fracture-responsibly/>.

38. David Brooks, "Shale Gas Revolution," *New York Times*, November 3, 2011, accessed October 1, 2016, <http://www.nytimes.com/2011/11/04/opinion/brooks-the -shale-gas-revolution.html>.

39. "George P. Mitchell Receives Lifetime Achievement Award from Gas Technology Institute," *PR Newswire*, accessed October 1, 2016, <http://www.prnewswire .com/news-releases/george-p-mitchell-receives-lifetime-achievement-award -from-gas-technology-institute-96564819.html>; Douglas Martin, "George Mitchell, a Pioneer in Hydraulic Fracturing, Dies at 94," *New York Times*, July 26, 2013, accessed October 1, 2016, <http://www.nytimes.com/2013/07/27/business /george-mitchell-a-pioneer-in-hydraulic-fracturing-dies-at-94.html>.

40. Russell Gold, "How Aubrey McClendon Led Today's Energy Revolution," *Wall Street Journal*, March 4, 2016, accessed October 1, 2016, <http://www.wsj.com /articles/how-aubrey-mcclendon-led-todays-energy-revolution-1457117234>.

41. Jacob Gronholt-Pedersen, "U.S. Exports First Shale Gas as LNG Tanker Sails from Sabine Pass Terminal," *Reuters*, February 24, 2016, accessed October 1, 2016, <http://www.reuters.com/article/us-shale-export-idUSKCN0VY08B>; Georgi Kantchev, "With U.S. Gas, Europe Seeks Escape from Russia's Energy Grip," *Wall Street Journal*, February 25, 2016, accessed October 1, 2016, <http://www.wsj.com/articles/europes-escape-from-russian-energy-grip-u-s -gas-1456456892>.

Chapter 19

1. Aldous Huxley, "Case of Voluntary Ignorance," in *Collected Essays* (New York: Harper Brothers, 1959).

2. Huxley, "Case of Voluntary Ignorance."

3. "History of Zip2," *Ecorner: Stanford University's Entrepreneurship Corner*, October 8, 2003, accessed October 1, 2016, <http://ecorner.stanford.edu /authorMaterialInfo.html?mid=397>; Ashlee Vance, *Elon Musk: Tesla, SpaceX, and the Quest for a Fantastic Future* (New York: HarperCollins, 2015), 72, 84–89; "Founding of PayPal," *Ecorner: Stanford University's Entrepreneurship Corner*. October 8, 2004, accessed October 1, 2016, <http://ecorner.stanford.edu/videos/378/ Founding-of-Paypal>.

4. "Completed Missions," *SpaceX*, accessed October 1, 2016, <http://www.spacex. com/missions>; Andrew Chaikin, "Is SpaceX Changing the Rocket Equation?," *Air & Space Smithsonian*, January 2012, accessed October 1, 2016, <http://www .airspacemag.com/space/is-spacex-changing-the-rocket-equation-132285884 /?no-ist>; "SpaceX Successfully Launches Falcon 1 to Orbit," *SpaceX*, September 28, 2008, accessed October 1, 2016, <http://www.spacex.com /press/2012/12/19/spacex-successfully-launches-falcon-1-orbit>; William Jacobs, "Launch Systems Rockets Priced to Move," *Popular Science* 267, no. 5, November 2004; William Harwood, "Experts Applaud SpaceX Rocket Landing, Potential Savings," *CBS News*, December 22, 2015, accessed October 1, 2016, <http:// www.cbsnews.com/news/experts-applaud-spacex-landing-cautious-about -outlook/>; Jeffrey Kluger, "Things to Know About SpaceX," *Time*, accessed October 1, 2016, <http://time.com/space-x-ten-things-to-know/>; Vance, *Elon Musk*, 5.

5. Fred Lambert, "SpaceX Transferred Novel Welding Techniques and Equipment to Tesla Motors," *electrek*, May 24, 2015, accessed October 1, 2016, <https://electrek.co/2015/05/24/spacex-transferred-novel-welding-techniques-and-equipment-to-tesla-motors/>; "Tesla Gigafactory," *Tesla*, accessed October 1, 2016, <https://www.tesla.com/gigafactory>; Jonathan O'Connell, "The Gamble on Tesla's Gigafactory in the Nevada Desert," *Washington Post*, April 10, 2015, accessed October 1, 2016, <http://www.washingtonpost.com/business/capitalbusiness/the-gamble-on-teslas-gigafactory-is-a-big-one--in-many-fashions/2015/04/10/50e9de40-d4c8-11e4-a62f-ee745911a4ff_story.html>.

6. Maria Galucci, "SolarCity to Build World's Largest Solar Panel Plant on U.S. Turf, Spurring Solar Manufacturing Revival," *International Business Times*, June 17, 2014, accessed October 1, 2016, <http://www.ibtimes.com/solarcity-build-worlds-largest-solar-panel-plant-us-turf-spurring-solar-manufacturing-1603556>; Apurba Sakti, Raanan Miller, and Fikile Brushett, "What's Cost Got to Do with It? An Assessment of Tesla's Powerwall," *MIT Energy Initiative*, June 12, 2015, accessed October 1, 2016, <http://energy.mit.edu/news/whats-cost-got-to-do-with-it/>; Dana Hull, "Tesla Powerwalls for Home Energy Storage Hit U.S. Market," *Bloomberg*, May 4, 2016, accessed October 1, 2016, <http://www.bloomberg.com/news/articles/2016-05-04/tesla-powerwalls-for-home-energy-storage-are-hitting-u-s-market>.

7. Vance, *Elon Musk*, 349; Jerry Hirsch, "Elon Musk's Growing Empire Is Fueled by $4.9 Billion in Government Subsidies," *LA Times*, June 1, 2015, accessed October 1, 2016, <http://www.latimes.com/business/la-fi-hy-musk-subsidies-20150531-story.html#page=1>; Daniel Wiesfield, "Peter Thiel at Yale: We Wanted Flying Cars, Instead We Got 140 Characters," *Yale School of Management*, April 27, 2013, accessed October 1, 2016, <http://som.yale.edu/peter-thiel-yale-we-wanted-flying-cars-instead-we-got-140-characters>; see, for example, Nicole Allan, "Who Will Tomorrow's Historians Consider Today's Greatest Inventors?," *Atlantic*, November 2013, accessed October 1, 2016, <http://www.theatlantic.com/magazine/archive/2013/11/the-inventors/309534/>.

8. Charley Grant, "More Than Money at Stake in Tesla's SolarCity Deal," *Wall Street Journal*, September 26, 2016, accessed October 1, 2016, <http://www.wsj.com/articles/more-than-money-at-stake-in-teslas-solarcity-deal-1474903454>; Adam Hartung, "Why Investors Should Support the Tesla-SolarCity Merger," *Forbes*, September 23, 2016, accessed October 1, 2016, <http://www.forbes.com/sites/adamhartung/2016/09/23/why-investors-should-support-the-tesla-solarcity-merger/#33263c784c32>.

9. Dana Hull, "Tesla Breakup with Mobileye Turns Ugly," *Bloomberg*, September 16, 2016, accessed October 1, 2016, <http://www.bloomberg.com/news/articles/2016-09-16/tesla-says-mobileye-tried-to-block-its-auto-vision-capability>.

10. Seth Fiegerman, "Samsung Won't Be the Last to Have Exploding Batteries," *CNN*, September 13, 2016, accessed October 1, 2016, <http://money.cnn.com/2016/09/13/technology/samsung-lithium-batteries/>; Katie Fehrenbacher, "Why Tesla's New Battery Pack Is Important," *Fortune*, August 24, 2016,

accessed October 1, 2016, <http://fortune.com/2016/08/24/tesla-100kwh
-battery-pack/>; Alex Knapp, "SpaceX Identifies Possible Cause of
Rocket Explosion, Anticipates November Launch," *Forbes*, September 23,
2016, accessed October 1, 2016, <http://www.forbes.com/sites/alexknapp
/2016/09/23/spacex-identifies-possible-cause-of-rocket-explosion-anticipates
-november-launch/#617c2aa27c62>.

11. Elon Musk, "Master Plan, Part Deux," July 20, 2016, accessed October
1, 2016, <https://www.tesla.com/blog/master-plan-part-deux>; Rebecca
Hersher and Camila Domonoske, "Elon Musk Unveils His Plan for Colonizing
Mars," *The Two-Way*, September 27, 2016, accessed October 1, 2016,
<http://www.npr.org/sections/thetwo-way/2016/09/27/495622695/this
-afternoon-elon-musk-unveils-his-plan-for-colonizing-mars>.

12. See, for example, Patrick J. Kiger, "U.S. Phase-Out of Incandescent Light
Bulbs Continues in 2014 with 40-, 60-Watt Varieties," *National Geographic Society*,
December 31, 2013, accessed October 1, 2016, <http://energyblog.national
geographic.com/2013/12/31/u-s-phase-out-of-incandescent-light-bulbs-continues
-in-2014-with-40-60-watt-varieties/>.

13. "Former SpaceX Employee Explains What It's Like to Work for Elon Musk,"
Business Insider, June 24, 2014, accessed October 1, 2016, <http://www
.businessinsider.com/what-its-like-to-work-for-elon-musk-2014-6>.

14. "Meet the Battery-Powered Home: Tesla Joins the Race to Help Homeowners
Unplug from the Grid," *Economist*, May 14, 2015, accessed October 1, 2016,
<http://www.economist.com/news/science-and-technology/21651106-tesla
-joins-race-help-homeowners-unplug-grid-dawn-battery-powered-home>;
Lambert, "SpaceX Transferred Novel Welding Techniques and Equipment to
Tesla Motors."

15. "Photovoltaics." *U.S. Department of Energy*, accessed October 1, 2016, <http://
energy.gov/sites/prod/files/2014/08/f18/2014SunShotPortfolio_PV.pdf>;
Eric Mack, "Why Elon Musk Spent $10 Million to Keep Artificial Intelligence
Friendly," *Forbes*, January 15, 2015, accessed October 1, 2016, <http://www
.forbes.com/sites/ericmack/2015/01/15/elon-musk-puts-down-10-million
-to-fight-skynet/>.

16. Vance, *Elon Musk*, 267–68.

17. Kim Bhasin, "Elon Musk: 'If We Published Patents, It Would Be Farcical,'"
Business Insider, November 9, 2012, accessed October 1, 2016, <http://www
.businessinsider.com/elon-musk-patents-2012-11#ixzz3ekSUvPCn>; Vance, *Elon
Musk*, 22.

18. "Utility Corporations: Summary Report of the Federal Trade Commission,"
38, 43.

19. Craig R. Roach, Frank Mossburg, and Vincent Musco, "Partnership, Not Preemption:
How State-Sponsored Planning Can Fit with FERC's Capacity Markets," *Public
Utilities Fortnightly*, December 2013, accessed October 1, 2016, <https://www.fort
nightly.com/fortnightly/2013/12/partnership-not-preemption>.

20. Craig R. Roach and Vincent Musco, "Federal Versus State Jurisdiction in the Electricity Business: Two Back-to-Back Decisions in 2016 Will Fundamentally Re-define the Jurisdictional Split," *Public Utilities Fortnightly*, May 2016, accessed October 1, 2016, <https://www.fortnightly.com/fortnightly/2016/05/federal-versus-state-jurisdiction-electricity-business>.

21. Hirsch, "Elon Musk's Growing Empire Is Fueled by $4.9 Billion in Government Subsidies"; Reem Nasr, "Elon Musk: Incentives Not Necessary, but Helpful," *CNBC*, June 1, 2015, accessed January 5, 2017, <http://www.cnbc.com/2015/06/01/elon-musk-we-are-not-getting-a-check-from-the-govt.html>.

22. Craig R. Roach and Vincent Musco, "Electric Vehicles (Update)," *Southwest Power Pool Annual Looking Forward Report: Strategic Issues Facing the Electricity Business*, April 15, 2016.

23. Brian Fung, "Why Tesla Keeps Fighting for Direct Sales When It Could Just Work with Dealers," *Washington Post*, October 22, 2014, accessed October 1, 2016, <https://www.washingtonpost.com/news/the-switch/wp/2014/10/22/why-tesla-keeps-fighting-for-direct-sales-when-it-could-just-work-with-dealers/>; Mike Ramsey, "Tesla Weighs New Challenge to State Direct-Sales Ban," *Wall Street Journal*, March 28, 2016, accessed October 1, 2016, <http://www.wsj.com/articles/tesla-weighs-new-challenge-to-state-direct-sales-bans-1459189069>; Mike Ramsey, "Tesla Presses Its Case on Fuel Standards," *Wall Street Journal*, August 2, 2015, accessed October 1, 2016, <http://www.wsj.com/articles/tesla-presses-its-case-on-fuel-standards-1438559469>; Henry Sanderson, "Tesla in Stand-Off over Lithium Supply," *Financial Times*, December 15, 2015, accessed October 1, 2016, <http://www.ft.com/cms/s/0/4a924a64-99df-11e5-987b-d6cdef1b205c.html#axzz47bErNWW4>; Susan Pulliam, Mike Ramsey, and Brody Mullins, "Elon Musk Supports His Business Empire with Unusual Financial Moves," *Wall Street Journal*, April 27, 2016, accessed October 1, 2016, <http://www.wsj.com/articles/elon-musk-supports-his-business-empire-with-unusual-financial-moves-1461781962>.

24. George Santayana, *The Life of Reason* (New York: Prometheus Books, 1998), 82; "A Likely Story . . . And That's Precisely the Problem," *Washington Post*, April 17, 2005, accessed October 1, 2016, <http://www.washingtonpost.com/wp-dyn/content/article/2005/04/16/AR2005041600154.html>.

25. "Greatest Engineering Achievements of the 20th Century," *National Academy of Engineering*, 2003, accessed October 1, 2016, <http://www.greatachievements.org/>.

26. Kayla Webley, "Hurricane Sandy by the Numbers: A Superstorm's Statistics, One Month Later," *Time*, November 26, 2012, accessed October 1, 2016, <http://nation.time.com/2012/11/26/hurricane-sandy-one-month-later/>; "PSE&G Reaches $1.22 Billion Settlement in Energy Strong Proceeding," *Transmission & Distribution World Magazine*, May 7, 2014, accessed October 1, 2016, <http://tdworld.com/distribution/pseg-reaches-122-billion-settlement-energy-strong-proceeding>; PG&E sought $3.9 billion and only received $1.22 billion; Rebecca Smith, "Assault on California Power Station Raises Alarm on Potential

for Terrorism," *Wall Street Journal*, February 5, 2014, accessed October 1, 2016, <http://www.wsj.com/articles/SB1000142405270230485110457935914194 1621778>; "Smart Grid Cyber Security Market & Electrical Energy Storage Technology in the Intelligent Grid Analyzed in New Market Research Reports," *PR Newswire*, September 17, 2012, accessed October 1, 2016, <http://www.prnews wire.com/news-releases/smart-grid-cyber-security-market--electrical-energy -storage-technology-in-the-intelligent-grid-analyzed-in-new-market-research -reports-170005586.html>; Roach and Musco, *Southwest Power Pool Annual Looking Forward Report* (Washington, DC: Boston Pacific Company, Inc., April 23, 2013) 62.

27. Morgan Kelly, "Two Years after Hurricane Sandy, Recognition of Princeton's Microgrid Still Surges," *Princeton University*, October 23, 2014, accessed October 1, 2016, <https://www.princeton.edu/main/news/archive/S41/40/10C78/index .xml?section=featured>.

28. "Renewable Portfolio Standard Policies," *Database of State Incentives for Renewables & Efficiency*, September 2014, accessed October 1, 2016, <http://www.dsireusa .org/resources/detailed-summary-maps/>.

29. James Bennet, "Bill Gates: 'We Need an Energy Miracle,'" *Atlantic*, November 16, 2015, accessed October 1, 2016, <http://www.theatlantic.com/magazine /archive/2015/11/we-need-an-energy-miracle/407881/>; Amy Harder and Bradley Olson, "Exxon Touts Carbon Tax to Oil Industry," *Wall Street Journal*, June 30, 2016, accessed October 1, 2016, <http://www.wsj.com/articles/exxon -touts-carbon-tax-to-oil-industry-1467279004>; George P. Shultz and James A. Baker III, "A Conservative Answer to Climate Change," *Wall Street Journal*, February 7, 2017, <https://www.wsj.com/articles/a-conservative-answer-to -climate-change-1486512334>.

30. Jack Stewart, "Tesla's Plan for World Domination Includes Buses and Semis," *Wired*, July 20, 2016. <https://www.wired.com/2016/07/teslas-plan-world -domination-includes-buses-semis/>.

Epilogue

1. Bill Gates, "We Need Energy Miracles," *Gatesnotes* (blog), June 25, 2014. <https://www.gatesnotes.com/Energy/Energy-Miracles>.

2. See, for example, Debbie Sniderman, "Harnessing Power from Slow Moving Currents," *ASME*, February 2012, accessed October 1, 2016, <https:// www.asme.org/engineering-topics/articles/mechanisms-systems-devices /harnessing-power-from-slow-moving-currents>.

3. Robert Rosner and Stephen Goldberg, "Small Modular Reactors—Key to Future Nuclear Power Generation in the U.S.," *Energy Policy Institute at Chicago*, November 2011, 1, accessed October 1, 2016, <http://www.eenews.net/assets /2013/03/13/document_gw_01.pdf>.

4. "How Do Photovoltaics Work?," *NASA Science*, accessed October 1, 2016, <http://science.nasa.gov/science-news/science-at-nasa/2002/solarcells/>.

5. "Hydrogen Fuel Cell Vehicles," *Center for Climate and Energy Solutions*, accessed October 1, 2016, <http://www.c2es.org/technology/factsheet/HydrogenFuel CellVehicles>.

6. Both Geim and Novoselov are at the University of Manchester, UK; "Scientific Background on the Nobel Prize in Physics 2010: Graphene," *The Royal Swedish Academy of Science*, revised November 29, 2010, 1, 2, accessed October 1, 2016, <https://www.nobelprize.org/nobel_prizes/physics/laureates/2010/advanced -physicsprize2010.pdf>; A. K. Geim and K. S. Novoselov, "The Rise of Graphene," *Nature Materials* 6 (2007): 183; "The Home of Graphene: Energy," *The University of Manchester*, accessed October 1, 2016, <http://www.graphene.manchester .ac.uk/explore/the-applications/energy/>; Peter Shadbolt, "Wonder Material Could Harvest Energy from Thin Air," *CNN*, November 3, 2015, accessed October 1, 2016, <http://www.cnn.com/2014/12/23/tech/innovation/tomorrow-transformed -graphene-battery>.

7. "Press Release: The Nobel Prize in Chemistry, 2016." *The Royal Swedish Academy of Sciences*, October 5, 2016, accessed December 31, 2016, <https://www.nobel prize.org/nobel_prizes/chemistry/laureates/2016/press.html>.

8. "What Is U.S. Electricity Generation by Energy Source?," *Energy Information Administration*, updated April 1, 2016, accessed October 1, 2016, <https://www .eia.gov/tools/faqs/faq.cfm?id=427&t=3>.

9. Bennet, "We Need an Energy Miracle"; Christopher Adams and John Thornhill, "Gates to Double Investment in Renewable Energy Projects," *Financial Times*, June 25, 2015, accessed October 1, 2016, <http://www.ft.com /cms/s/2/4f66ff5c-1a47-11e5-a130-2e7db721f996.html#axzz4FiVaQ67b>; Christopher Martin, "Google Is Making Its Biggest Ever Bet on Renewable Energy," *Bloomberg*, February 26, 2015, accessed October 1, 2016, <http:// www.bloomberg.com/news/articles/2015-02-26/google-makes-biggest-bet-on -renewables-to-fund-solarcity>.

10. Bennet, "We Need an Energy Miracle."

11. Bennet, "We Need an Energy Miracle"; Adams, "Gates to Double Investment in Renewable Energy Projects."

12. Adams, "Gates to Double Investment in Renewable Energy Projects."

13. "Introducing the Breakthrough Energy Coalition," *Breakthrough Energy Coalition*, accessed October 1, 2016, <http://www.breakthroughenergycoalition.com/en /index.html>.

INDEX